Designers
The Key to Successful Product Development

Springer
London
Berlin
Heidelberg
New York
Barcelona
Budapest
Hong Kong
Milan
Paris
Santa Clara
Singapore
Tokyo

Eckart Frankenberger, Petra Badke-Schaub
and Herbert Birkhofer (Eds)

Designers

The Key to Successful Product Development

With 90 Figures

Springer

Dr.-Ing. Eckart Frankenberger
Professor Herbert Birkhofer

Technische Universität Darmstadt, Fachgebiet Maschinenelemente und Konstruktionslehre, Magdalenstrasse 4, D-64289 Darmstadt, Germany

Dr. Petra Badke-Schaub

Universität Bamberg, Psychologie II, Markusplatz 3, D-96045 Bamberg, Germany

ISBN-13: 978-1-4471-1270-9 e-ISBN-13: 978-1-4471-1268-6
DOI: 10.1007/978-1-4471-1268-6

British Library Cataloguing in Publication Data
Designers : the key to successful product development
Birkhofer
1. Engineering design 2. Engineering design - Research
I. Frankenberger, Eckart II. Badke-Schaub, Petra
III. Birkhofer, Herbert
620'.0042
ISBN-13: 978-1-4471-1270-9

Library of Congress Cataloging-in-Publication Data
Designers, the key to successful product development / Eckart Frankenberger, Petra Badke-Schaub, Herbert Birkhofer, eds.
p. cm.
Proceedings of a conference held in Darmstadt, Germany, December 1997.
Includes bibliographical references (p.).
ISBN-13: 978-1-4471-1270-9
1. Design, Industrial--Congresses. 2. New Products--Congresses.
3. Engineering design--Research--Congresses. I. Frankenberger, Eckart, 1966- .
II. Badke-Schaub, Petra, 1960- . III. Birkhofer, Herbert, 1945- .
TS171.A1D477 1998 98-7548
658.5'752--dc21 CIP

Softcover reprint of the hardcover 1st edition 1998

Typesetting: Camera-ready by editors
Printed and bound at the Athenæum Press Ltd., Gateshead, Tyne & Wear
69/3830-543210 Printed on acid-free paper

Contents

Preface

Designers – The Key to Successful Product Development

At first glance, the goal of engineering design seems to be very easy: engineering design aims to create and develop competitive technical products, and design research should ensure that this goal is reached.

However, remembering the wide variety of technical products and the multiple aspects that need to be considered in their development, we realise the complexity of engineering design and consequently of design research. Questions arise such as: 'what is a 'good' product?', and 'how do we create 'good' products?'. By posing these questions, we have broadened the subject, and different approaches suggest themselves:

We can start to analyse technical products in an effort to understand their composition on a general level. Knowing the logic behind the make-up of technical systems, we should then be able to optimise the methods of making them. This traditional *product oriented* view by engineers provides valuable rules and guidelines for the assembly of components. Over decades, engineers have collected masses of information in thousands of standards to avoid technical failures. Nevertheless, failures originating from poor engineering design become apparent almost daily in the trivial, dangerous or even fatal incidents of everyday life.

So other factors must be involved. Obviously, the development of technical products is more than just following technical rules. In the light of the vast array of activities involved in engineering, designers considered another approach to engineering design research: the *process oriented* view, which approaches the development of technical products as a complex technical and social process with influences from several different fields.

First of all, each designer has *individual* prerequisites such as experience of the problem or priorities reinforced by the individual's education, abilities, capacities and aptitudes.

But engineering designers are not only working in isolation: the cooperation essential in modern design teams causes additional influences according to the prerequisites of the *group*, such as the character of the group or the ways of conflict solving within the group. Moreover, the design work takes place under certain *organisational* conditions, such as the operational aspects or the financial situation of the company.

Last but not least, design work is determined by the characteristics of the *task*, which can vary greatly. All these influences on the process together lead to the *result* of the design work.

This overall view highlights the designer as an individual, a group member, a manager and an organiser of a network of influencing factors that he experiences in a subjective way:

The designer is thus revealed as the *key* to successful product development!

The engineering perspective on product development is consequently influenced by both psychological and organisational aspects. This point of view can be observed internationally in several research groups. As an example, the *keynote* contribution of this book introduces the motivation and stages of the design research developed by the collaboration of engineers and psychologists from the universities of Darmstadt, Bamberg and Munich in Germany.

The structure of this book

During recent years much scientific knowledge has been accumulated on factors influencing the design process, including the individual, the team and the organisational context. The aim of this book is to provide a survey of the state of this scientifically based knowledge and the trends of engineering design research concerning the influences that lead to successful product development.

This book is based on the proceedings of the symposium 'Designers – The Key to Successful Product Development', held in December 1997 in Darmstadt, Germany. During this meeting exponents from important research groups in Engineering Design came together to present and discuss their work.

Discussing research on engineering design processes and bringing together the work on influencing factors from different studies demands first of all a methodical presentation of the problem: Thus, the book starts with a section on *design research methods*.

Moreover, the contributions focus on different aspects of design work. Therefore, the subsequent sections are directed to the main fields of influence on the engineering design process and the result: the *task* itself, the *individual* priorities of each designer, the prerequisites of the designers as a *group*, and the *organisational* conditions under which the designers work.

Figure 1 illustrates the interconnections between the main fields of influence on the design process, to show both the basic model and the structure of this book.

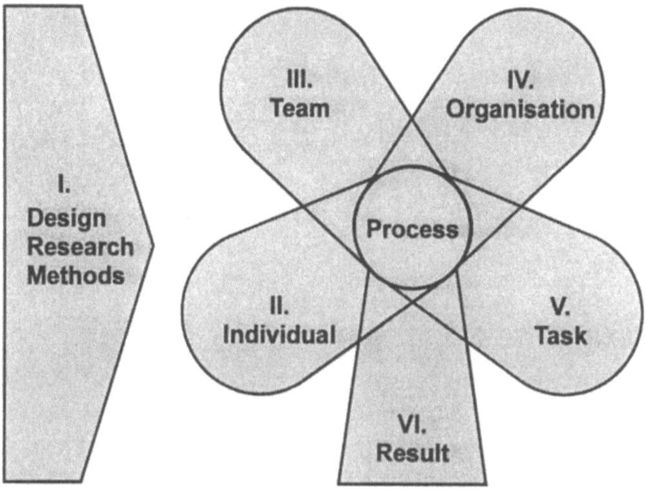

Figure 1: The structure of this book based on the main fields of influence on the design process and the result (showing numbers of the relevant sections).

Together with the first section on design research methods, this book is structured according to the basic model of influence into six sections:

I. The section on *design research methods* provides an overview of the main approaches of design research and their different perspectives on the design process.

II. The section on *individual* influences presents results concerning experience, methodical knowledge and the internal processes of the individual designer.

III. The section on *team* influences deals with the basic processes of verbal communication between designers, the question of team-building and team-improvement in engineering education, and the integration of the group, the individual, and the external influences in the design process.

IV. The section on *organisational* influences focuses on management issues, which determine how proceedings are handled both in the project itself and in the company.

V. The section on *task and process variables* discusses the influences of different types of design tasks and important process variables, such as information availability.

VI. The section on *evaluation of result and design work* approaches the difficult question of how to compare and measure design performance in terms of product and process.

Each section is introduced by a short preface and the main results and factors are summarised in a résumé at the end. The book closes with an overall discussion on the state of engineering design research and the trends and open questions in this fascinating and important issue.

Keynote

Historical Background and Selected Results of Interdisciplinary Research between the Universities of Bamberg, Darmstadt and Munich

Gerhard Pahl

Abstract

With the shift from judging engineering design as an artistic activity to the idea that designing can be taught [1], methodical design research was established., During the last few years, especially engineering design research has been instituted as an important discipline with broad expectations from theory and practice. This paper will focus on two main aspects: firstly, a short survey is given on the historical development of design research in Germany, with special emphasis on the interdisciplinary research with psychology between the universities of Bamberg, Darmstadt and Munich; secondly, the main results of this research will be summarised.

1. Historical Background

At first glance, collaborative output from design researchers and cognitive psychologists seems a surprising combination. As a keynote to this volume, this contribution considers how and why collaboration with cognitive psychologists from Bamberg, Darmstadt and Munich began, with a brief historical excursion.

The Ilmenauer school in the former East Germany first demonstrated methodical work in the area of precision mechanics with the fundamental work of Hansen, Bock and Bischoff in the 1950s [2, 3], and this was further developed in the following years.

From about 1965 to 1980, design research worked out a more general design methodology in accordance with systems technology and by observing successful as well as unsuccessful procedures while developing technical products. It was an answer to the problem which was designated in an article in 1967 as 'Bottleneck Engineering Design' [4], coined by university lecturers who had, without exception, previously experienced successful and responsible design activities in industry.

The methodology was based on their own observations, from which a more or less standard procedure was abstracted. Within the framework of the German Society of Engineers (VDI), professors came together and worked as a team under the direction of F. Kesselring, and as a result, in 1973 this group drew up the VDI Guideline 2222 in draft form [5]. In addition, design method-oriented works of Rodenacker (1970) [6], Hubka (1973) [7], Koller (1976) [8], Pahl & Beitz (1977) [9], Schregenberger (1981) [10] and Roth (1982) [11] emerged and preceded well in advance by numerous lectures and magazine publications.

At the International Conferences on Engineering Design 1981 and 1983, criticism was expressed, mostly by Anglo-American scientists, concerning two aspects:

- The methods presented in VDI 2222 and those of the individual schools were too rigid in their requirements.
- Empirical investigations confirming the efficiency of the design methods were lacking.

Further, the publications were written in German and received little international attention, or were only partially recognised. In addition, poor translations and several extreme theoretical presentations at conferences contributed to the misunderstandings.

With the onset of the 1980s, in spite of continued intensive work on design methods, no essentially new insights into methodology appeared. Instead, so-called 'white fields' were worked on, such as the completion of catalogues, the creation of systematic representations of narrower fields of knowledge, and the refinement of evaluation methods, etc. Early cost recognition, which was promoted especially by Ehrlenspiel in his design book 'Constructing Cost-effectively' (1985) [12], proved a valuable addition to the field. In this area, productive cooperation developed between the institutes in Munich and Darmstadt in the early 1980s.

This was the situation when, during an excursion from the Machine Elements Conference in Stuttgart, 1983, Prof. Ehrlenspiel and I agreed that new impulses were necessary in order to further develop the rather stagnant design methodology. What had been achieved up to this point had clearly shown us the technical processes and partial goals necessary for the development of a technical product. It also produced a series of useful singular methods. However, we did not recognise how humans, with their creative capabilities, could optimally contribute to such a process. We came to the conclusion that new impulses and insights into the further development of design methods could only be obtained if we escaped the boundaries of our profession and sought collaboration with cognitive psychologists.

In 1983 Prof. Ehrlenspiel, who initiated Rutz' work on 'Design as an Intellectual Process', (completed in 1985 [13]) began to take note of the 'Lohhausen' project by Professor Dörner in Bamberg [14], thanks to comments by Dr. Schregenberger. Independently of this, Professor Weinert, with whom I cooperated during that time as a Vice President of DFG (German Research Council), advised me of a new DFG report from Prof. Dörner on the project entitled 'Lohhausen: Dealing with Vagueness and Complexity'. I realised that very essential statements were made in this report. Considerable parts were directly transferable to the work and thought patterns of design activity.

Together with Prof. Ehrlenspiel and Dr. Schregenberger we arranged a first meeting with Prof. Dörner in Bamberg in 1984. This meeting proved to be unusually interesting and valuable. Prof. Dörner expressed a strong interest in collaborating with engineers. He thought, among other things, that the results of actions in the development of technical products were more concrete and hence could be better judged and tested than those in the political arena – which was one of his areas of interest at that time.

We arranged a cooperative research project on processes of thinking and reasoning in design between the Universities of Bamberg, Darmstadt and Munich, within the framework of the German Research Society (DFG).

The year 1984/85 focused on the clarification of suitable forms of cooperation and the preparation of research proposals. In August 1985 our new proposals were submitted. The Munich project was primarily interested in general design processes, while in Darmstadt the main concern was design procedure in comparison with the targets of design methodology.

Initially we received quite massive objections from referees, who were sceptical about the prospects for success of our unconventional collaboration. Two months later, in March 1986, in a joint statement we were able to clarify these reservations. We referred to the fact that processes in the designer's head could not be grasped and understood simply by focusing on the technical system.

The much cited, but not very well understood process of synthesis during design requires close investigation of the designer's behaviour in thought and action. The ability to solve problems is not determined by knowledge of the technical system alone but also by experience of problem-solving strategies and by personal traits, for example, by individual heuristic competence.

In an attempt to analyse the design process from a psychological point of view, the following goals emerged for collaboration:

- Clarification of the process of thinking, i.e. how do designers think during problem-solving? How do they arrive at a good solution?
- What intellectual and procedural barriers exist and how are they overcome?
- What mistakes in thought and action dominate?

- Can a successful process of thinking and reasoning be described and how can it be supported?
- Are the proposed procedures and methods the right ones and can improvements be made?
- How should design procedures and methods be fashioned, adapted or changed to realise a more effective creative process?

With these stirring questions in mind, we made a conscious effort to move beyond the design methodology as it existed at the time. It was also clear that in doing so, the design methodology and our own methodical attempts could come under fire. But this risk had to be taken if we were to further develop design methodology.

When Prof. Ehrlenspiel used the occasion of the Machine Elements Conference in Munich to present our intentions and mentioned the planned set of investigation instruments (video camera, thinking aloud etc.), we were met with criticism and lack of understanding. In particular, our colleagues who concentrated specifically on systematic methods could not understand how this could lead to new knowledge on physical effects, catalogue contents and systematic methods.

The aspect of a better understanding of the designer as the person who instigates action, and the idea of supporting the process of thinking and reasoning of the designer seemed to be suspect and unnecessary.

At the ICED 85 in Hamburg we set out clear markers for the first time. In a section specifically created for this purpose, entitled 'New Impulses for Systematic Engineering Design', we created the opportunity to discuss and consider the necessity and involvement of psychology. This resulted in 17 contributions at an international level by renowned scientists such as Archer, Bromme, Ehrlenspiel, Franconian, Gregory, Holliger, Hongo, Jorden, Klimek, van den Kroonenberg, Müller, Pahl, Riehm, Schregenberger, Tayfeh-Emamverdi, Wallace and Wingert [15]. It was to our great regret that Professor Dörner could not be present at that time because of other professional obligations. A workshop provided the opportunity for gaining a deeper understanding of the presentations and for drawings conclusions for future work. The results were encouraging and confirmed the route we set out.

One year later in August 1986, the projects were sanctioned by the DFG without essential cutbacks. Now, also in practice, the road was clear for collaboration. The first research team consisted of the engineers Norbert Dylla (Munich) and Gerd Fricke (Darmstadt) and the psychologists Walburga Preussler and Rüdiger von der Weth (Bamberg). Twice a year the whole working-team met to discuss the results obtained in the interval. These meetings always took place in a very congenial and friendly atmosphere. Initially we needed time to clarify the technical terms that each of us used. Soon, however, very fruitful insights emerged, which we perceived as a strong enrichment to our own ideas. The concerns initially expressed by the referees that psychologists and engineers would have nothing to say to one another were quickly dispelled, and none of today's participants would like to miss the constant exchange of thoughts and ideas.

The colleagues from Bamberg collaborated vigorously in the actual investigations that took place, often spending several weeks in Darmstadt and Munich. They influenced the course of events in a very productive collaboration. Also, Dr. Petra Badke-Schaub, Renate Eisentraut, Kerstin Härer and Peter Auer from Bamberg later joined the group.

In the subsequent period a multitude of publications [for example, 16–21], symposia and the dissertations of Dylla (1991) [22] and Fricke (1993) [23] emerged. Their work has been continued by Joachim Günther from Munich and Eckart Frankenberger from Darmstadt (dissertation 1997) [24]. Most of these works are now internationally known. A chronological table is added at the end of this paper.

2. Selected Results

As discussed in the previous section, the main issue of our interdisciplinary collaboration was the attempt to understand the design process from a psychological point of view. Knowledge about the designer's process of thinking should disclose barriers and failures and, thus, should allow methods to be adapted or changed in order to improve the design process in various ways.

The results presented in the following are necessarily reduced to a few statements and cannot cover the broad area of the different projects. Nevertheless, some of the most important results should be presented and briefly discussed.

Can a successful process of thinking and reasoning be described and how can it be supported?

First of all, in different settings as well as in different studies, it can be stated that there are various ways to arrive at a good result. There is no one unique solution – furthermore, by observing designers' strategies in great detail we can state that each designer shows a very individual pattern of designing.

Dylla [22], Fricke [23] as well as von der Weth [21] and Günther & Ehrlenspiel (in this volume) illustrate two contrasting design procedures – the strict *design phase-oriented procedure* with sequential passing of the different phases (task clarification, conceptual, rough and embodiment design) vs. the *sub-problem-oriented procedure* without the phase-oriented micro-structure – and both may produce good results.

However, it is necessary to mention that the procedure following the work on different sub-problems without any subsequent design phases was especially used by designers without methodological education. And another important difference was found in regard to the generating of solutions: non-methodological educated designers worked mainly according to a *'corrective solution generation'* [17, 19]), which means only one solution was produced and if proved unsatisfactory, this solution was adjusted in order to correct the fault as far as possible; whereas methodological educated designers created more than one equally valid solutions and kept an overview of several suitable variants – as proposed by design methodology – this procedure being called *'productive solution generating'*.

Without regard to methodological education, it was found that successful designers showed the following action strategies and procedures during the design process:

1. A detailed goal analysis, which means that the task clarification is characterised by an elaborate information search – especially where imprecise problem formulation exists.
2. A conceptual phase and a subsequent concrete realisation phase.
3. A divergent and then a convergent solution search, with adequate resolution level switching between abstract and concrete levels of reasoning.
4. A permanent evaluation of solutions.
5. Constant reflection on their own action and thinking strategies.

It is not surprising that all these elements reflect the strategies of a 'good designer' and are simultaneously constituents of design methodology, because general design methodology was worked out by observing successful as well as unsuccessful procedures in the practice of product development. On the other hand, it was the aim of design research to establish whether the proposed procedures and methods are the right ones and what improvements can be made.

We learned during many empirical research projects that the proposed plans of design methodology are useful, but it seems necessary to adjust these general strategies to the particular problem. Sometimes it may be reasonable to leave out one or two steps, whereas in another case it may be necessary to put special emphasis on a design step – in the words of Fricke [20, 23] 'successful designers produce a flexible solution-related tactic'.

What are the main individual prerequisites in a successful design process?

There are many different individual characteristics possible that could influence the design process in a positive way. Of course, first of all we would assume that *intelligence* must be a very important prerequisite of a successful designer. However, comparing designers producing good results with those producing poor results, there is no difference in regard to intelligence measured by CFT [19, 27].

Moreover, aspects of *spatial ability*, which should be an important prerequisite in design work, are not of major importance in each investigation. In one study it was found that the most successful designers came out of the group of subjects with the highest scores in the assessment of spatial ability (measured with the 'Schlauchfiguren-Test', which only recognises one aspect of spatial ability, the 2D/3D relation); in another study with various tests which yielded an assessment of different aspects of spatial ability, no relations with regard to the quality of design work were detected [28]. It can be stated that the problems of this research are connected with the emphasis on the product and the neglect of the strategies during the process. It seems vital to establish further research on the question of the function of visual representations and spatial ability during the design process, especially during different phases of the process.

In several other results, *'heuristic competence'* was proved to be a sensible personality trait, which allows individual strategies to be organised throughout the whole design process and to be combined with a more adequate search for solutions – and thus to yield a better overall result [21, 23, 27].

Experience: a guarantee of successful results?

Of course, *experience* and *knowledge* are essential individual cognitive elements which seem to assure a good result. However, this statement has to be modified. Defining experience as assimilated and applied knowledge about facts and methods adjusted by an individual's own practice means that there is no such thing as 'the experienced designer': Each designer has a particular quantity and structure of experience according to the differing participation in methods and types of knowledge. Until now, we have only very global knowledge about the different aspects and consequences of the design process.

Designers with experience often state that they possess a special intuition that tell them how to proceed [29] and how to develop an adequate solution, and they mostly deny a systematic procedure.

On the one hand, results confirm that experienced designers are fast workers – faster than non-experienced students with methodological education [27]. On the other hand, it was found by Frankenberger and Badke-Schaub (in this volume) that experience which is not flexible and open to corrections may underlie manifold problems and cause long-term problems in the design process.

However, a positive result has to be stressed: designers with methodological education arrive at better results in new situations where experience alone is not an advantage. It seems to be necessary to teach experienced designers how to use and adjust strategies of design methodology [26, 30, 31].

Moreover, by investigating the individual designer in laboratory settings as well as in practice over several years, it became obvious that an important situation in practice may alter the whole individual design process: that is the work of individuals in a *team* [30,31]. Therefore, a section of the empirical interdisciplinary research group is now occupied with design team processes in practice. According to the results of this project, it seems necessary to develop design methodology as an instrument for providing design teams with information about the availability of adequate information and to give guidance on the organisation and acceptance of responsibilities in the course of decisions. Some of the results of this project can be found in the contribution of Frankenberger and Badke-Schaub in this volume and also in [24].

During this short outline of the results of our interdisciplinary collaboration it has been shown that much empirical research has been done that clarifies delivers important aspects of the designer's thinking and acting, but that many questions still remain. It may be the task of the other contributors to highlight the necessities for further research in order to sharpen theories of design processes and thus to improve the quality of products and processes.

Chronological Table

Listed here are only the main events and completed works. The latter have many precursors in journals and public lectures.

1956	Hansen: Systematic Design [2]
1967	Bottleneck 'Engineering Design' [4]
1970	Rodenacker: Methodical Design [6]
1972–74	Pahl/Beitz: For the Practice of Design. Journal Series Design
1973	VDI – Guideline 2222, Part 1 (draft) Conception of Technical Products [5]
1973	Hubka: Theory of Technical Systems [7]
1974	Hansen: Design Science [3]
1976	Koller: Design Methods for Machine, Instrument and Apparatus Construction [8]
1977	Pahl/Beitz: Engineering Design [9]
1977	VDI – Guideline 2222 Part 1, Conception of Technical Products
1981	Schregenberger: Methodical Conscious Problem Solving [10]
1981	Dörner *et al.*: Lohhausen. About Dealing with Uncertainty and Complexity [14]
1982	Roth: Designing with Design Catalogues [11]
1984	First Meeting in Bamberg between Professors Dörner, Ehrlenspiel and Pahl – in the presence of Dr. Schregenberger
1985	VDI – Guideline 2221 (draft) Methodology for the Development and Design of Technical Systems and Products
1985	Dissertation of Rutz: Design as an Intellectual Process [13]
1985	ICED 85 in Hamburg with section 'New Impulses for Systematic Engineering Design' [15]
1985	Joint DFG Proposals for the project: Processes of Reasoning during Design
1986	Acceptance of the Proposals and Start of Cooperation on a Project
1990	Müller: Working Methods of the Technical Sciences [25]
1991	Dissertation of Dylla: Thought and Action during Design [22]
1993	Strohschneider/ v.d. Weth (eds.): 'Yes, Just Make a Plan....'. Symposium [18]
1993	Dissertation of Fricke: Design as a Flexible Problem-solving Process [23]
1994	Pahl (ed.): Psychological and Pedagogical Questions in Methodical Design. Ladenburger Diskurs [26]
1997	Dissertation of Frankenberger: Product Development by Teams: Empirical Investigations and Conclusions for Teamwork in Engineering Design [24]
1998	Dissertation of Günther – in preparation [27]

References

1. Kesselring F. 1942: Die 'starke' Konstruktion. Gedanken zu einer Gestaltungslehre. *VDI-Zeitschrift* 86, 21/22, 321–330.
2. Hansen F. 1956: *Konstruktionssystematik.* VEB-Verlag Technik, Berlin.
3. Hansen F. 1974: *Konstruktionswissenschaft – Grundlagen und Methoden.* Hanser, München.
4. VDI-Fachgruppe Konstution (ADKI) 1967: Engpaß Konstruktion, *Konstruktion* 19, 192–195.
5. VDI-Richtlinie 2222 Blatt 1 (Draft) 1973: *Konzipieren technischer Produkte.* VDI-Verlag, Düsseldorf.
6. Rodenacker W.G. 1970: *Methodisches Konstruieren.* Konstruktionsbücher Bd. 27. Springer, Berlin.
7. Hubka V. 1973: *Theorie der Maschinensysteme.* Springer, Berlin.
8. Koller R. 1976: *Konstruktionsmethode für den Maschinen-, Geräte- und Apparatebau.* Springer, Berlin.
9. Pahl G., Beitz W. 1977: *Konstruktionslehre – Handbuch für Studium und Praxis.* Springer, Berlin.
10. Schregenberger J.W. 1981: *Methodenbewußtes Problemlösen – Ein Beitrag zur Ausbildung von Konstrukteuren.* Haupt, Bern.
11. Roth K. 1982: *Konstruieren mit Konstruktionskatalogen.* Springer, Berlin.
12. Ehrlenspiel K. 1985: *Kostengünstig Konstruieren.* Springer, Berlin.
13. Rutz A. 1985: *Konstruieren als gedanklicher Prozeß.* PhD-Thesis, TU München.
14. Dörner D., Kreutzig H.W., Reither F., Stäudel T. (eds) 1983: *Lohhausen. Vom Umgang mit Unbestimmtheit und Komplexität.* Hans Huber, Bern.
15. Archer, Bromme, Ehrlenspiel etc. (17 contributions in the section: 'New Impulses for Systematic Engineering Design'). In: V. Hubka (ed.) *Theory and Practice of Engineering Design in International Comparison. Proceedings of ICED 85.* Edition Heurista, Zürich.
16. Fricke G., Pahl G. 1991: Zusammenhang zwischen personenbedingtem Vorgehen und Lösungsgüte. In: V. Hubka (ed.) *Proceedings of ICED 91.* Edition Heurista, Zürich, pp. 331–341.
17. Ehrlenspiel K., Dylla N. 1991: Untersuchung des individuellen Vorgehens beim Konstruieren. *Konstruktion* 43, 43–51.
18. Pahl G., Fricke G. 1993: Vorgehenspläne beim Methodischen Konstruieren und Vermeidung von Anwendungsfehlern. In: S. Strohschneider & R.v.d. Weth (eds) *Ja, mach nur einen Plan, Pannen und Fehlschläge – Ursachen, Beispiele, Lösungen.* Huber, Bern.
19. Ehrlenspiel K., Dylla N. 1993: Experimental Investigation of Designer's Thinking Way and Design Procedure. *Journal of Engineering Design* 4, 234–248.
20. Fricke G. 1994: Erfolgreiches individuelles Vorgehen beim Konstruieren – Ergebnisse einer empirischen Untersuchung. *Konstruktion* 46, 181–189.
21. Von der Weth R. 1994: Konstruieren – Heuristische Kompetenz, Vorerfahrung und individuelles Vorgehen. *Zeitschrift für Arbeits- und Organisationspsychologie* 38, 102–111.
22. Dylla N. 1991: *Denk- und Handlungsabläufe beim Konstruieren.* Hanser, München.
23. Fricke G. 1993: *Konstruieren als flexibler Problemlöseprozeß – Empirische Untersuchung über erfolgreiche Strategien und methodische Vorgehensweisen.* VDI-Verlag, Düsseldorf (Fortschrittberichte VDI-Reihe 1, Nr. 227).

24. Frankenberger E. 1997: *Arbeitsteilige Produktentwicklung – Empirische Untersuchung und Empfehlungen zur Gruppenarbeit in der Konstruktion.* VDI-Verlag, Düsseldorf (Fortschrittberichte VDI-Reihe 1, Nr. 291).
25. Müller J. 1990: *Arbeitsmethoden der Technikwissenschaften – Systematik, Heuristik, Kreativität.* Springer, Berlin.
26. Pahl G. (ed.) 1994: *Psychologische und pädagogische Fragen beim methodischen Konstruieren. Ergebnisse des Ladenburger Diskurses.* Verlag TÜV Rheinland, Köln.
27. Günther J. in preparation: *Individuelle Einflüsse auf den Konstruktionsprozeß.* Dissertation TU München.
28. Eisentraut R. 1994: *Das Bild im Kopf. Verarbeitung räumlicher Informationen beim Konstruieren.* unpublished diploma-thesis. Otto-Friedrich-Universität, Lehrstuhl Psychologie II, Bamberg.
29. Hubka V., Eder W.E. 1992: *Einführung in die Konstruktionswissenschaft.* Springer, Berlin.
30. Pahl G., Grote K.-H. 1996: Interdisciplinary Design – Knowledge and Ability Needed. *Interdisciplinary Science Reviews* 21, 4.
31. Pahl G. 1997: Wissen und Können in einem interdisziplinären Konstruktionsprozeß. In: G. Frhr. zu Putlitz & D. Schade (eds) *Wechselbeziehungen Mensch, Umwelt, Technik.* Schäffer-Poeschel Verlag, Stuttgart.

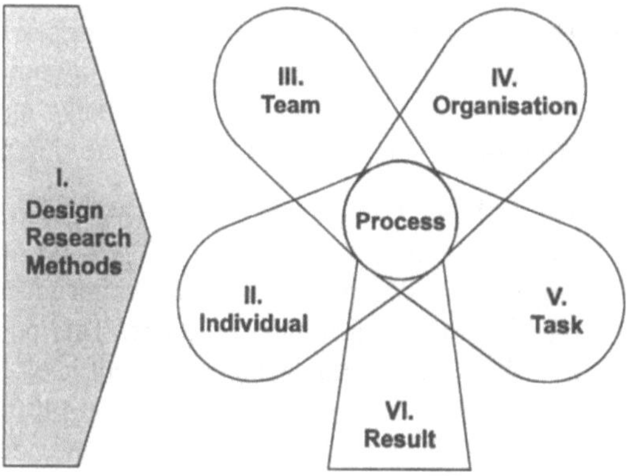

Section I:

Design Research Methods

Engineering design research embraces a broad area of approaches and methods in order extend knowledge about the influences on the processes and results in design.

Section I deals with the challenge of how to establish design research and design research methods. Five contributions address the following questions:

How to gain knowledge about complex problem-solving processes in the fields of engineering design and product development?

What are the different perspectives and approaches of the engineering design process?

What problems occur when comparing results in design research?

How should we proceed in design research and in developing design methods?

The contributions follow from a general discussion on how to proceed in investigating design processes to illustrate different approaches, and the question of future visions in design research.

- The first contribution by *Dörner* addresses in a general manner the cognitive processes involved in designing. The other essential aspect of the paper discusses research strategies, especially in complex environments, and the question of *why* and *how* to investigate single cases.

- The following two papers by Hubka & Eder and Roozenburg & Doorst present two different approaches that address the question of *what* to investigate in engineering design. *Hubka & Eder* introduce a traditional approach based on the structure of technical products. This perspective basically takes the view that the nature of the technical system itself determines the process of creating this system ('from the technical system to the method'). The authors support their plea for a theoretically founded methodology by an example of a design process.

- *Roozenburg & Doorst* take another view of design activities by means of a critical analysis of Schön's characterisation of designing. The focus is on the basic moves of the designer in the process ('reflection in action'), which thus creates a better understanding of the design process ('from human activity to the method').

- *Blessing, Chakrabarti & Wallace* discuss and evaluate the papers of this volume. They illustrate the difficulties of comparing results because of the different approaches used, and make a plea for a common research methodology in engineering design.

- Starting from a common research methodology with the validation of methods (comparable to Blessing *et al.), Schregenberger* assesses the development of design research and proposes further collaboration with other disciplines in order to obtain better performing design methodologies in the future.

Thought and Design – Research Strategies, Single-case Approach and Methods of Validation

Dietrich Dörner

Abstract

Designing a piece of machinery requires thought, at least in most case. For although sometimes one meets machines or technical devices where not very much thought seems to have been invested, it is normally necessary to think when facing the problem of designing a new machine.

In this paper I shall try to identify the different forms of thinking which can be observed in designing (I shall not consider prescriptions about how designing should proceed, see [1]) and I shall also try to answer the question of how a theory about thought in designing may be constructed.

1. From a Cloud to a Machine

In the beginning there is a cloud! When designing begins there exists a more or less cloudy idea about how the machine should look like and how it should work. This cloudy idea 'crystallises' in the course of time and is transformed into a clear and complete image of the machine in the form of an exact drawing. Figure 1 shows some of the stages of this transformation in the course of designing of an adjustable slide projector support.

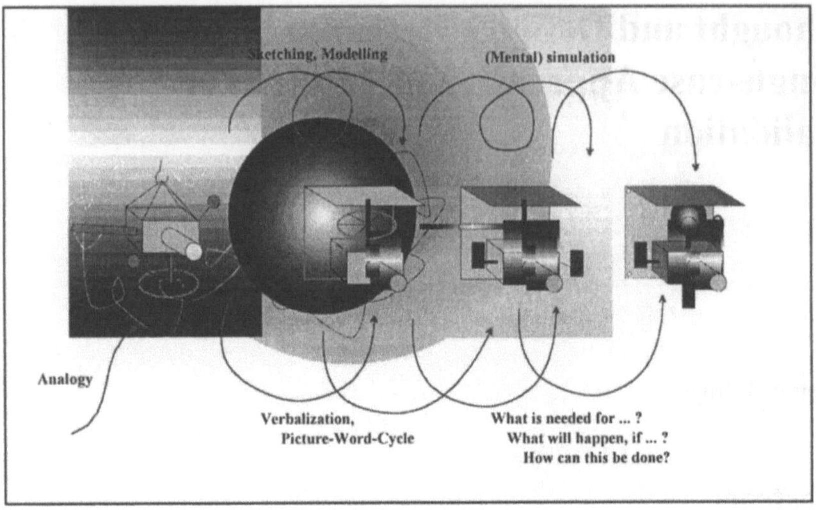

Figure 1: The designing of an adjustable slide projector support.

Through what stages does this process of crystallisation of the cloud run?

First we need the 'cloud' itself. Where does it come from? One source of ideas about a new piece of machinery is the use of what one already knows about machines. This knowledge can be a source of analogies.

'Do I already know machines that serve similar functions?' If this is the case, then these machines are used as models for the new machine. This method of producing ideas about new machines in terms of similar machines is helpful — but it is dangerous! Experience generally encourages conservatism and may sometimes fence the design process in the (more or less) narrow bounds of the past, thus inhibiting the formation of new ideas. Experience can be a great help, but it can also be a barrier to new developments.

Now we have a more or less vague idea about the machine. In general the initial idea is rough, incomplete and abstract – the details are missing and the idea is more a cloud than a definite outline. Elaboration is necessary, in two respects. The idea needs to become more concrete and more complete. Elaboration of the initial design means setting the idea in a more concrete form and completing it. And to make the idea more concrete is a prerequisite for the completion of the idea, for only in a concrete picture of an idea it becomes apparent what is missing. — How is a rough idea developed into a more concrete form and how is its completion accomplished?

2. Sketches and Models

How can a cloudy idea be brought in a more concrete form? Very often by means of sketches, using a drawing board or CAD. But other methods are used too. For instance some designers make *models* of a machine, using paper, wood, plasticine, wire or even staples. In all cases, these two- or three-dimensional 'pictures' of a machine serve the purpose of making the idea of the machine more concrete. Sketches or models, however, do not just serve the purpose of saliency enhancement. According to observations they fulfil three functions:

1. They clarify the characteristics of the machine. The initial 'cloud' partially crystallises during the sketching or modelling process, and this makes the idea amenable to critical consideration. Now becomes apparent what will work and what will not, what will fit together and what will not. 'Is this case really strong enough as a bearing for this axis? Or is it too weak? And wouldn't it be better to weld this case rather then to connect it by screws?'
2. Sketches and models may additionally serve the purpose of forming a *logbook* of the whole design process, so that it is easier to appreciate the earlier stages of the development of one's ideas. Often it is necessary to abandon a specific line of thought and to start anew, although it might be reasonable to reuse abandoned ideas. They might prove to be useful in the new approach. Therefore it is reasonable to store the sketches and models as a stockpile of raw material for new approaches.
3. The sequence of sketches and models is not only a stockpile of ideas, but a good basis to reveal the mechanics of one's own thinking. This sequence shows how one has proceeded and therefore shows up the strengths and weaknesses of one's strategies, and reveals fixations and conservative thinking. It is therefore a good basis for reforming one's own strategies of thought in self-reflective thinking.

3. The Picture–Word Cycle

Another method to make ideas clear might be to put them into words. If an idea is unclear, explain it to somebody! It may be that your partner won't understand anything. But *you* will! You will appreciate what is clear to you and what is not, which of your ideas can be easily put into words and which cannot. To force oneself to express ideas means a separation of the unclear ideas from clear ones. And it is very helpful to know which of your ideas need further elaboration and which do not. Verbalisation helps in finding the weak parts of your ideas.

But this is not the only way in which verbalisation helps. Even if everything is quite clear, 'rephrasing' can alter your perception of a problem. A very important system for new thoughts is the 'picture–word cycle'. You can describe a machine in this way and in many other ways. Try for instance to describe your machine in a very abstract manner. For example instead of 'soldered joint', 'screw' or 'weld' use the word 'connection'. With some certainty this will open your eyes to alternative realisations of parts of your machine. The following sentences have nearly the same meaning but they evoke quite different images (see Figure 2):

'The axis should be attached to the case!'

'Axis and case should fit each other!'

Or look at these sentences:

'The projector should be adjustable!'

'It should be possible to project to different spots in the room!'

Superficially the sentences in each of these two pairs mean the same thing. But it is obvious that they do not have the same meaning in terms of the images they evoke! (To project to different spots in a room, for instance, is possible without adjusting the position of the projector! Use glass fibre? Or an appropriate lens system?)

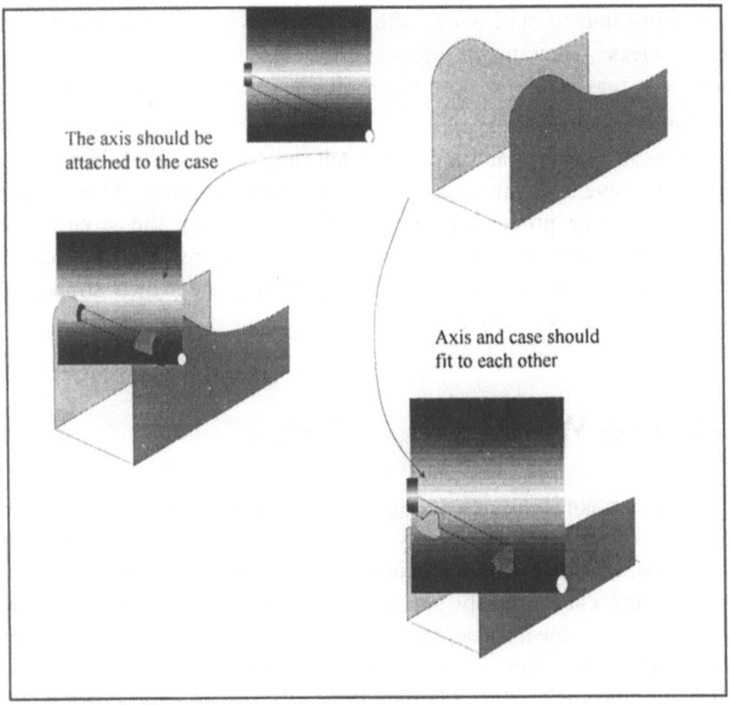

Figure 2: The effects of verbalisation.

Rephrasing clarifies your ideas, because putting images into words does not just mean this literally. Words evoke new and different images and therefore alter the original image. If you tell your partner 'and then this has to be fixed to this ...' the word 'fix' will evoke a number of other methods of fixation than you have thought about before. You will have a certain image in your mind. By putting it into words you will almost certainly produce a lot of alternative pictures. The 'benefits of chatting' and generally the benefits of teamwork are based partially on the inherent necessity to communicate one's ideas and thus unconsciously to launch the 'picture–word cycle'.

However it is not always beneficial to put ideas into words. Germinal processes exist in human thought, in which casting the ideas into verbal form is premature because it would destroy the dynamics of thought. These germinal phases, which can be regarded as rapid and partially unconscious recombinations of images, may result in 'sudden insights' and shouldn't be disturbed by attempts to verbalise.

4. Elaboration

To have a clear idea about a machine is not at all sufficient. In terms of the psychology of problem-solving, a designing problem is dialectical. Normally the *main* functions that the machinery should serve will be known, otherwise nobody would start a design process. But it is not known how these functions could be realised and this has to be found out during the process of designing. The phase of elaboration will often assume (and should assume) the form of a *dialectical process*. The designing engineer will look for 'contradictions' and try to remove them. This 'looking for contradictions' is advanced by posing a certain type of question.

'To attach this axis to the machine we need bearings. Bearings for this kind of an axis will have a weight of x. Is the case to which the bearings will be attached strong enough to carry them? No, it is not!' A contradiction is detected which can now be removed by changing the case, the axis, the bearings or all three of them.

'Will this axis exhibit strong imbalances when rotating? Can the bearings stand this?' — sometimes a *mental simulation* of the running machine is the appropriate method of identifying deficiencies and contradictions within the structure of the machine. Elaboration takes place by asking questions such as

- What is needed for proper functioning?
- What are the conditions for the proper attachment of this part?
- What will happen when ... /if ... ?
- What side effects will result from this change?
- What will happen to this structure in the future?

Even though the need to ask (oneself) such questions sounds trivial, a 'natural' demand, many people seek to avoid such questions. Sometimes it seems that designers are so happy to have arrived at a solution that seems to work, that they do not want to face such questions. They side-step them because they don't want to be confronted with the uneasy experience that they have not yet arrived at an appropriate solution to the problem. (This kind of thinking we called 'ballistic', because a decision is 'fired' without much monitoring and control in order to preserve the illusion of a final solution. Such tendencies may be less frequent with engineers compared with, say, politicians; however one could find them even in the engineering community. In computer programming – a kind of designing task, too – this kind of faulty thinking is not rare at all.)

Designing a machine means the assembly, combination and recombination of individual parts. Designing is not a progression of subsequent, different stages, but a process of 'amplification' of one structure. More and more parts are added to a basic structure; the whole structure grows and becomes larger and larger. But the whole machine is still one system. This means that the addition of one single element to one part of the machine could affect the function of the whole system. Thus it is not only necessary to analyse the functions of single parts, but to analyse the machine as a coherent system. We can observe weaknesses in this respect when we see teams designing big machines. Normally the work is divided in such a way that each team specialises in one part of the machine, and here the inherent danger is to lose the overview.

We know from computer programming that one of the most prominent and frequent mistakes consists in changing one part of the program without considering the side-effects caused by this change in other units of the program. We have no data about the frequency of errors of this type in the field of designing work, but from case studies we know that this kind of mistake seems to be here very common.

Additionally, questions about the 'environment' of the machine have to be raised. A plan for a wonderful machine is simply senseless if certain parts cannot be manufactured because the appropriate facilities are missing. — How will the machine perform if it is put in an oven? And how will this brewery-machine work when you put it into a fridge? (A Bamberg brewery-machine factory faced this problem *after* it had delivered a complete brewery to a Russian town in Siberia.)

5. Towards a Theory of Thought and Design

We have now described some forms of thinking and reasoning that can be observed in the process of designing. However it is clear that these ingredients of the process do not constitute a theory of thought in design. *This* engineer designs in this way, uses no sketches at all, but likes to chat a lot with his colleagues. Another one doesn't like teamwork at all, but likes to bury himself in his thoughts and reappears after days with brilliant ideas. Another one likes to produce sketches, but hates CAD. Is there ***one*** (optimal) process of designing? Absolutely not! How designing work is done, how thinking proceeds from a cloudy idea to the clear picture of a machine is dependent on a lot of personal and environmental conditions and on the characteristics of the task.

One person might possess a kind of inner projection screen of high quality and hence is able to produce vivid and clear images of a machine or its parts, and so doesn't need a larger quantity of sketches. But as he doesn't produce many sketches he has no accurate record of his own thinking and hence gets into difficulties in explaining clearly how his thinking proceeds. As he has no accurate logbook of his thinking, no accurate record is available for review, or to rethink his ideas or change his strategies.

Another person might be very sensitive to stress, losing all calmness and reasoning abilities when put under time pressure, but an excellent problem-solver when not under pressure. A third person might need stress to get the right 'kick' to work systematically and in a coherent manner. When there is no time pressure such a person could get lost in side routes and irrelevant activities.

Many of the strategies just mentioned (the list is far from being complete) may diminish efficiency, others increase. And some may be equivalent although different. It is not necessarily good to use CAD instead of a drawing board. For some users these two devices are not alternatives, but complement each other.

We need much more knowledge about how design work happens 'in the real world' before we are able to give advice on how to train thinking and reasoning abilities according to the strengths and weaknesses of individuals.

Thinking while designing is a heterogeneous process, composed of very different elements. We have to collect material about thinking in design by studying single cases very carefully. We should look at excellent (and less excellent) engineers when they are designing machinery of the most diverse kind. We should observe their behaviour and we should try to extract from the different patterns of behaviour a general theory about the process of designing tasks.

But how can this be achieved? How is it possible to arrive at a general theory by studying a few individuals in different situations? We assume that the following research strategy will work:

1. Look at the initial stages of the thought process and try to get a very precise image about the kind and sequence of the different processes of thinking, reasoning, sketching, communication and chatting, looking at catalogues, handbooks, etc.
2. Try to attribute these processes and the sequence of its elements to the conditions of the person, the environment, the task. Try to understand the 'machinery of thought' of the individual; why doesn't this person use sketches, why does that person never look at handbooks, why does this person prefer to work alone?
3. Try to *predict* the behaviour of the person under investigation in the next stages of problem-solving. What kind of product will eventually result? How long will the process continue? What will be the elements of the solving process? Which questions will the person ask himself in the elaboration phase of the process? Which methods will the person use to put his ideas into concrete forms?

What could be predicted? This is a crucial question because to predict the behaviour of a person when designing a piece of machinery means to test the validity of a theory. It is very simple to predict trivial aspects of the behaviour of a subject which are obvious. Certainly this person will wear shoes when he comes to work tomorrow. Certainly he will open the door to his room and look for messages. Certainly he will switch on the computer. It is silly to predict such forms of behaviour. It is not a test of the theory, it is no test at all.

But if you predict that the subject under observation will abandon his approaches completely and start anew, or that a person who likes to work alone will tomorrow try to find opportunities to chat with his colleagues, or that a person who has never used CAD tomorrow will ask someone to explain him the way it could be used, then your predictions are risky because they are far from being natural or obvious. Is it possible to predict such forms of behaviour, such ruptures in the stream of behaviour? If this were not the case it would mean that a psychological theory of the designing process would not be possible. But I believe that such predictions are possible. For instance, if it could be observed that the feeling of competence of an individual because of unsuccessful attempts to solve a problem diminishes continually, then a 'rupture' in behaviour becomes (according to the theory of the role of feeling of competence in the organisation of behaviour) probable. Or if the capability of a person for imaging is somewhat restricted, then it will be predictable that this person will use the drawing board more often than other designers. The requirements to make such predictions are

- a general theory of thinking, remembering and imaging in the course of the design process,
- a general theory about the role of such parameters as the 'feeling of competence' and other parameters which are of importance for the regulation of emotions,
- observation of a person's behaviour during a design process to assess the individual parameters.

If the prediction of the aspects of an individual's behaviour on the basis of a general theory and on the basis of the assessment of the individual's behaviour proves to be correct, this can be considered as a validation of the theory. If however the predictions prove to be wrong, the theory or assumptions about the parameters should be changed.

Proceeding in this way means using single case studies to construct *valid* theories. (As an example for such a strategy of theory building, see the work of Badke-Schaub and Frankenberger [2] and Frankenberger [3].) Normally the role of single case studies is viewed differently in psychology; case studies serve in the initial phase of orientation as a method of obtaining ideas about the phenomena in the domain under investigation. But single case studies in the above mentioned form include theory building *and* theory testing. We formulate theories for individuals which are validated as we do not only identify the behaviour of these individuals, but use what we observe to construct theories about the reasons why an individual behaves in one way and not in another. And we test this theory by prediction.

Using single case studies in this form we first arrive at theories for the individuals. But a closer look to these theories will reveal what is specific for one individual and what is not. A small number of such single case studies will hence lead rather quickly to general theories.

References

1. Pahl G., Beitz W. 1997: *Konstruktionslehre*, 4th edn. Springer, Berlin.
2. Badke-Schaub P., Frankenberger E. 1997: Modelling Design Processes of Groups in Industry: An Empirical Investigation of Cooperative Design Work. In: D. Harris (ed.), *Engineering Psychology and Cognitive Ergonomics. Job Design and Product Design*. Ashgate, Aldershot.
3. Frankenberger E. 1997: *Arbeitsteilige Produktentwicklung. Empirische Untersuchung und Empfehlungen zur Gruppenarbeit in der Konstruktion*. VDI Verlag, Düsseldorf (Fortschr.-Ber. VDI Reihe 1 Nr. 291).

Theoretical Approach in Design Methodology

Vladimir Hubka & W. Ernst Eder

Abstract

The goal of rationalising engineering design evoked tasks of defining design methodology. Design methodologies were formulated by 'good' engineering designers, but lacked a theoretical foundation. Experience of teaching such approaches has been good, but not fully satisfactory. Our approach uses theoretical insights for scientifically generating methods based on relationships between theory, method and object as formulated by Klaus. The basic knowledge was collected and extended to formulate Design Science. A brief overview of its structure is given. The relevant design methodology as the strategy for designing uses various models derived from the theory. An example of an engineering design problem is used to illustrate the application of these models and its tactical tools.

1. Introduction

In the 1950s it was recognised that engineering design was deficient; both the products and the design processes needed to be improved and rationalised. Approaches to design methodology were formulated by 'good' engineering designers, but lacked a theoretical foundation. Experience of teaching such approaches has been good, but not fully satisfactory.

The goal for setting up a coherent theory was and is to help engineering designers in practice, by rationalising engineering design. Defining these goals evoked consequent tasks, namely of defining the design methodology. Whilst acknowledging the needs, possibilities and advantages of intuitive, idiosyncratic and opportunistic procedures, only a planned and conscious (but iterative and recursive) designing procedure can ensure that an optimal solution to a problem can be approached in an effective process.

But how does a 'good' designer really work? And how can a 'good' designer be supported to improve his or her capability? These questions can lead to very different answers.

Abstracting by competent people from self-observation, as was done for the earlier methodologies, is one essential approach, but it is not the most reliable way of formulating good design processes. An alternative is based on observation by independent external observers, who analyse what they see and record. However, they can miss the essence of the design task, because human thought is faster than the capability of demonstrating what was done, for example by talking aloud. The hope of deriving a theory of designing from external observations is likely to take time, if it can ever be achieved.

A different approach has utilised both experience and theoretical insights that are necessary for appropriately, holistically and scientifically generating methods. This paper reports on this theoretical approach, and uses a recent engineering design problem to illustrate the role played by models and tactical tools of this approach.

2. Science and Design Science

Science, as a codification of knowledge, can never be *true* in any formal sense. As a means of exploring by research to obtain knowledge, it cannot provide *proofs*. Its investigations and results (as in mathematics) depend on unprovable axioms to justify the existence of that science. Science and its theories should provide a reasonable and rational structure for an area of knowledge, and therefore form that knowledge into a system. For engineering design, the possibilities of research can be divided into two classes:

- with respect to its *aims*: (1) to formulate Design Science as holistic and complete way as possible; or (2) to investigate individual regions of interest;
- with respect to its *methods* of research investigations.

According to Cross [1], three forms of design research are useful (our comments added to Cross' keywords):

- research *into* design, by various kinds of independent observation, descriptive studies such as narratives, anecdotes and stories, e.g. protocols, pragmatic or theoretical developments, and their testing;
- research *for* design, to create tools (especially computer-resident ones), design methods and forms of modelling, preferably based on theory;
- research *through* design, e.g. abstraction from self-observation and other observations during designing, hypothesising and testing, and setting out (formal descriptive/declarative) theories.

For validity, this research should be purposeful, inquisitive, informed, methodical and communicable. The following reflect the research methods [2]:

(a) self-observation whilst designing and directing a design team, or participating in third-party observation [3], and abstracting from this experience; active industrial supervision and tutoring can be useful;
(b) third-party observing, protocol analysis, etc. are apparently scientific, independent and relatively objective, and can detect the observable;
(c) formalising, theorising, proposing models, etc. is capable of logical coherence, and should result in holistic or detail generalisations.

These are almost always used in combination with each other. Research and knowledge must also be interpreted for use in engineering design.

If the theory is represented in schemes, figurative models or diagrams, with which their contents can be visualised and memorised, the effect of *topical learning* (Aristotle) can prove very effective for engineers.

Engineering design, a sub-section of integrated product development, is complex. It is an essential activity within the socio-technical system [4], and is the basis (via manufacturing) for a large part of the gross national product (GNP) of any developed country. It is thus useful to investigate the mixture of factors that influence the product, the designing, and their contents.

The basic knowledge of this kind was gathered for technical artefacts in the theories of technical systems [5, 6] and of design processes [7]. This knowledge was summarised for engineering design [8, 9] with examples [10]. Teaching [11] and creativity [12] were also considered. The knowledge was extended into a formulation of Design Science [13]. A brief overview of the structure of Design Science and its constituents is part of this paper.

As G. Klaus [14] stated in cybernetics (see also [15]), relationships exist between the object under consideration (its nature as a product or process), the theory and the method. The theory should describe and provide a foundation for both the behaviour of the object (with adequate and sufficient precision), and the utilised methods (for using and/or operating the object, and for designing the object). The method should be adapted to the object. These three phenomena are of equivalent status. An interplay between object (phenomenon), theory and method, refined and examined on each other, characterises normal human and social progress. Quoting Klaus:

'Both method and theory emerge from the phenomenon of the object.'

If the theory of an object-region is mature, then the method is founded on the theory. The theory describes reality, the method prescribes how the scientific and practical activity and behaviour of the humans should occur.

Where no comprehensive theory is available, methods can still be proposed, even where the structure or behaviour of the objects is not completely known (cybernetic interpretation). The method can be characterised as an input–output relationship ('black-box' principle, first formulated by Ashby in 1956). We know that corresponding results will be generated when we act on a system in a certain fashion. The theory will then – and often after a lengthy delay – give an explanation of why this is so.

Design Science is built on the systems (or systemic [16]) view of cybernetics. Regular and law-like relationships exist between the functions and structures of a system. The structure determines the function (and/or way of functioning, mode of action). The function does not uniquely determine the structure, a function can be realised by several different structures.

For the designers' work we need accurate knowledge (of objects and processes, but also of design processes) that corresponds to reality. Practically all these considerations are realised in the products and accordingly demonstrated in reality. This happens gradually, stepwise, so that the theory soon contains almost no hypothetical elements and will exist only as a logical consequence of the facts – according to Bavink [17], the need is for an *executing theory*, not *only* an *explaining theory*.

For technology, and in Design Science, two classes of system are used:

1. a *process system* as a set, and unity of phases, partial processes and operations, which represent a process, course of action or development;
2. a *real object*, thing, real system, artefact, formation or technical system as a set, and unity (relationships) of the elements (e.g. parts, components, organs, functions, processes) from which this object is assembled.

Both of these terms are relative: a partial system is itself a system, and a system is a partial system of a larger formation. Our starting point is the general model of the transformation system, Figure 1 [6, 9, 13]. From this we can derive several more specific models for the main constituent elements of a general system (and product) and for engineering design.

3. Process Systems – Process Theory

The word *process* is applied to a procedure or operation, including its method, technique or practice. Technology (in the wider sense) sees countless applications: manufacturing process, assembly process, etc. Usually these processes are only designated by a participle (used as a noun or an adjective), e.g. hardening, measuring, turning, preparing. If processes are analysed (Figure 1, upper diagram), it can be seen that various *objects* (also animal and human), material, energy and/or information are changed. If one defines the change of state of these four classes as a transformation, then the *transformation process* can serve to formulate a process theory. A method is then the instructions (prescription) for implementing a process.

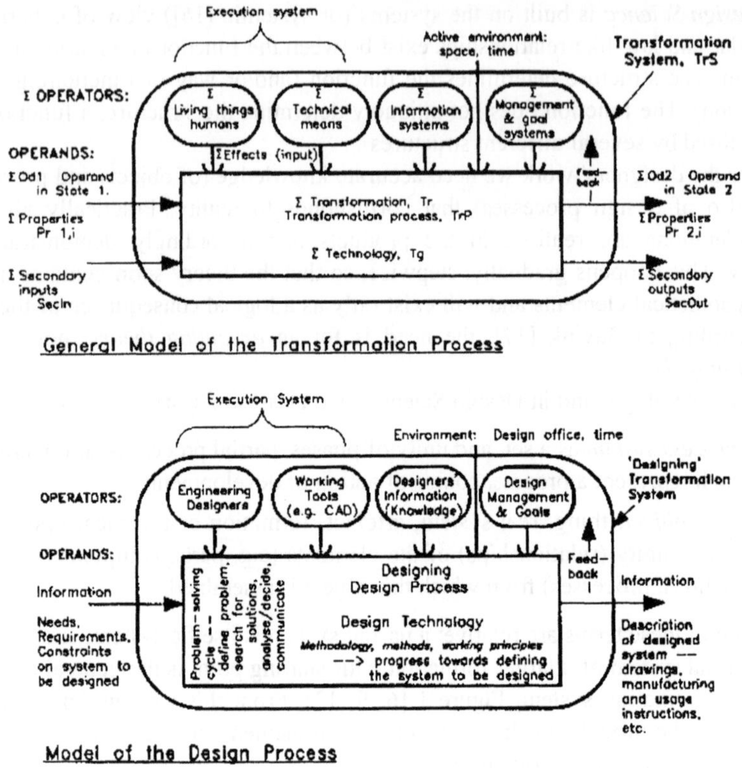

Figure 1: General model of transformation system and design process.

Figure 1 (upper diagram) describes the transformation as follows: an operand in state 1 is transformed into state 2 by applying a suitable *technology*, by means of *effects* delivered by the operators (humans, technical systems, information and knowledge, management, environment).

In this transformation, the operand passes through several intermediate states, in space and time. Apart from the operand, each transformation process model shows secondary inputs and secondary outputs. The model also shows the fundamental relationships between the elements. In designing such a system, the state 2 of the operand is usually the final desired outcome (e.g. by the customer) of the transformation.

Investigating the elements of the model yields their description and relationships, and a classification (taxonomy) of the kinds of process elements. These results, with the principles of their regularity and their explanation, form the process theory, which is valid for all transformation processes.

The large number of existing processes can be divided and categorised from several points of view – features, characteristics, etc. One of the tasks of the theory is to accomplish a suitable classification, by investigating the individual classes in detail and thereby bring them closer to the user.

An important class is formed by those processes in which technical, artificial means are employed. We usually place these *technical processes* at the centre of our interest. The existence of *natural processes* (those that proceed without the assistance of technical means) should not be forgotten.

The transformation process can be used as a basis for designing:

a) As soon as designers have understood the transformation process (as a black box, as an object and its theory), they can apply the appropriate method. This consists of: determining the essential tasks and operations of the transformation process, choosing a favourable technology, establishing all necessary output effects of the operators (input effects needed to transform the operands) and distributing these optimally among the humans and the technical system in the existing situation. This method is preferably used in the phase *clarification of the design task*, although it already belongs to the solution process (*conceptualising*) phases. It should be obvious that this procedure cannot be completed in a linear-sequential fashion, since iterative and recursive working is essential, and feedback from later stages to earlier ones will progressively drive the solution proposals towards an optimal state. If any opportunistic and intuitive step is taken outside this procedure, a check should at least be made to ensure that the results do not violate the procedural considerations and outcomes.

b) The origination and life of the technical system (its life cycle) can be clearly presented with the help of several models of the *transformation system*, see Figure 2 [6, 9, 13]. A matrix formed from the life phases and operators of the processes directly involving the technical system presents the search field for properties (characteristics) of the technical system to be designed, and categorises the areas of *design for X* – DFX [18].

c) The general transformation process model serves as starting point for the construction of the models for specific areas. This includes the design process and its model, Figure 1, lower diagram [9, 13], as discussed in section 5 below. Information appears as the operand of the design process, and this is transformed from the state of established needs, requirements and constraints to a description of the component structure of the technical system. The procedure of the transformation (technology) of the design process can be very different. The design strategy is based on the knowledge of the finality and causality connection. The model serves as the structural elements of the theory of design [7] along with the appropriate concretisation, e.g. human designer, etc.

Several kinds of process (operations) always appear as part of each transformation process: the working transformation process itself; auxiliary processes; propelling (driving, energy delivering) processes; regulating and controlling processes; connecting and supporting processes; and preparing, implementing/executing and finishing (checking, verifying, record-keeping) processes. This can constitute a check-list for a process structure.

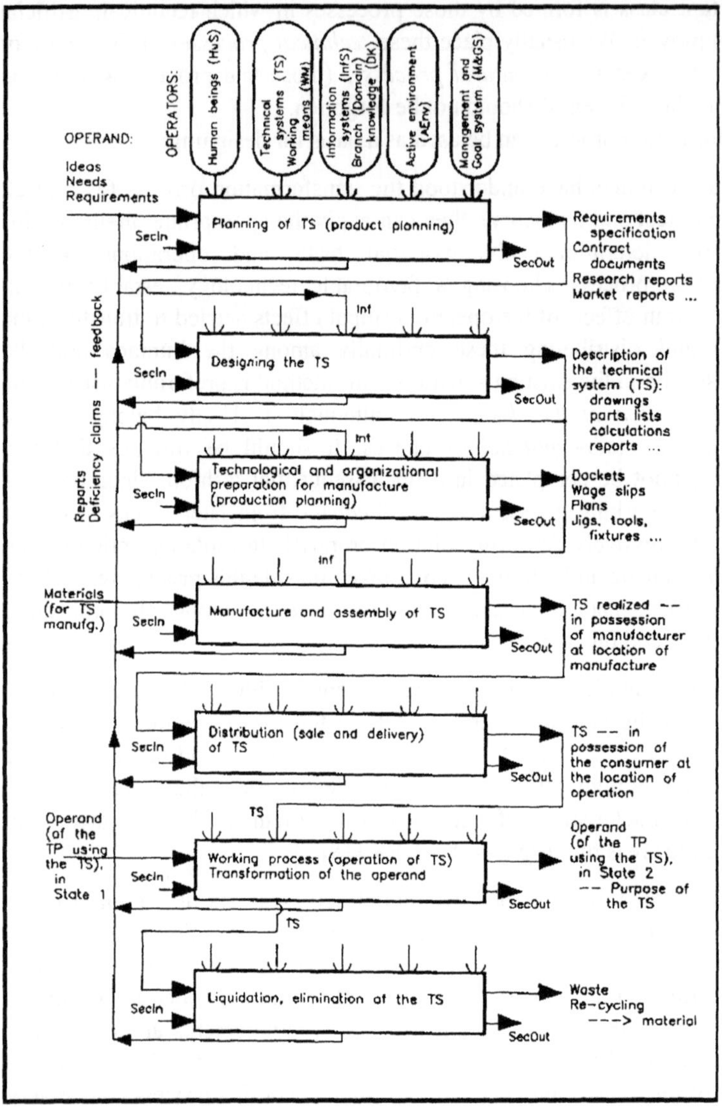

Figure 2: General model of the life cycle of technical systems.

4. Object Systems – TS Object Theory

In the transformation process, the role of the technical system (technical means – object system) is that it should achieve the desired transformation (of the operand) through certain effects on the operand (in addition to the effects delivered by human operators). The output, the effects and the task of the technical system are therefore given. The model of the object system, the technical system, and its various structures (using various elements) are shown in Figure 3 [6, 9, 13].

Cybernetics, and the generalisation of human experiences with the technical systems, have resulted in the hypothesis giving the inputs to technical systems as *material - M, energy - E and signal/information - S* (Figure 3, level I). The (internal) task of the technical system is now the transformation of its inputs (M, E, S) into the sum of the needed output effects. For this purpose several internal transformations (capabilities, functions) are necessary (Figure 3, level II). These internal transformations demand certain effects, which are executed via suitable means within the technical system. Further capabilities are demanded from the elements, the means.

This process, a chain (scheme) of *effects → means* (or *goals → means*), guides and steers (not determines) the work of the designer. The task is to search for (appropriate) means, in various modelling steps and levels of abstraction, e.g. function, organ and component structures. The final goal is to find the elementary design properties: the component structure and, for each component the form, dimensions, raw materials, production process, etc. The main possibilities for optimisation are contained here. An expert in logic (philosophy) sees this as the role of *finality* (synthesis, moving towards the intended goal), which stands in contrast to *causality* (analysis).

The number of steps (effects → means) depends on the complexity of the task, the means being selected from the repertoire of those systems elements (organs, components) whose capabilities are known. The main group of these means in engineering form the *machine elements*. This group contains not only the elementary level of complexity (i.e. machine elements in the narrower sense, e.g. shafts, screws, gear wheels, etc.), but also more complex technical systems such as gear boxes, couplings and fittings, which themselves are composed of the elementary elements. For a technical system of high complexity, e.g. machine tools, the necessary structure must be established recursively, by subdividing the system in several steps.

After the TS internal transformations are established (Figure 3, level II), the technology (the mode of internal action) can be established (selected from among the alternatives), and therefore the capabilities of the technical system can be shown. We speak here about TS functions, and if these functions and their relationships are presented, the function structure (Figure 3, level III) emerges. The search for suitable means (e.g. by the method of morphology) leads to establishing the organs that will execute the functions (but usually this process evokes further functions). The terminology organ – organism has been chosen in analogy with biological systems, and designates the abstract classes of means of the technical system, e.g. connections, guidances, drives (Figure 3, level IV). The steps effects/functions → organs are repeated down to the elementary levels of complexity. The elementary organ structure serves as the starting point for development of the component structure (layouts and details), i.e. for the choice of the components (Figure 3, level V). The assignment of components to organs is not unique, some components can participate in realising several organs.

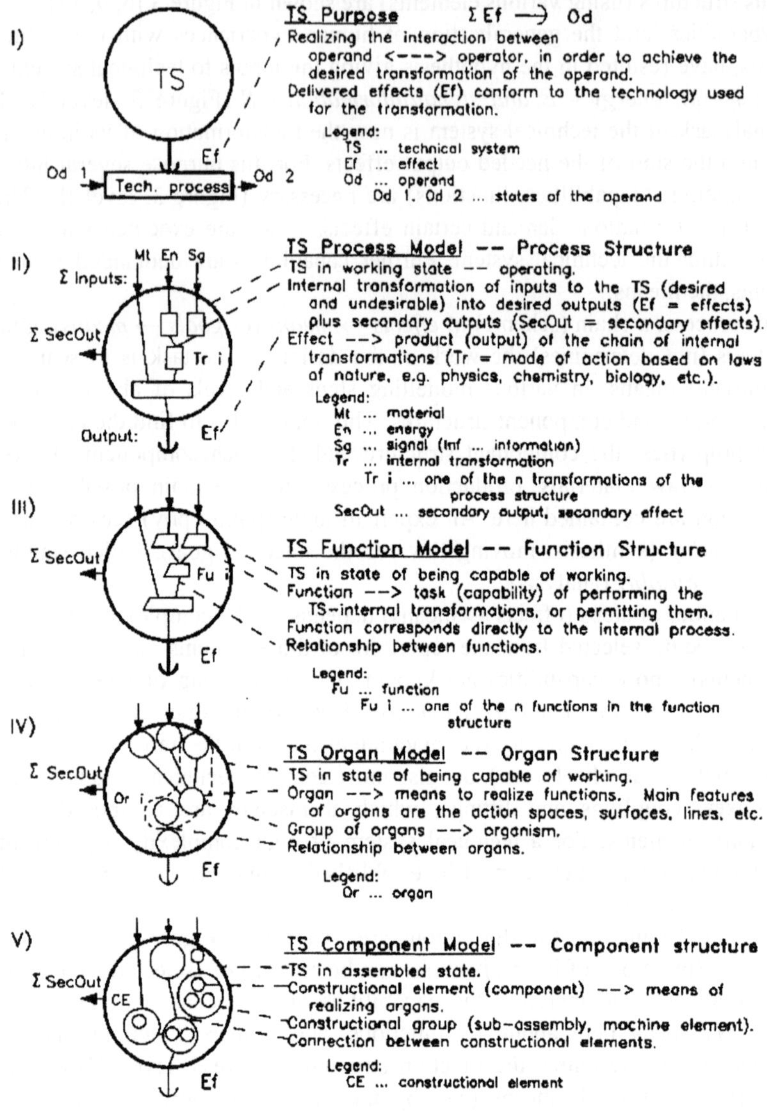

Figure 3: General models of technical systems.

Establishing these (function, organ and component) structures has an important influence on our understanding of methodical procedures. There is a fundamental difference between whether the goal for a design phase is determined, i.e. *establish the function, organ or component structure*, or whether we only speak of conceptualising, laying out (embodying) or detailing.

5. Characterising Designing

Designing an engineering system (following Figure 1, lower diagram) is a process that should start with information delivered by customers (or is assumed to exist) about the needs, requirements and constraints they place on the transformation and/or technical system (process and/or product) to be designed – as input *operand* in state 1. The output (operand in state 2) should usually be a full description of the designed system, such that it can be manufactured and/or implemented and used. This should also (e.g. under concurrent engineering) include a description of the manufacturing system (process and devices) needed for the product.

The *process of designing* involves repeated application of a cycle of problem-solving steps to achieve progress in transforming the input into the output. This transformation is stepwise, iterative and recursive, each step achieves a measure of progress on a portion of the problem and its proposed solution. The technology of designing consists of applying one or more of the principles, methods and operating tools (including judgement, intuition, reflection [19, 20], opportunism and creativity [12]), and information (recorded and tacit knowledge, heuristic data and guidelines) in each part of a step. Performing the design process is the task of its *operators*:

- the *human*, engineering designers who sometimes act alone, but more frequently as members of a design team. Each designer must have some knowledge about the systems being designed, and about the processes and methods of designing. If these two aspects of knowledge are largely internalised (by a human), we call it *expertise* applicable to that range of products. Progress in designing (solving a design problem) involves establishing *shared meaning* within the team about the problem and the possibilities of solving it. If team-building is too successful, a situation of *group-think* may arise, when faults in the shared meaning are no longer critically detected [21]. Whether designing is done by an individual or by a team, some *systematisation* (codifying the knowledge by abstracting, categorising, using methods, etc.) is usually beneficial.
- the *technical system*, set of tools, equipment and facilities that designers can use for performing steps in the design process. This includes pencils, drafting machines, CAD workstations (and programs and peripherals), desks, etc. that are available to designers during their work.
- the *information system*, which consists of data, scientific and heuristic models, tables, catalogues, lists, textbooks, etc. from which relevant items may be selected (and referenced) to provide added information for designing. Aspects of this information system are:
  - *knowledge related to objects*, dealing with physical phenomena, their mathematical models, etc., usually explicit and recorded. Especially important are those items of knowledge that have been scientifically investigated and abstracted – the engineering sciences.

– *knowledge related to procedures*, design process knowledge, including theories and methods of designing and processing the object knowledge [7, 9], see the previous section of this paper.

In addition, the types of knowledge that exist in Design Science can be classified as:

– *descriptive* (declarative) in the form of a coherent theory – as distinct from the more usual usage of the word *descriptive* to imply a narrative, an anecdote, an essay, a story or a metaphor [24];
– *prescriptive* as advice, directives, instructions, heuristics or guidelines for practice and application (e.g. of a method);

It is obvious that designers need to coordinate these two aspects of knowledge and information, and adapt them to the specific design situation (company, product group, problem, etc.) in order to be able to design a system (product).

- the management system, managing the product (in relation to the customers and market), its development process (including intermediate states of the product) of which designing is a vital part, managing the design process itself, and managing the company.
- the environment in which designing takes place also affects the design process, and consequently the designed quality of the future product.

What happens with the model of the technical system? What is the appropriate method based on the theory? In order to generate the *design methodology* as the strategy of designing, various basic models from the theory have been used, e.g.:

- the model of transformation processes, Figure 1
- the model of the life cycle of technical systems, Figure 2;
- the models of structures of technical systems, Figure 3;
- the model of properties of technical systems, Figure 4; etc.

In addition, several sections form a theoretical basis for tactical methods, e.g. the theory of properties of technical systems as a basis for generating a design specification (list of requirements), with possible support from QFD, and its enhancements [23].

6. A Design Task

The following description of a task shows a pattern of systematic (and methodical) designing. The process occurred in November and December 1996, but was not played as formally as shown, since there was an element of chance in the sequence, which we modified in retrospect to clarify the remembered thought processes. In particular, we omitted the few iterative cycles, and the discussions with the customer. We also added comments to the figures to clarify the theoretical context. We are indebted to Mr. Orest Koroluk, Technical Officer in Mechanical Engineering. He attended the RMC design courses in mechanical engineering, and then participated in the discussions on conceptualising the apparatus, and to consciously following the steps. He also assessed the layouts and offered many suggestions.

6.1 Design Task as Given, and Clarification

Each *design specification* (list of requirements) should be compiled in some detail. All statements should relate to performance and capabilities, but where possible should not refer to the actual configuration of the apparatus itself. Guidelines (a method) for establishing these lists of requirements are available [9], in particular the classes of properties of technical systems [6, 9, 13]. It is useful to list these statements, and note a priority level for each, and the classes of properties to which it refers.

The problem: Recently a gymnastics apparatus was ordered *informally*. The Display Team of cadets at the Royal Military College of Canada (RMC) wishes to display static exercises (without swinging) on three trapeze drops (six ropes side by side) with six gymnasts. They have used an apparatus that was easily attached to beams in a gymnasium building. The display needs now to take place on the parade square, in free air, and an apparatus should be newly built to suit these conditions. It should be possible to erect and dismantle the apparatus and it should be easily transported. Size, dimensional and weight limitations are prescribed. Simple, standardised and easily assembled and released connections should be used. Good appearance is desired, but safety, permanence, etc. are especially important.

Searching the market, the team was offered a stand for one trapeze (really intended for gymnastic rings) at a price of $10,000. Both the cost and the ballast needs were disadvantages, but there were advantages in other respects.

6.2 Conceptualising – Establish Process, and Function and Organ Structures

The trapezes and climbing ropes should hang freely from a beam, which is held at the correct height by two uprights. A releasable connection between the beam and the subdividable uprights is anticipated.

The transformation process and the technologies (according to Figure 1) for quick erection and disassembly, and the preparation for transporting this apparatus, are vital for solving the task. For erection, the process can be performed in various stages:

(b1) the whole apparatus is assembled flat on the ground, and then erected; (b2) the uprights and beam are assembled flat, and erected on to the base; (b3) only the uprights are assembled flat, erected on to the base, and then the beam (with the ropes and trapezes) is hoisted into place; (b4) the uprights are progressively assembled vertically on to the base, and the beam is hoisted into place. The vertical position of the apparatus can be maintained by ballast weight, by tensioning with guy ropes, or by a sufficiently broad standing area on the ground. Erection by hand in sequence (b4) on a broad standing area seems an initial best choice. However, this raises several other possibilities:

- use of a ground beam with cantilevered uprights (stiff in bending); or
- construction of the uprights as three-legged stands with ground width in two directions, e.g. in the beam direction and transverse to it as a 45 degree triangle.

Such an upright can be assembled flat on the ground, and then lifted either around the long side or one of the short sides of the base. A lift rope must be run over the top, so that the beam can be hoisted.

With such considerations about the transformation process and technology, and by producing some sketches (see Figure 4), we conceptualised the apparatus to a sufficient degree so that the next step could be started. The functions (Figure 3, level III) for the technical system (like 'join the constructional elements'; 'fasten and guide the climbing, trapeze and hoisting ropes', etc.) were derived from the current considerations. They now have to be solved in principle by particular organs (Figure 3, level IV). For instance, an easily fitted and released connection of the elements can be achieved through a transverse screw and nut, a transverse pin, a bayonet plug and rotate connection, or a friction collet. The second alternative was selected, suitable spring-ball retained pins being commercially available. The function of the beam can also be explored: it must transmit the vertical forces of the trapezes on to the uprights. The principle can be of pure bending in a beam or of resolving the forces into tension and compression members (ties and struts). The latter has smaller weight, but higher costs. These considerations were fairly extensive, have explored the solution field quite well, but have caused almost no expenditure. The resulting decisions have predetermined many of the costs for material and manufacture in orders of magnitude. Any changes at this point would be easily possible, without actual expense. In the stages of layout (embodiment) and elaboration, manufacturing planning, etc., the cost loading that a change causes will increase progressively. Cautious conceptualising, where possible using systematic methods, gives the possibility of approaching an optimal solution with least cost and time.

6.3 Embodiment, Layouts, Elaborating, Detailing – Establish Final Component Structure

Preliminary layouts (the first entry into the component structure, Figure 3, level V) for important sections of the apparatus were captured in sketch form, e.g. see Figure 4. From these, with the right sizes shown in the design specification, the first dimensional layout was produced (Figure 5), which shows a tentative arrangement of the members in the beam and the uprights. Some *quick and dirty* (order-of-magnitude) calculations showed the approximate sizes of members to resist the forces and buckling effects. A truss program [25] was used to verify the member sizes in the beam, and calculate the expected stiffness and deflection under load, which caused a major change in the arrangement of the beam members. Further details were also investigated.

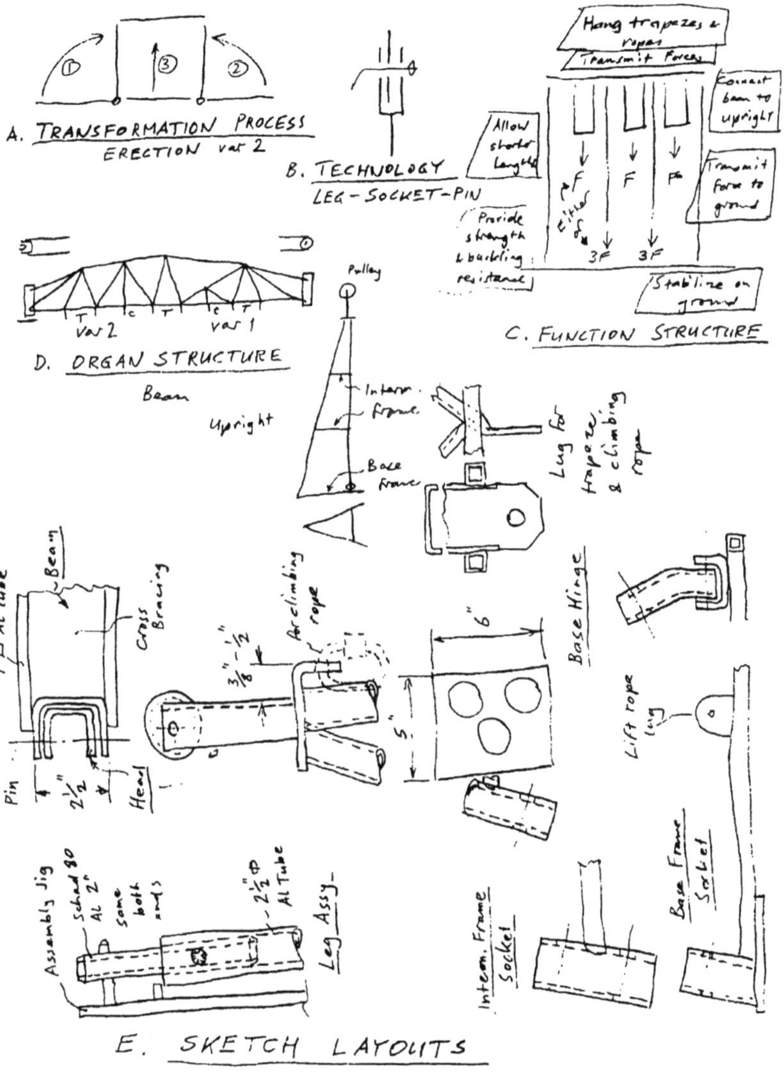

Figure 4: Sketches of transformation, technologies and layout.

Detail drawings of components were produced, using commercially available raw materials, with particular attention to manufacturing processes. The welded groups were shown in subassembly drawings.

This step is so well known from experience (although a theoretical illumination is still largely missing) that it needs almost no discussion. Computer-aided drafting could have been used. For the manufacture of the components, some discussions with the technical personnel were still necessary, so that different tools and fixtures could be made, e.g. for cutting the tubes for the uprights, especially the connection angle at the corners of the intermediate frames. The apparatus was built according to the drawings. Some adjustments were needed to accommodate welding distortion. The apparatus was used in May 1997 for the RMC Graduation Tattoo.

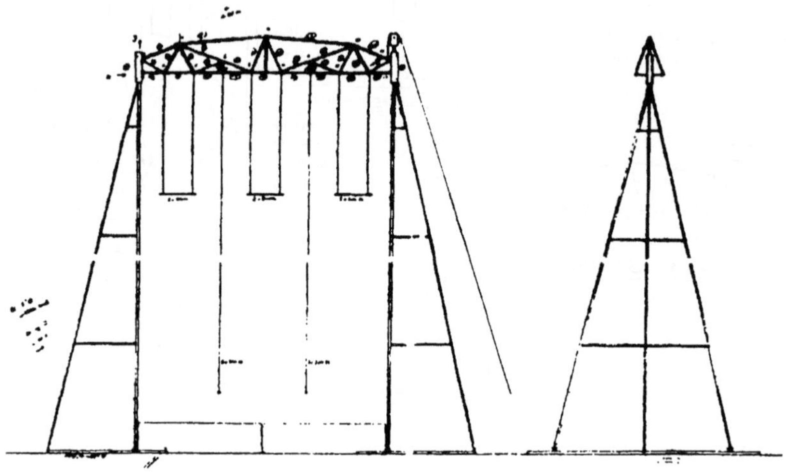

Figure 5: Dimensional layout.

7. Closure

A *natural* design process, learned by previous experience (even for 'good' designers), frequently involves a quick progression to the details. Usual advice is that designers should define the problem by first writing a design specification [26], should search for alternatives, should delay giving details for as long as possible, etc.

With suitable knowledge of Design Science, improvements over the *natural* design processes are possible: write the design specification using the classes of properties of technical systems (Figure 4), and of their life cycle (Figure 2); search for alternatives with function structures, morphology, organ structures and component structures. The 'systemic view' [16] can therefore be made practical. Many conflicts [27] can be avoided by using clearer models. Designing is both a broad and diffuse progress through different stages, and an amplification of several structures (compare [28]).

Searching for alternatives can easily lead to an exponential increase in the number of available combinations – *combinatorial complexity*. Periodic evaluation of the alternatives must be used at suitable stages of designing (e.g. for each of the named structures) to control this increase. A return to a previous position is possible, because all alternatives and the reasons for choices are recorded – which is not usual in *natural* design processes.

There are no restrictions on creativity, intuition, opportunism, reflection, incubation, iteration, recursion, decision-making, teamwork, etc. All of these are essential for designing, but none of them by themselves describe the essence of designing. The myth that *system and method destroys creativity* can finally be laid to rest.

The close connection of a sound theory (Design Science) with a workable and coordinated method (of designing) to produce an object (the trapeze apparatus) should also be obvious.

It is usual to find that experienced people will resist changing their proven procedures (see also [28, 29]). Any attempt to introduce systematic methods into engineering design practice depends on many factors:

(a) designers must be aware of the knowledge of Design Science;
(b) design managers must give designers the time and encouragement to learn this knowledge and the appropriate methods;
(c) design managers must show that they accept that the performance of designers will be reduced during learning – and no known metric can measure by how much the performance will have increased (despite e.g. [30]), or how many errors (and consequent costs) have been avoided by using the improved design processes;
(d) design managers must require that the intermediate results from design steps based on Design Science should be on record, they should be audited with the final design results – and designers must be given the time to produce these records.

This last point can provide the necessary documentation for defence in any subsequent liability case.

References

1. Cross N. 1995: Editorial. *Design Studies* 16, 2–3.
2. Eder W.E. 1994: Developments in Education for Engineering Design – Some Results of 15 Years of WDK Activity in the Context of Design Research. *Journal of Engineering Design* 5, 135-144.
3. Hales C. 1991: *Analysis of the Engineering Design Process in an Industrial Context, 2nd edn.* Gants Hill Pub., Eastleigh, Hants.
4. Ropohl G. 1979: *Eine Systemtheorie der Technik (A Systems Theory of Technology).* Carl Hanser Verlag, München.
5. Hubka V. 1974: *Theorie der Maschinensysteme;* and 1984: *Theorie technischer Systeme (2nd edn, Theory of Technical Systems).* Springer, Berlin.
6. Hubka V., Eder W.E. 1988: *Theory of Technical Systems.* Springer, New York.
7. Hubka V. 1976: *Theorie der Konstruktionsprozesse (Theory of Design Processes).* Springer, Berlin.
8. Hubka V. 1980: *WDK 1: Allgemeines Vorgehensmodell des Konstruierens.* Fachpresse, Goldach.
9. Hubka V., Eder W.E. 1992: *Engineering Design.* Heurista, Zürich.

10. Hubka V., Andreasen M.M., Eder W.E. 1988: *Practical Studies in Systematic Design.* Butterworths, London.
11. Hubka V. 1978: *Konstruktionsunterricht an Technischen Hochschulen (Design Teaching at Technical Universities).* Leuchtturm, Konstanz.
12. Eder W.E. (ed.) 1996: *WDK 24 – EDC – Engineering Design and Creativity – Proceedings of the Workshop EDC.* Heurista, Zürich.
13. Hubka V., Eder W.E. 1992: *Einführung in die Konstruktionswissenschaft (Introduction to Design Science).* Springer, Berlin; and 1996: *Design Science: Introduction to the Needs, Scope and Organization of Engineering Design Knowledge.* Springer, London.
14. Klaus G. 1965: *Kybernetik in philosophischer Sicht (Cybernetics in the Philosophical View), 4th edn.* Dietz Verlag, Berlin.
15. Ashby W.R. 1968: *An Introduction to Cybernetics.* Methuen Univ. Paperbacks, London.
16. Cross N., Clayburn Cross A.C. 1998: *This volume,* pp. 71–84.
17. Bavink B. 1947: *Das Weltbild der heutigen Wissenschaften und seine Beziehung zu Philosophie und Religion (The World Picture of Current Science and its Relationship to Philosophy and Religion*). Silva-Verlag, Iserlohn.
18. Hubka V. 1996: *Design for ...: Inhalt und Struktur (Contents and Structure).* In: 8. *Symposium Egloffstein, Allgemeine Systematik der DF (General Systematics of Design for Classes),* TU, Erlangen.
19. Schön D.A. 1983: *The Reflective Practitioner: How Professionals Think in Action.* Basic Books, New York.
20. Roozenburg N., Dorst K. 1998: *This volume*, pp. 29–41.
21. Dörner D. 1989: *Die Logik des Mißlingens: Strategisches Denken in komplexen Situationen.* Rowohlt, Reinbek; and 1996: *The Logic of Failure: Strategic Thinking in Complex Situations.* Metropolitan Books, New York.
22. Hales C. 1993: *Managing Engineering Design.* Longman, Harlow, Essex.
23. Hosnedl S., Borusikova I., Wilhelm W. 1997: TQM Methods from the Point of View of Design Science. In: A. Riitahuhta (ed.) *World Class Design by World Class Methods. WDK 25: Proceedings ICED 1997 Tampere* Vol. 1, pp. 391–394. Heurista, Zürich.
24. Lloyd P. 1997: *This volume*, pp. 113–124.
25. Akhras G. 1996: *Simple Computer Programs for Truss Analysis.* RMC, Kingston, ON.
26. Hacker W., Sachse P., Schroda F. 1997: *This volume,* pp. 205–216.
27. Frankenberger E., Badke-Schaub P. 1997: *This volume*, pp. 149–164.
28. Dörner D. 1997: *This volume*, pp. 3–11.
29. Weth R. v. d. 1997: *This volume*, pp. 98–108.
30. Duffy A., O'Donnell P. 1997: *This volume*, pp. 269–283.

Describing Design as a Reflective Practice: Observations on Schön's Theory of Practice

Norbert F. M. Roozenburg & Kees Dorst

Abstract

In this paper we shall discuss Schön's views on design, in particular his model of the structure of design as reflection-in-action. Our assessment of Schön's ideas takes account of both theoretical and empirical considerations. In our view, the chief importance of Schön's work is his plea for the emancipation of design in overly knowledge-oriented universities. He effectively criticises the rather simplistic view of professional practice as little more than the application of scientific knowledge, and his descriptions of design practice stand as a vindication of what practitioners really do. His concept of framing has proved fruitful for describing conceptual design activities. However, the Schön model of the structure of reflection-in-action does not present a radical alternative to traditional models of problem solving in design and, taken as an empirical theory, Schön's theory of reflective practice is, admittedly, weak and fuzzy.

1. Introduction

In recent years rational, systematic approaches to designing in architecture, engineering and product design have been challenged by design theorists who approach design from 'post-modern', i.e. relativist, constructivist and hermeneutic points of view. Among the exponents of this approach are Bucciarelli [1], Coyne [2] and Schön. In their view, traditional models of design are rooted in outdated, positivistic idealisations of science and technology, and fail to recognise the ambiguities and complexities of real design practice. They claim that design research can be relevant for practice only if it succeeds in developing a better understanding of designing as it occurs in everyday design situations. They see it as their primary aim not to prescribe but 'to understand and, thereby, assist what already works' [3]. To that end they present alternative models of designing, which they prefer to see as metaphors. In these models, or metaphors, they emphasise the role of tacit knowledge, experience, skill, interpretation, reflection and socio-cultural practices.

Donald A. Schön contributed a great deal to the advancement of this new understanding of designing. More than anyone else, he showed what a radical alternative for traditional ways of seeing designing could look like. For Schön designing is not primarily a form of problem solving, information processing or search, but a 'reflective dialogue' between the designer and the materials of the situation that he or she is in.

In this paper we shall assess Schön's views on design, in particular his model of the structure of design as reflection-in-action. Our assessment of Schön's ideas takes account of theoretical and empirical considerations, the latter being based upon the PhD thesis-project of the second author [4]. One of the aims of this project was to compare the descriptive potential of Schön's model with those of the model of design as rational problem solving; in the thesis project these models were considered as representing different paradigms for design research.

The layout of this paper is as follows. First, we shall recapitulate Schön's ideas on designing, mainly as set forth in his books *The Reflective Practitioner*, *Educating the Reflective Practitioner*, and *Frame Reflection,* mostly in his own words [5, 6, 7]. Then we shall introduce the thesis project and briefly discuss some results of the empirical comparison of the two paradigms. In conclusion we shall make a number of observations, based on the empirical study and on theoretical reflections on Schön's theory of design.

2. Schön's Stance on Technical Rationality

In *The Reflective Practitioner* Schön argues for 'a new epistemology of practice' that takes as its point of departure the competence and artistry already embedded in the thinking and doing of skilful practitioners. Schön contrasts this new epistemology with what he calls 'the model of Technical Rationality'. For Schön, Technical Rationality does not appreciate the design-like character of professional competence. It portrays professional activity as 'instrumental problem solving, made rigorous by the application of scientific theory and technique' [5, p. 21]. In this view, professionals are thought to apply general principles and standardised knowledge, preferably validated by rigorous research, to concrete, well-formed problems [5, p. 24].

For Schön the model of Technical Rationality is the heritage of positivist doctrines, in which practical knowledge was construed in terms of relationships of means to ends [5, p. 33]. Because of the successes of engineering and medicine in the late 19th and early 20th centuries, it became the dominant model of professional practice. In academia, Technical Rationality still pervades thinking about practice. The model is implicit in the institutionalised relations between research and practice, and in the normative curricula of professional education [5, p. 26]. Its concept of 'application' of scientific knowledge entails a hierarchy of professional knowledge: first – and highest in rank – comes the underlying basic science, then applied science and the 'engineering' component (e.g. diagnostic and problem solving techniques), and lastly professional skills and attitudes [5, p. 24].

However, Schön contends, problems of real-world practice do not present themselves well-formed but as messy, indeterminate situations. If practitioners are to get a well-formed problem matched to their familiar theories and techniques, they must *construct* it from the materials of a 'problematic' situation [6, p. 4]. Schön sees this constructing largely proceeding through complementary acts of 'naming' and 'framing'. In Schön's words, 'the practitioner selects things for attention and organises them, guided by an appreciation of the situation that gives it coherence and sets a direction for action. So problem setting is an ontological process . . . a form of world making' [6, p. 4]. Practitioners may frame problematic situations in different ways and debates arising from conflicting frames are not easily resolvable by appeal to facts. Yet, it is through naming and framing that technical problem solving becomes possible [6, p. 5].

Even when constructed, a problem may escape the categories of applied science because it may present itself as unique or unstable. In order to solve a problem by the application of existing theory or technique, a practitioner must be able to map those categories onto features of the practice situation. But, according to Schön, 'a unique case falls outside the categories of applied theory and an unstable situation slips out from under them' [5, p. 41]. Moreover, problematic situations often are situations of conflict among different professional views and different requirements and values. So, practitioners must also reconcile, integrate or choose among conflicting appreciations of a situation so as to construct a *coherent* problem worth solving [6, p. 6].

In Schön's view, these 'indeterminate zones of practice' cannot be dealt with within the model of Technical Rationality; yet, he considers them central to professional practice. Therefore he turns the problem of professional knowledge and competence upside down: 'We should start not by asking how to make better use of research based knowledge but by asking what we can learn from a careful examination of 'artistry', that is, the competence by which practitioners actually handle indeterminate zones of practice' [6, p. 13].

3. Schön's Alternative: Reflective Practice

According to Schön, inherent in the practice of competent professionals is a core of artistry. Artistry is a kind of knowing that is rigorous in its own terms, and we can learn a great deal about it by carefully studying the performance of unusually competent performers. And, although applied science and research based technique have an important role to play, their use is bounded on several sides and mediated by artistry [6, p. 13]. Central to the artistry by which proficient professionals deal with indeterminate situations Schön postulates a process which he calls 'reflection-in-action'. Reflection-in-action is thinking, in the midst of action, about the largely tacit knowing-in-action on which the workaday life of professionals – just like other people – depends to a large measure. Schön's knowing-in-action refers to the sorts of know-how that are involved in physical activities like riding a bike and mental activities like instant analysis of a balance sheet. In both cases, says Schön, following Ryle and Polanyi, 'the knowing is *in* the action. We reveal it by our spontaneous, skilful execution of the performance; and we are characteristically unable to make it verbally explicit' [5, p. 50; 6, p. 25].

Indeed, common sense acknowledges knowing-in-action, but common sense also recognises that professionals think more or less consciously about what they are doing, while they are doing it; they can 'reflect-in-action'. But, as far as intuitive, spontaneous performance produces expected results, we tend not to think very much about it. Hence, reflection-in-action hinges on the experience of surprise. Starting from here, Schön characterises the process of reflection-in-action as follows [6, p. 28]:

- There is, to begin with, a situation of action to which we bring spontaneous, routine responses. These reveal a knowing-in-action that may be described in terms of strategies, understandings of phenomena, and ways of framing a task or problem appropriate to the situation . . .
- Routine responses produce a surprise – an unexpected outcome, pleasant or unpleasant, that does not fit the categories of our knowing-in-action . . .
- Surprise leads to reflection . . . Reflection is at least in some measure conscious . . .
- Reflection-in-action . . . questions the assumptional structure of the knowing-in-action. We think critically about the thinking that got us into this fix or this opportunity; and we may, in the process, restructure strategies of action, understandings of phenomena, or ways of framing problems . . .
- Reflection-in-action gives rise to on-the-spot experiment. We think up and try out new actions intended to explore the newly observed phenomena, test our tentative understandings of them, or affirm the moves we have invented to change things for the better.

Thus, says Schön, reflection-in-action tends to focus interactively on the outcomes of action, the action itself, and the intuitive knowing implicit in the action [5, p. 56]. Reflection-in-action is a kind of *inquiry* that is neither dependent on the categories of established theory and technique nor on prior agreement about ends. In this form of inquiry means and ends are not kept separate, and thinking is not separated from doing, as implementation is built into the inquiry. Reflection-in-action can proceed, even in situations of uncertainty or uniqueness, because it is not bounded by the dichotomies of Technical Rationality [5, p. 69].[1]

[1] According to Schön the positivist epistemology of practice rests on three dichotomies. Given the separation of means from ends, instrumental problem solving can be seen as a technical procedure to be measured by its effectiveness in achieving a pre-established objective. Given the separation of research from practice, rigorous practice can be seen as an application to instrumental problems of research-based theories and techniques whose objectivity and generality derive from the method of controlled experiment. Given the separation of knowing from doing, action is only an implementation and a test of technical decisions [5, p. 165].

4. The Structure of Design as Reflection-in-action

In Schön's view, designing, broadly conceived, is an exemplar of the process fundamental to the exercise of artistry in all professions [6, p. 41]; see also [8]. In a search for what he calls 'the structure of reflection-in-action', he analysed protocols of architectural designing and demonstrated that a similar 'skeleton' can be found in cases of psychotherapy, engineering design, policy making and, even, in scientific investigation. The reconstruction Schön arrived at can be briefly summarised as follows [5, p. 130]:

- Initially, the practitioner is unable to solve a problem as set. He responds to this unsatisfactory situation by restructuring or reframing the problem. Implicitly reframing suggests a direction for reshaping the situation;
- The practitioner then takes the reframed problem and conducts an 'on-the-spot' experiment to discover what consequences and implications can be made to follow from it. . . this he does through a web of moves, discovered consequences, implications, appreciations and further moves. Within the larger web, individual moves yield phenomena to be understood, problems to be solved, or opportunities to be exploited;
- Moves also produce unintended changes which give situations new meanings . . . the situation 'talks back', causing the practitioner eventually to reframe the situation once again;
- In this reflective conversation, the practitioner's effort to solve the reframed problem yields new discoveries which call for new reflection-in-action.

Thus, says Schön, 'the reflective process spirals through stages of appreciation, action and reappreciation. The unique and uncertain situation comes to be understood through the attempt to change it, and changed through the attempt to understand it' [5, p. 132].

5. Describing Design as Reflection-in-action; a Protocol Anaysis

Undoubtedly Schön's analysis of design is very compelling. Yet, the question arises to what extent his model accurately and fruitfully describes what occurs and is experienced in designing by designers themselves. In his thesis-project the second author looked for an answer to this question [4]. In this project Schön's constructionist view of designing was looked at as a new paradigm for design research, and theoretically and empirically compared with the paradigm of design seen as rational problem solving.[2] The comparison was focused on the strengths,

[2] In the thesis-project a paradigm is defined as 'a fundamental image that ... defines and interrelates the exemplars, theories, methods and instruments ... within a science' [9].

weaknesses and biases of the two ways of describing design activities that are implied by these different paradigms.

The study of design as a form of rational problem solving, started in the early 1970s, was mainly inspired by Simon's ideas on a science of design [10]. Along the lines of Newell and Simon's human problem solving theory, design is modelled as a search process; the design problem defines a 'problem space' which has to be explored in search of an optimum or satisfying solution. This model of design is bound to a positivist stance on science and a striving for rigour of design research: objective observation and logical analysis should lead to general, formal models of the design process. Simon quotes 'hard' models and optimisation methods, as being used in statistical decision theory and management science, as prime examples of what a real science of design could and should be [10, p. 60]. Simon's proposal for a 'science of design' did not withstand the test of time in all respects. Yet, his ideas have broadly been taken up by many design researchers, and, in one form or the other, they are still a dominant force within the field.[3]

In the thesis-project the descriptive potential of this paradigm was contrasted with the descriptive potential of the paradigm of design as reflection-in-action. To that end, for each paradigm an encoding system was developed which reflected the corresponding paradigm as faithfully as possible. With both encoding systems the data of empirical protocols were analysed. The final assessment of the paradigms was based on the comparison of the performance of the encoding systems in capturing design-as-experienced. To provide some focus, the analyses were directed to the paradigms' ability to account for the phenomenon of 'integration' in design.[4] This put them to the test on neutral territory, as integration has not yet been analysed in either of the paradigms.

In order to develop an encoding system in the spirit of 'reflective practice', Schön's wide-ranging and rather general characterisation of designing had to be operationalised. Now, Schön's model of design as reflection-in-action does not describe an overall design process from 'problem' to final design – or episodes thereof – but presents a pattern or 'mechanism' of design thinking, which pertains to design activities at a much more detailed, albeit rather abstract, level. For Schön the basic elements of design activities are actions, and the kernel of the design ability is to make intelligent decisions about those actions. Consequently, Schön's reflection-in-action was idealised as a process of *naming*, *framing*, making *moves* and *evaluating* them (Figure 1).

---

[3] As an illustration: in the two Research in Design Thinking workshops organised in 1992 and 1994 by the Delft Faculty of Industrial Design Engineering, visited by a fair cross-section of design researchers from Product Design, Engineering and Architecture, Simon was referred to more than anyone else: 31 direct references and goodness knows how many indirect ones in 32 papers.

[4] In the thesis-project, integration was implicitly defined: someone is designing in an integrated manner when he/she displays a reasoning process building up a network of decisions concerning a topic (part of the problem or solution), while taking account of different contexts (distinct ways of looking at the problem or solution).

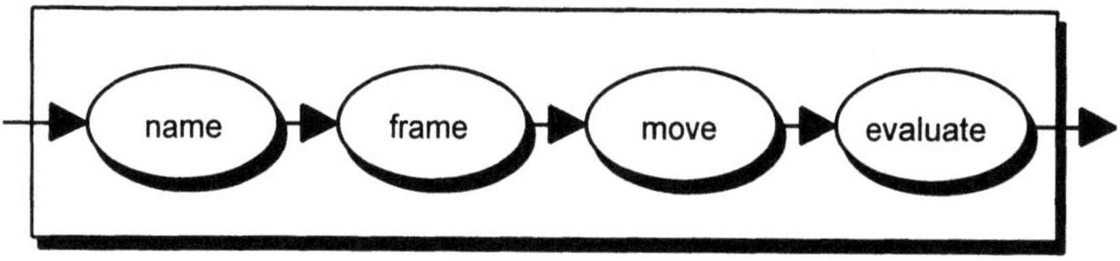

Figure 1: The model of reflective practice.

In the 'naming' step the objects or aspects to be considered in the design situation are selected by naming them. In the 'framing' step these named entities are put into a context, and an overall coherent perspective on the design task is constructed. In making a 'move' the designer takes a tentative action towards the solution of the problem, based on the naming and framing of the design task. The evaluation of a move leads to either satisfaction, the making of new moves, or the reframing of the problem. Evaluation could also lead to a complete new view of the design task, causing the designer to start naming new entities in the design situation.

Schön suggests three general criteria for evaluation of the moves [5, p. 136]; (a) coherence: is this move well-directed in the light of the goal of the design activity, and is it consistent with earlier moves? (b) accordance with the performance specification: is this move within the bounds of an acceptable solution to the problem? and (c) problem-solving value: has this move succeeded in reducing the problems, and did this move keep inquiry going?

A complete design project consists of many of these reflective processes.

The scheme of Figure 1 provided the framework for the encoding of nine protocols of experienced designers. They were given the task of creating one or more concepts for a litter disposal system in a new Dutch train, on which they worked for 2.5 hours. This task is typical for a product development assignment, in that it calls for the integration of ergonomics, construction, form-giving, manufacturing and business aspects.[5]

A main conclusion of the empirical study was that modelling design as reflection-in-action works particularly well – i.e. is accurate and fruitful – for conceptual stages of a design process, but less so for the activities preceding conceptual design. The analysis of the nine protocols showed that in the conceptual phase designers indeed actively frame and reframe problems and solutions until a satisfying and promising combination is constructed. Moreover, in this phase the designers' actions seem largely to be controlled by the peculiarities of the assignment and very little by some sort of preconceived plan or procedure. This behaviour is more accurately described as reflection-in-action than as problem solving, as the description method for capturing design as reflective practice tries to preserve the links between the design process, the content of the task and designers' views on it. The concept of 'frame' turned out to be fruitful in capturing these links. Moreover, although activities like

[5] This study was reported upon earlier in [11, 12].

integration, in which the process and content of design activities are intricately connected, could be *described* to some extent within the rational problem solving paradigm, yet, they were more precisely *understood* using the reflective practice paradigm.

In contrast, the model of reflection-in-action captured little of the rationale behind the many naming-actions and moves that were found in the parts of the protocols preceding conceptual design, such as gathering and selecting information. Designers seem to refrain from associating with a particular frame in this early stage, as a result of which the model of reflection-in-action leads to a description that does not express much of what actually occurs. In this study the designers did not go into detailed design very much, because their task was to develop a concept. Yet, we expect that for detail design reflection-in-action will also lose some of its descriptive power, as in this part of the design process both the problem frame and the solution frame (i.e. the chosen concept) are usually fixed and stable.

In the course of this empirical study it was also found to be difficult to draw general conclusions from the description of design as a reflective practice. The treatment of design as reflection-in-action lacked the clarity and rigour that was achieved by the rational problem solving paradigm. Because of the weakness of the underlying theory, e.g. the problem of finding clear empirical references for Schön's general concepts, it was deemed premature to draw general conclusions from this description of design.

6. Some Observations on Schön's Theory of Practice

In this section we shall comment on two topics that are central to Schön's views of designing, i.e. the structure of reflection-in-action and the role of framing. But we shall begin with a remark on Schön's critique of Technical Rationality.

Indeed, the view that designers and other practitioners merely apply (or should apply) general principles and standardised knowledge is a fiction.[6] For Schön, Technical Rationality is a misconception because the problematic situations of real practice are unique, unstable, indeterminate and the like, and therefore do not fit the categories of existing theory or technique. No doubt this is an important reason for the limited use of research based theory and technique in practice, as all theories are idealisations. But, there is, we think, a more fundamental reason, which is that neither the design of a new product nor its functions can be deductively inferred from scientific theory. Hence, even when a practitioner succeeds in constructing a well-formed problem that matches familiar theories, the latter do not guarantee that he will find a solution for the problem as he has come

[6] For an overview of different positions with respect to the thesis that technology is (not) applied science, see [14]. It is also to be noted that much of Schön's critique on the shortcomings of Technical Rationality had already been raised within the design methodology community a decade earlier, especially by Rittel and Webber [15] and Hillier, Musgrove and O'Sullivan [16].

to see it. Scientific theories can provide powerful heuristics and strategies for solving practical problems, but, ultimately, new solutions to design problems have to be conceived of by a creative act of invention [13]. Apart from that, we feel that Schön exaggerates the uniqueness of practical problems. It is true, no two practical problems are identical but that does not make them unique in every respect, and inaccessible to research based theory and technique.

The kernel of Schön's theory of practice is the process of reflection-in-action. For Schön, reflection-in-action is the key to the artistry by which competent designers deal with situations of uncertainty, uniqueness and conflict. And, in as far as designing is a rational activity, he sees this rationality hinging on the designer's ability to reflect on the frames they construct and reconstruct in the process [7, p. 165]. In several places in his books Schön gives somewhat different descriptions of the structure of reflection-in-action. In the preceding section, we have summed up his ideas as: naming – framing – move(s) – evaluation, which is to be taken as a cycle of activities that repeats itself and leads to a typical spiral-shaped development of problems and solutions. If this interpretation of Schön is right, then it seems to us that the cycle of reflection-in-action is very similar to the cycle of 'perception' – 'trying' – 'evaluation', which was already known for many years, from psychological studies in problem solving. With regard to the latter cycle the eminent Dutch methodologist of behavioural sciences, A.D. De Groot, stated that we have to do with a *necessary* construct, a logically indispensable model [17]. 'This construct obtrudes itself on our every attempt to come to grips with such phenomena as experience, goal-striving behaviour, learning, problem solving and directed and creative thinking. If we are at all to understand the interaction of a subject and the external world by which experience is acquired, we cannot do without the assumption of some process of 'observation' or 'perception', followed by one of 'trying' (in the sense of tentatively (re-)acting on the external world or on a representation of it), and one in which the information is 'digested', i.e. in which it is determined, whether the results of the trial are good or ill, and what this means to the subject – what it has learned from it – namely the potential influence of the experience on future behaviour.' Given the clear similarity between these two cycles, we do not consider Schön's model of the structure of reflection-in-action to be a radical new view on designing. The 'perception' – 'trying' – 'evaluation' cycle has been embedded in many different models of practical problem solving, and notably in the problem solving model of Systems Engineering, which has become part and parcel of traditional systematic approaches to designing.[7]

As was said, Schön's model of designing does not describe an overall design process from 'problem' to final design. It describes in very general terms design

[7] Here we refer to problem solving cycles like the one in the VDI Guideline 2221 which comprises problem analysis, problem definition, system synthesis, system analysis, evaluation and decision making. Schön's model of reflection-in-action also shows similarities with Dörner's schema of the steps in the organisation of complex action [18] and with Hacker's *et al.* characterisation of designing as interaction of internal and external processing [19].

activities as being controlled locally: recurringly people frame problems, make moves and evaluate these on their *immediate* problem solving value. So, Schön describes designing as being in large measure an opportunistic activity. However, designers not only think about the possibilities for a next move, opened up by the results of previous ones, but they also (have to) think ahead. They usually are under the pressure of external goals and constraints over which they have little control, which forces them to anticipate the things still to be done in order to reach the end of a project. This demands a kind of strategic thinking which is for instance reflected in the phase models or procedural models of designing and product development. Schön's theory of reflective practice pays little or no attention to these 'higher level' strategies.

In Schön's view framing and reframing are the *sine qua non* of reflection-in-action. Frames are 'underlying structures of belief, perception, and appreciation' [7, p. 23] and framing is 'the operation of selectivity and organisation' by which people make sense of complex, information-rich situations. Frames determine what counts as fact and what arguments are taken to be relevant and compelling, as there is no way of perceiving and making sense of social reality except through a frame [7, p. 30]. Moreover, frames have both descriptive (diagnostic) and prescriptive force: frames determine the features one attends to, the order one attempts to impose on the situation and the direction in which one will try to change it. In the process of naming and framing, both the ends to be sought and the means to be employed are identified [5, p. 165]. Schön's idea of framing is clearly constructivist [7, p. 215]. Constructivists share the belief that the external world is, at least to some extent, a 'product' of the human mind and knowledge of that world is always a construct of individuals, or communities and institutions, or cultures. Constructivists, thus, deny that people can have objective (subject-independent) knowledge of a reality that exists separately and independently of themselves. All knowledge is subjective as it depends intimately on individuals' perceptions and preconceptions and the socio-cultural context in which it is being acquired. Understandably, in the context of scientific research, constructivist ideas – especially the more radical versions of it – must have been and still are difficult to accept for those philosophers and scientists who adhere to the canons of logical-positivism. Constructivism directly challenges the idea of universally true statements, laws and theories, which is, in positivistic views, what scientific research is about. However, we feel that designers and other practical 'problem-solvers' will have fewer problems in accepting the subjectivity and relativism that go with constructivist views. We think so because designing and practical problem solving in general are not aimed at producing new knowledge – although new knowledge can and often is produced in the process – but at solutions for practical problems. Not many practitioners, if any, claim objectivity and universality, for the problems they work on, nor for the solutions they develop. It seems to us that practitioners themselves have always properly understood that the concept of 'problem' presupposes perceptions, expectations and valuations of people that try to intervene in an unsatisfactory situation. Practical problems and solutions are by there very nature, so to say, individual or socio-cultural constructs. Hence, Schön's plea for framing and frame reflection might look to many practitioners as hammering at an open door. But, this detracts nothing from his claim and we are in agreement with Schön that designing involves framing. However, we think that Schön overstates the role of framing in the context of design. He seems to ignore the dual nature of material products; products are both socio-cultural and physical constructs. Products are socio-cultural constructs in the sense that the way they are perceived, used and valued is largely socially constituted. But products are also physical objects that are submitted to the laws of nature. Now, the physical world seems to offer much less room for different subjective knowledge constructs than the socio-cultural world. Thus, with regard to the physical side of product designing, framing might play a much more limited role than Schön suggests.

Given the importance of framing in Schön's theory, practical questions arise, such as: on what grounds can 'good', useful or fruitful frames be recognised and how can inter-personal frame conflicts be settled? In *The Reflective Practitioner* and *Educating the Reflective Practitioner* Schön leaves questions like these largely unanswered, but not so in *Frame Reflection*, although this work does not address the designing of material artefacts but the resolution of policy controversies. On constructivist views, frames are inherently subjective and cannot be falsified, as there are no objective, frame-neutral observers; everyone's frame is seen by others through their respective frames [7, p. 30]. But if this is so, Schön and Rein wonder, on what basis do we assume that frame conflicts are resolvable? Is there an objective basis for choosing among frames, or reframing policy issues, in one way rather than another? How can we possibly resolve frame conflicts when frames themselves determine what counts as evidence and what interpretations of evidence are acceptable? [7, p. 41]. The radical relativist answer to this question is that such a basis cannot be found, but Schön and Rein, as we do, consider this position clearly as an uncomfortable and unacceptable trap. Therefore, in *Frame Reflection* they elaborate on the possibility of a *pragmatic* frame-critical approach, which accounts for the fact that policies *are* sometimes reframed in action, and that their reframing *does* sometimes result from the actors' reflection on frame conflicts that arise in the evolving, politically coloured process of policy design. Unfortunately, the conclusions at which Schön and Rein arrive are rather general and obvious: participants in a policy conversation must be able to put themselves in the shoes of other actors in the environment, and they must have a complementary ability to consider how their own action frames may contribute to the problematic situations in which they find themselves. To that end participants must be able to reflect on their own action frames, on the action frames held by their antagonists, and on the action frames that underlie the transactions through which they may have helped to promote miscommunication or exacerbate design flaws. Now, these may be vital abilities, but it leaves practitioners with the problem of how to achieve them.

7. Conclusions

In this paper we have set out Schön's approach towards practice, and in particular his reconstruction of reflection-in-action, which he considers the essence of designers' and other practitioners' competence. We have also introduced some results of an empirical study in which Schön's model of the process of reflection-in-act was put to an empirical test and we ended this paper with some critical observations on Schön's ideas, in an effort to contribute to a proper understanding of the paradigm of design as a reflective conversation. In our view, the chief importance of Schön's work is his plea for the emancipation of design in overly knowledge-oriented universities. He effectively criticises the rather simplistic view of professional practice as little more than the application of scientific knowledge and his descriptions of design practice stand as a vindication of what practitioners really do. Schön turns away from Technical Rationality, but he addresses design practice without crossing into irrationality: there is a structure in design activities which can be captured from the study of what proficient practitioners do. Schön presents this structure as being person-dependent in a rather implicit but surely valid way: the frames that a designer imposes on a situation are, at least to a considerable extent, part of the personal heritage of the designer. With all this we are very much in agreement. However, taken as an empirical theory, Schön's theory of reflective practice, as it stands, is admittedly weak and fuzzy. One looks in vain for explicit definitions of his central concepts, criteria for 'good' frames are largely missing, and Schön does not come very far in explicating 'the rigour in its own terms' that he ascribes to the process. But, maybe we, as design methodologists, have been unjust and over-eager in trying to read into Schon's books a new 'practical' theory of designing, while Schön has just presented what he calls a 'primer' for a new theory of problem solving in practice.

References

1. Bucciarelli L.L. 1994: *Designing Engineers.* MIT Press, Cambridge, Mass.
2. Coyne R. 1995: *Designing Information Technology in the Postmodern Age.* MIT Press, Cambridge, Mass.
3. Snodgrass A., Coyne R. 1992: Models, Metaphors and the Hermeneutics of Designing. *Design Issues* Vol. IX, No. 1.
4. Dorst K. 1997: *Describing Design: A Comparison of Paradigms.* PhD thesis, Delft University of Technology.
5. Schön D.A. 1983: *The Reflective Practitioner: How Professionals Think in Action.* Basic Books, New York.
6. Schön D.A. 1987: *Educating the Reflective Practitioner: Toward a New Design for Teaching and Learning in the Professions.* Jossey-Bass, San Francisco.
7. Schön D.A., Rein M. 1994: *Frame Reflection: Toward the Resolution of Intractable Policy Controversies.* Basic Books, New York.

8. Schön D.A. 1988: Designing: Rules, Types and Worlds. *Design Studies* Vol. 9, No. 3.
9. Ritzer G. 1975: *Sociology: a Multiple Paradigm Science.* Allyn and Bacon, Boston, Mass.
10. Simon H. A. 1969: *The Sciences of the Artificial.* MIT Press. Cambridge, Mass.
11. Christiaans H.H.C.M., Dorst C.H., Cross N. 1993: Levels of Competence in Product Designing. In: N.F.M. Roozenburg (ed.): *Proceedings International Conference on Engineering Design ICED 93.* Heurista, Zürich.
12. Dorst C.H. 1995: Analysing Design Activity: New Directions in Protocol Analysis. *Design Studies* Vol. 16, No. 2.
13. Roozenburg N.F.M. 1993: On the Pattern of Reasoning in Innovative Design. *Design Studies* 14, 4–18.
14. Staudenmaier J. 1985: *Technology's Storytellers: Reweaving the Human Fabric.* MIT Press, Cambridge, Mass.
15. Rittel W.J., Webber M.M. 1973: Dilemmas in a General Theory of Planning. Reprinted in: N. Cross (ed.) 1984: *Developments in Design Methodology.* Wiley, Chichester, pp. 135–144.
16. Hillier B., Musgrove J., O'Sullivan P. 1972: Knowledge and Design. Reprinted in: N. Cross (ed.) 1984: *Developments in Design Methodology.* Wiley, Chichester, pp. 245–264.
17. De Groot A. D. 1969: *Methodology: Foundations of Inference and Research in the Behavioral Sciences.* Mouton, The Hague.
18. Dörner D. 1996: *The Logic of Failure; Why Things Go Wrong and What We Can Do to Make Them Right.* Metropolitan Books, New York, p. 43.
19. Hacker W., Sachse P., Schroda, F. 1998: *This volume*, pp. 205–216.

An Overview of Descriptive Studies in Relation to a General Design Research Methodology

L.T.M. Blessing, A. Chakrabarti & K.M. Wallace

Abstract

This paper describes a general design research methodology that aims at a more rigorous approach to design research by piecing together the various existing design research areas and by encouraging collaboration with other disciplines. It discusses the challenges involved in doing design research, such as the large number of factors that influence design, and highlights the importance of descriptive studies. The papers presented in this book are used throughout as examples, and set in the context of the proposed methodology in the Conclusion and Outlook at the end of this book. These papers and others revealed some of the key issues that need addressing: the formulation of criteria for success, the use of results from descriptive studies in the development of design methods and tools, and the validation of such methods and tools against these results.

1. Introduction

Technical products and the processes of their creation have undergone considerable changes over the last decades. Products have become more complex using new technological developments and integrating knowledge of various disciplines. Increasing competition, stronger customer awareness and stricter legislation have resulted in shorter product life-cycles and tighter requirements. Products have to be technically as well as commercially successful. As a consequence of product changes, the product development process has changed. Complexity, quality pressure and time pressure have increased.

New approaches to improve the effectiveness and efficiency of the product development process are needed in order to cope with these changes and remain competitive. The overall aim of engineering design research is to support industry by developing knowledge, methods and tools that improve the chances of producing a successful product.

This aim raises a number of important questions:

1. What do we mean by a *successful* product?
2. How is a successful product *created*?
3. How do we *improve the chances* of being successful?

The first question leads to issues such as what *criteria* should be used to judge success, that is, what measures will determine whether our research has been successful? The second question leads to issues such as what are the *influences* on success, how do these influences interact and how can they be measured? Investigating these issues will increase our understanding of design, which is needed to improve the design process. The third question gives rise to issues related to the translation of this understanding into *design methods and tools,* and to the validation of these methods. Validation is needed to determine whether the application of these methods indeed leads to more successful products (as determined by the criteria).

A pure product-focused research effort cannot resolve these issues. That this has been recognised is shown by the increasing number of descriptive studies – that is, studies of the way in which a design process actually takes place – to increase understanding of this process both as a cognitive and a social process, and within the organisation (for an overview see [1] and [2]). Traditionally this is not the type of research conducted within engineering and it is not possible to transfer research methods directly from other disciplines. A new approach is required.

2. A Design Research Methodology

To address these issues in an integrated and systematic way, a research methodology is needed. Such a methodology should help in identifying *research areas* and projects within those areas, and selecting specific *research methods* to address the issues.

Two characteristics of design research require the development of a specific research methodology. First, the selection of research areas is not straightforward because of the numerous influences (e.g. in [3]) and the interconnectivity between the influences. Design involves, among other things, people, products, tools and organisations. Each of these is the focus of a particular discipline with its own research methodology and methods, such as social science, engineering science, computer science and management science. Design research is therefore bound to be multidisciplinary [see also Schregenberger [4] who pleads for a holistic approach). An additional complication is the uniqueness of every design project. This particularly affects the repeatibility issue in scientific research.

The second characteristic of design research is that it not only aims at understanding the phenomenon of design, but also at using this understanding in order to change the way the design process is carried out. The latter requires more than a theory of what is; it also requires a theory of what would be desirable and how the existing situation could be changed into the desired. Because this cannot be predicted, design research involves design – the creation of methods and tools – and requires proper validation, just as any design needs testing. Again, methods from a variety of disciplines are needed.

Many different methods can be, and have been, used to address the various issues involved in design research. The authors propose to piece these research areas together into a generic design research methodology that links the research questions together and addresses them in a systematic way. The methodology has been tried and evaluated as a framework. Specific guidelines regarding the set of research methods that are suitable for each question are currently being developed. The aims are: to provide guidelines for better, more rigorous design research; to help identify the focus of the research; to structure the research; to make it more efficient; and to provide a context in which to position a particular research project relative to other design research. We consider a commonly accepted research methodology as one of the main characteristics of an established area of scientific research.

Figure 1 shows the overall methodology [5]. A simple example will help clarify the terms that are used. Each step and some of the research methods that have been applied will be discussed in more detail in the following sections, with an emphasis on descriptive design studies.

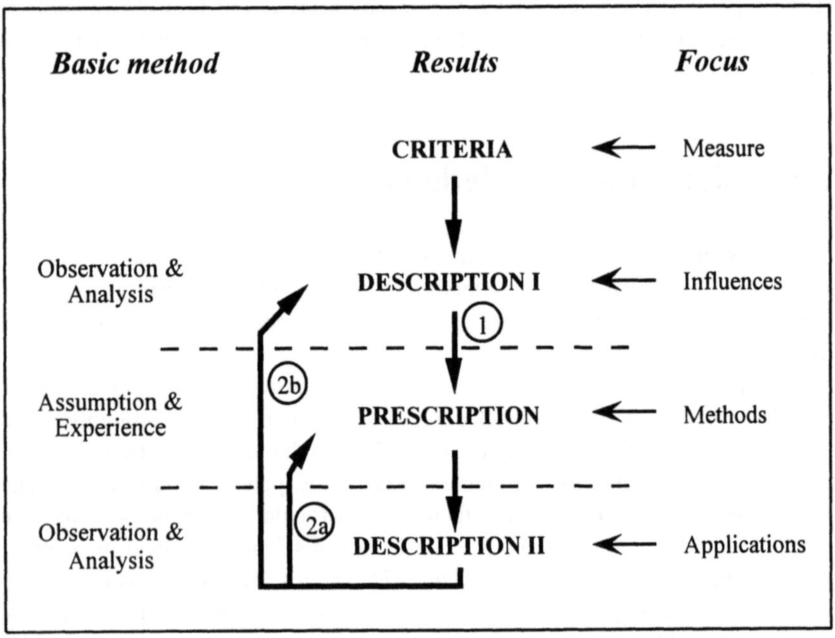

Figure 1: Design research methodology framework and links largely missing in current research (1, 2a and 2b).

Example:

Criteria (section 3):	A reduction in time-to-market is identified as a criterion for success. This provides the focus for the next step and is the measure against which to judge a design method or tool.
Description I (section 4):	A descriptive study, involving observation and analysis, shows that insufficient problem definition relates to high percentages of time spent on modifications, in accordance with VDI 2210 [6], which is assumed to increase time-to-market. This description provides an understanding of the various factors that influence, directly or indirectly, the main criterion, in this case time-to-market.
Prescription (section 5):	Based on the outcome of the descriptive study (link 1) and introducing assumptions and experience about an improved situation, a method or tool is developed to encourage and support problem definition. Developing methods and tools is a design process in itself.
Description II (section 6):	The method is applied and a descriptive study is executed to validate the method or tool. This includes two tests. The first test is whether problem definition is supported (a comparison with Prescription, link 2a). The second test is whether less time was spent on modifications, and whether this in turn reduced the time-to-market (a comparison with Description I, link 2b). There might be reasons why the second test fails, such as side-effects of the method.

Descriptive studies are likely to reveal a chain or a network of causes and effects, connecting influencing factors with the criterion. Methods developed in the Prescription step directly address one or more influencing factors in this chain, which are then expected to affect the rest of the chain. However, interference in this chain or network, as intended by the method or tool, may result in a different network of causes and effects such that the effects are not as predicted, that is, side-effects may occur. Validation (Description II) should take this into account and at least test all causal links between the influencing factors addressed by the method and the success criteria, and not limit itself only to the factors that the method addresses explicitly.

Description I includes the validation of the findings. The same could have been done for the validation of the results of Prescription as validation can be considered part of the proper development of methods and tools. However, it has been mentioned as a separate step (Description II) to highlight its importance and to make explicit the two types of validation (links 2a and 2b). The need to highlight this step emerges from the lack, in most design research, of the type of validation suggested here.

It cannot be expected that each of the steps of the methodology are executed in depth in every single project. Many projects, such as PhD projects, have severe time limits. Besides, where possible one should build upon existing research. Nevertheless, the overall methodology should be taken into account and the link between the step which is the focus of the research, and the related steps, should at least be addressed. Thus, a study of particular influences on the design process (Description I) should at least indicate how this can be used to improve the design process. We have found that using this framework from the beginning of a research project helps provide a direction to the research at hand and identify what areas need addressing.

As the following sections will show, answering the research questions is a challenging task because design research deals with, and tries to improve activities in an industrial context that provides a large number of the influencing factors. These sections highlight the importance of descriptive design studies as part of design research, to increase our understanding and to verify the methods and tools developed.

3. Success Criteria

Defining success criteria plays three roles. The first is to identify the focus of the research, that is the aim this research is ultimately trying to fulfil. The second role is to be able to determine the factors that contribute to or prohibit success (Description I). The third is to use the criteria as measures against which to judge any design method or tool proposed on the basis of the identified factors for success (Description II).

Two types of criteria are used by design researchers who focus on the factors that influence success (see Table 2). A common criterion used in research in an industrial context is success in the market (sales, profit, return on investment), as used by Cooper [7] and Dwyer & Mellor [8]. In laboratory research a common criterion is the fulfilment of technical requirements (a good design), as used by Fricke [9] and Dylla [10]. An interesting criterion is the one used by Hales in this volume [11]: reduction of cost of insurance and product liability. Although this criterion does not reveal the ultimate criterion – which could be increased sales through reduced price, or increased profit by maintaining the price, or something else – it is specific enough to be used.

Most studies, however, do not state their criteria explicitly. For example, Hales & Wallace [3, 12] tried to identify the factors that influenced the design process of a gasifier test rig in an industrial context. Often criteria are hidden in the overall aim of the research, such as improving the effectiveness of the process. Unless followed by a measurable definition, effectiveness cannot be used as a criterion because it does not support the three roles identified earlier. If a process is to be assessed rather than only described, or if support is to be developed, the definition of criteria is of crucial importance. The paper of Duffy & O'Donnell in this volume [13] shows the difficulties in criteria formulation for product performance.

4. Influences, Interrelations and Measures

Success criteria permit the identification of factors that influence success and the way in which they influence success. This in turn helps in identifying the characteristics of successful and non-successful design processes. This knowledge contributes strongly to the development of effective methods and tools.

The identification of factors influencing the design process is part of descriptive studies. These studies aim at a better understanding of design. Because of the complexity of design, the factors are strongly linked. These networks of influencing factors are difficult to analyse, in particular because of the human element involved and the uniqueness of each design project.

Descriptive studies can be characterised by the options that the researcher selects for the various aspects that have to be determined in the design of the study, such as the setting, the subjects and the data collection methods (see Table 1). To choose the most suitable options it is necessary to first pay close attention to the formulation of the focus of the study, that is the formulation of: the aim of the research; the specific research questions, hypotheses, model or theory; and the various constraints that are outside the researcher's control. Alternative, contradictory hypotheses, models or theories can be included if the aim is to find the best amongst them.

Table 1: Characteristics of descriptive studies and their options.

Aspects	**Options**
Environment	Where the study took place: lab or industry
Nature of the study	(Comparative) exploratory, (comparative) action research
Data collection methods	Pure observation, thinking aloud, introspection, participant observation, diary keeping, archival research, questionnaire, interviews
Subjects	Level of experience
Number of cases	The number of data sets collected, e.g. the number of experiments or interviews conducted
Team size	Number of subjects involved in each case
Time constraint	Any limits to the design time imposed on designer
Duration	Length of process studied
Continuation	Is the process observed continuously?
Role of researcher	Contribution of the researcher to the process and the outcome
Required results	Results of the observed part of the process: e.g. specification, layout drawing, prototype, product
Design object type	Original, variant, redesign
Topic	Object to be designed
Batch size	Mass, large batch, small batch, one-off

Many of the options are interrelated, e.g. the decision to go into industry will reduce the number of options available for data collection techniques, and a particular data collection technique is likely to affect the number of cases that can be investigated. The options shown in Table 1 were identified in [1, 2]. In combination with the aim of the research, they are expected to cause differences in findings.

It is important to make these options explicit so that the context of the study can be determined by researchers who wish to make use of the findings. An example of the confusion that can arise can be found in this volume. Cross & Clayburn Cross [14] conclude that expert an designer 'devoted a major amount of his available time to reaching an understanding of the problem and the potential solution space before creating a concept [...]'. Von der Weth [15] found in this and earlier work that 'experienced designers' make 'important decisions early in the design process without careful analysis of the task, solution principles and concepts [...] and produce good solutions in this way'. A very similar finding is reported in Günther & Ehrlenspiel's paper [16].

At face value these results contradict one another, but there may be one or more reasons why both could be valid. First, it is possible that we are comparing apples with pears. The characteristics of the studies may be completely different because different combinations of options have been chosen. Second, one or more influencing factors, not identified yet, may have had an overruling effect. A situation is seldom determined by one factor only, as illustrated in earlier studies of Dörner [17]. Third, statistical chance may have caused the difference. Most detailed descriptive studies base their results on a small number of cases. The effect of this is aggravated by the numerous influencing factors that play a role. Fourth, different definitions for identical terms may have been used, e.g. the definition of what makes a designer an expert.

In this section we assumed that the data have been analysed and the results have been presented in a manner that is in line with the characteristics of the descriptive study and addresses the stated aim, research questions, hypothesis or theory/model. Collected data can be analysed in a variety of ways. Analysis of the collected data is an important step that can seriously affect the generality of the findings. Therefore it is important to document how data were analysed. This includes the definition of the analysis categories (coding scheme), the analysis process (encoder reliability) and any assumptions made. The latter reflects our view on design. In the paper of Günther, for instance, the assumption is that methodical design is the approach to recommend.

In Table 2 three different studies and their results (not from this volume) have been summarised to illustrate the diversity of descriptive studies.

Table 2: Different descriptive studies: research methods and results.

Research methods			
	Hales / Wallace [3, 12]	Cooper / Dwyer / Mellor [7, 8]	Fricke / Dylla [9, 10]
Aim	Measure influences	Influences on success	Influences on success
Criteria for success	?	Profit Sales	Fulfilment of requirements
Options	Participant observation Industry Team 1 case	Questionnaire, Interview Industry Company 200+ cases	Thinking aloud, video taped Laboratory Individual 15 cases
Analysis	Classification	?	Classification
Results	List of influences and their measures	Success relates to proficiency at early stages	Success relates to methodical, flexible, goal-oriented approach

Two characteristics are discussed in this paper: the nature of the study and the data collection method.

4.1 Nature of the Study

A distinction can be made between exploratory research and action research, each of which can be comparative. In exploratory research no design methods or tools are introduced by the researcher. The objective of exploratory research is to identify the influencing factors and their links by studying how a particular process takes (took) place. The process may include the use of design methods and tools, but it is not the objective to validate these. Exploratory research is typical for Description I. Exploratory research is assumed to have little effect on the process being observed even though the chosen options might influence the process. Typical examples in this volume are the papers of Frankenberger & Badke-Schaub [18], von der Weth [15] and Lloyd [19].

Action research deliberately affects the process by introducing a design method or tool. Action research is typical for Description II, the step in which methods and tools are validated. This type of research is discussed in section 6.

Comparative exploratory research is exploratory research in which the cases are divided into groups with different characteristics, for example different tasks or environments for different groups or different types of background as used by Günther & Ehrlenspiel [16]. The objective is to find the effect of the different characteristics so as to better identify the influencing factors and their links. Similarly, comparative action research involves a division into groups with the objective to better identify the effects of a method or tool.

4.2 Data Collection Method

Several methods have been used for collecting data, many of which were originally developed for use in various other disciplines, such as psychology and sociology. In many cases multiple methods are needed to be able to answer all research questions, or verify all hypotheses.

The diversity of data collection methods used for design research is illustrated by the list in Figure 2, based on an investigation of 66 descriptive studies in the area of mechanical engineering design [1].

- simultaneous verbalisation, either video or audio taped
    - thinking aloud (individuals)
    - recording team discussions (talking aloud)
    - introspection (commenting on one's own mental activity)
    - recording questions asked
- recording drawings
    - photographs
    - video
    - collecting / intermediate copies
- case history compilation / archive analysis / document analysis / reports
- task analysis (in hours) / time-writing
- observation by taking notes / observational notes
- diary keeping / subject as observer / activity recording
    - subject is not the researcher
    - subject is researcher (describing own process)
- retrospection
    - reports by subject
    - interview / cross-examination
    - questionnaire

Figure 2: Data collection methods used in design research [after [1]].

4.3 Observations

Using our methodological framework as guidance, the following observations could be made with respect to the publications on descriptive studies, including those in this volume.

Although most papers describe how data were collected, this hardly ever covers all the above listed aspects.

Most publications do not describe the data analysis methods in sufficient detail to be able to verify the results.

Only a few publications reveal the network of influencing factors characteristic for design. Most focus on the link between two factors or on a (short) chain of effects.

Most researchers use one main data collection method, such as protocol analysis or observation, and complement this with exit questionnaires or some summarising interviews. For an example of the use of several methods in this volume, see [18]. Hardly any studies mention the use of multiple methods for triangulation to validate the findings.

A variety of influencing factors is studied with a variety of aims. This emphasises the complexity of design as a research area, but also reveals, as a consequence of the low number of studies (and of cases) that deal with the same topic, the limited understanding we still have of design. A side effect may be that hardly any contradictions in findings were found.

Most importantly in view of our methodology, few studies focus on the explicit link between success (or quality) of the product and the way the process was executed, that is the influencing factors. Only this type of research will be able to give clear indications as to the support required to improve design.

5. Design Methods Development

The most common starting point for the development of a method or tool is reasoning based on experience and logical argumentation, rather than using results of descriptive studies. An example is the assumption that searching a wider solution space improves the quality of the result.[1] Based on this assumption and others, extensive methodical approaches have been proposed by design researchers such as Pahl & Beitz [20], Hubka [21] and Andreasen [22]. The same assumption was also used in developing design support tools such as the Cambridge Materials Selector by Ashby & Cebon [23], and the FuncSION software for synthesising concepts to fulfil intended functions by Chakrabarti [24]. Experience and logical argumentation can be used, so long as this is done consciously and made explicit.

However, we argue that it is important, not only for the development of methods and tools, but also for their dissemination and acceptance in industry, that the actual design processes are taken into consideration. Few methods and tools are explicitly based on descriptive studies. Possible reasons are the recency of descriptive studies in the field of mechanical engineering design, and the fact that most studies do not aim at identifying the factors that influence success. We expect that the increasing number of descriptive studies will give a new push to the development, improvement and implementation of prescriptive methods and approaches.

The results of descriptive studies can be used in industry directly as guidelines or examples, e.g. the findings described in [14]. Using the results for the development of methods and tools can be expected to add more value to the results and lead to more useful methods and tools.

---

[1] The descriptive study by Dylla [10] supports this assumption, provided that the generation of solutions is linked to timely evaluation of the solutions.

The problem is that the transformation of the findings of descriptive studies into methods and tools is not straightforward. Assuming that the aim of these methods and tools is to *improve* the design process, it is necessary to determine what the future, improved situation with support would look like. Only seldom can this be derived directly from a description of the current situation, as this description provides the characteristics of *existing* processes, that is, those we aim to improve, unless we know the links between these characteristics and the success of the process. Even then, the development of methods and tools remains a creative activity in which experience and assumptions have to be used. Design methods and tools should support the positive, success-bringing characteristics, prevent the characteristics that have a negative impact, and take into account those negative characteristics that cannot be prevented [1, 25]. As mentioned before, most studies do not aim at identifying the factors that influence success. Assumptions have to be introduced to permit this transformation.

An attempt has been made to formulate a practical method that helps summarise and illustrate an envisaged design method or tool by identifying its scope and the underlying assumptions [26]. The method consists of a checklist and a supporting technique for those planning or just starting to develop a design method or tool. The resulting description clarifies: the problem that the method or tool addresses; the research findings or assumptions about design upon which it is based; the approach used; and the possible implications. The proposed method can thus be used by a variety of people involved in assessing a tool or method in an early stage of its development. It also helps determine the boundaries of a method or tool. This is important because design support may not be suitable in all cases – design processes are unique, and companies and their situations are very different. For the developers themselves the method reveals how realistic the envisaged method or tool is, and whether the scope has to be narrowed. The method links back to the findings of descriptive studies.

Although this method clarifies much of the transformation process from descriptive studies into a method or tool (prescription), more guidance for the design researcher/developer is needed.

6. Validation of Methods

Once new methods and tools have been developed, they need to be validated or at least tested and evaluated. The validation of design methods and tools is one of the most difficult research areas. The two main issues are: (i) to identify whether the method or tool has the expected effect on the influencing factors that are addressed directly; and (ii) to identify whether this indeed contributes to success (see the feedback arrows in Figure 1). As illustrated earlier, unexpected side-effects may occur.

The name of this step (Description II) indicates that the research methods applied are very similar to those in the initial descriptive study (Description I). The crucial difference is that for validation, action research or comparative action research is needed, rather than exploratory research (see section 4). Action research studies how a process takes place in which a design method or tool, developed either by the researcher or others, is introduced specifically with the objective of studying its effects in order to validate the method or tool. Action research needs more planning than exploratory research and the subjects may need to be trained in the use of the method or tool.

Comparative action research involves a division of the cases in the study into groups depending on whether or not they were introduced to the method or tool. Introducing more than two groups is possible in cases where particular aspects of the tool are being investigated. Comparison of the data collected in the various groups allows the drawing of more precise conclusions about the effects of the method or tool.

Identifying the effect of a method or tool on the factors that are directly addressed, e.g. improving problem specification, can be relatively straightforward. However, identifying whether this indeed contributes to success is far more difficult and the results are not easy to generalise. Success is difficult to measure other than in a real, industrial situation, and action research in an industrial situation is notoriously difficult, let alone comparative action research. Furthermore, the success of a method or tool depends on the context in which it is being used. This context is different for every design process, because every design project is unique. An increased understanding of the effects and interrelationships of the different influencing factors (the focus of Description I) can contribute strongly to the interpretation of the findings resulting from validation studies.

One method of validation that has been applied is the non-comparative observation and analysis of experienced designers using the method or tool under investigation, in an industrial context. Birkhofer [27], for instance, studied the application of the methodical approach proposed by Pahl & Beitz [20] for the design of a textile machine. Hales [3] did the same for the design of a gasifier test rig. Only very few studies involve a comparative analysis. Examples of comparative analyses in a laboratory context have been described by Blessing [1], Ottoson [28] and Christiaans & van Andel [29].

In general, however, systematic testing of methods and tools has not received much attention in design research, despite its importance for the introduction and acceptance of methods and tools in industry. The highlighted difficulties should not be a reason for avoiding validation. Various levels of validation are possible, and creativity in devising validation methods based on the various descriptive methods should be encouraged.

An attempt has been made to create an overview of the descriptive studies in this volume in relation to the proposed design research methodology. The overview is based on the analysis of the papers and the presentations during the seminar only. The results are given in the Conclusions and Outlook at the end of the book.

7. Conclusions

It is remarkable that for so long the investigation of design processes has been neglected. In order to significantly improve the design process, it is necessary to understand this process and for this, proper studies of the process and its influencing factors are essential. Fortunately, this has been recognised by several research groups and an increasing number of publications over the last decade have been the result. The availability of techniques developed in other disciplines for analysing human behaviour, and technological developments such as video recording, contributed to this by providing the methods and tools for detailed observation and analysis. This not only increased knowledge of the design process but also knowledge of doing design research and the various methods that can be applied.

Initially many design studies were very explorative and anecdotal. The time has now come to demand a rigorous approach from everyone working in this area. This rigour applies to the formulation of criteria, the collection of data, the transcription of data, the analysis of data, the drawing and phrasing of conclusions, and the documentation of the study. The involvement of other disciplines with more established research approaches, in particular the social sciences, has strongly contributed to a more rigorous approach, but their help is still needed as we go beyond the boundaries of our current focus of research.

This paper proposes a methodology for engineering design research that underlines the importance of descriptive studies for the development and validation of methods and tools for design. Most of the steps in this methodology have been addressed in design research, in various areas, using many different methods.

The proposed methodology also allows the comparison, classification and assessment of design research. Using the methodology to analyse various design research publications, including those in this volume, revealed some key issues that need to be addressed:

- identification of subsets within each research area that can be investigated scientifically, and that will provide information for the development of practical design methods;
- establishment of a link (link '1' in Figure 1) between the results of descriptive studies (Description I) and the development of design methods (Prescription);
- proper validation of developed methods through the establishment of links between Description II and Prescription (link '2a' in Figure 1), and between Description II and Criteria/Description I (link '2b' in Figure 1).
- development of new research methods, and the adaptation of existing research methods from various other disciplines such as computer science, sociology, psychology and management studies for each step in the methodology, in collaboration with researchers from these disciplines.

It is hoped that the proposed methodology contributes to a more rigorous approach to design research, pieces together the various existing design research areas and encourages collaboration with other disciplines. This is expected to lead to the improvement of design methods and tools, which, in turn, is expected to improve the chances of producing a successful product – the ultimate aim of design research.

References

1. Blessing L.T.M. 1994: *A Process-based Approach to Computer Supported Engineering Design.* PhD thesis, University of Twente.
2. Dwarakanath S., Blessing L.T.M., Wallace K.M. 1996: Descriptive Studies: A Starting Point for Research in Engineering Design. In: T. S. Mruthyunjaya (ed.): *International Conference on Advances in Mechanical Engineering,* 1996. Naros Publishing House, New Delhi, pp. 341–361.
3. Hales C. 1987: *Analysis of the engineering design process in an industrial context.* PhD thesis, University of Cambridge. Gants Hill Publ., Hampshire U.K.
4. Schregenberger J. 1998: *This volume*, pp. 57–67.
5. Blessing L.T.M., Chakrabarti A., Wallace K.M. 1995: A Design Research Methodology. In: V. Hubka (ed.) 1995: *International Conference on Engineering Design, WDK 23.* Heurista, Zürich, pp. 50–55.
6. Verein Deutscher Ingenieure 1975: *Datenverarbeitung in der Konstruktion, Analyse des Konstruktionsprozesses im Hinblick auf den EDV-Einsatz.* VDI 2210 Entwurf Nov. '75. VDI Verlag, Düsseldorf.
7. Cooper R.G. 1990: New Products: What Distinguishes the Winners? *Research-Technology Management* 27–31.
8. Dwyer M.G., Mellor R. 1991: New Product Process Activities and Project Outcomes. *R&D Development* 21, 31–42.
9. Fricke G. 1993: *Konstruieren als flexibler Problemlöseprozeß: empirische Untersuchung über erfolgreiche Strategien und methodische Vorgehensweisen beim Konstruieren.* VDI-Verlag, Düsseldorf (Fortschrittberichte VDI 1/227).
10. Dylla N. 1990: *Denk- und Handlungsabläufe beim Konstruieren.* Hanser, München.
11. Hales C. 1998: *This volume*, pp. 137–148.
12. Wallace K.M., Hales C. 1987: Detailed Analysis of an Engineering Design Project. In: V. Hubka (ed.) *International Conference on Engineering Design, WDK 13.* Heurista, Zürich, pp. 94–101.
13. Duffy A.H.B., O'Donnell F. J. 1998: *This volume*, pp. 269–283.
14. Cross N., Clayburn Cross A. 1998: *This volume*, pp. 71–84.
15. Von der Weth R. 1998: *This volume*, pp. 98–108.
16. Günther J., Ehrlenspiel K. 1998: *This volume*, pp. 85–97.
17. Dörner D 1989: *Die Logik des Mißlingens*. Rowohlt Verlag, Hamburg.
18. Frankenberger E., Badke-Schaub P. 1998: *This volume*, pp. 149–164.
19. Lloyd P. 1998: *This volume*, pp. 113–123.
20. Pahl G., Beitz W. 1994: *Konstruktionslehre*, 3rd edn, Springer Verlag, Berlin.
21. Hubka V. 1982: *Principles of Engineering Design.* Butterworth Scientific, London.
22. Andreasen M.M. 1987: Design Strategy. In: V. Hubka (ed.): *International Conference on Engineering Design, WDK 13.* Heurista, Zürich, pp. 171–178.
23. Ashby M.F., Cebon D. 1993: Materials Selection in Mechanical Design. *Journal de*

Physique IV 3, 1–9, No Colloque C7, supplement to Journal de Physique III.
24. Chakrabarti A. 1991: *Designing by Functions.* PhD thesis, University of Cambridge.
25. Blessing L.T.M. 1994: From Design Characteristics to System Requirements. *Technical Report CUED/C-EDC/TR16,* Cambridge University Engineering Department.
26. Blessing L.T.M. 1997: A Method for Identifying Scope and Assumptions in Developing and Selecting Methods and Tools. *Technical Report CUED/C-EDC/TR*, Cambridge University Engineering Department.
27. Birkhofer H. 1979: Methodisches Entwickeln von Druckknopfansetzmaschinen. *VDI-Berichte* Nr. 347, pp. 51–59.
28. Ottosson S. 1998: *This volume*, pp. 196–201.
29. Christiaans H., van Andel J. 1998: *This volume*, pp. 233–248.

The Further Development of Design Methodologies

Johann W. Schregenberger

Abstract

After some personal remarks about the previous empirical investigations into design processes I make a plea for a holistic understanding of these processes (because, in the end, they have to contribute to the stakeholder value of business enterprises) and for an interdisciplinary style of research that is more broadly orchestrated to obtain better performing design methodologies in the future.

1. Modern Empirical Research on Design Processes, Especially on the Behaviour of Designers

'Before further contrasting elaborated models, it is essential to know what people actually do when they are free to choose, not just what scientists think they should do' (von Cranach & Harré [1, p. 389]). It was with this rousing appeal to action researchers that I opened the memorable Bamberg meeting of 23th March 1984. At that time I was keen to animate D. Dörner, K. Ehrlenspiel and G. Pahl to launch interdisciplinary empirical research into the behaviour of designers. In my contributions for ICED 83 [2], ICED 85 [3] and on other occasions [4, 5] I proposed a research paradigm and strategy for the further development of design methodologies (Figure 1).

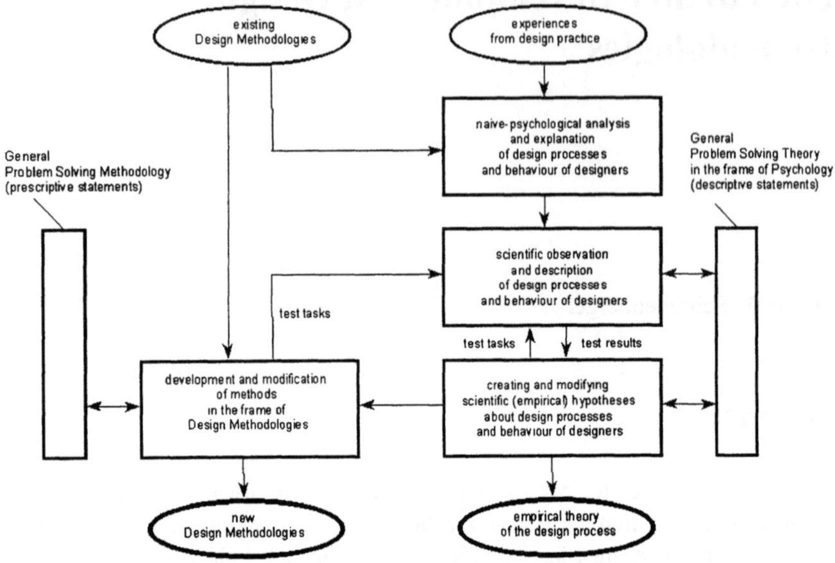

Figure 1: Research activities on the way to empirically better based design methodologies (literal translation of [5], Figure 1).

Since the early 1980s, empirical research on the behaviour of designers has evolved in many places round the world and is now showing its first fruits. We know of several circumspect and sensitive reviews of this research, such as the comments of Cross *et al.* [6, 7] and of Blessing [8]. In the following, I will not pen another review but merely record some lasting, subjective impressions about recent research.

Acceptance, Platforms and Coherence of Empirical Research

- It seems that, at least in mechanical engineering circles (who set the tone of design research), empirical research is nowadays regarded as a promising way of giving new impulses to the further development of design methodologies.
- The design research community has become a remarkably coherent one. We have an 'International Network of Engineering Design Researchers' and researchers can rely on the service of specialised technical journals.
- Over the years, interest groups have taken the initiative in organising several workshops where a hard core of researchers has discussed the state of the art. The periodic ICED and other professional conferences have also been important round-tables for design research.

Interdisciplinary Cooperation

- There are still few engineering research centres and groups engaged in empirical research. Up to now cognitive scientists have been the major research partners. Experience from joint projects (e.g. Frankenberger & Badke-Schaub [9]) seems to be encouraging further cooperation.

Project Definitions and Objectives

- Some projects have obviously been driven by the urge to validate methods already known, such as the core concepts of the VDI 2221 guideline. In these cases the engineering partner has dominated the research objectives and even the methods.
- In recent studies on the behaviour of designers and design teams, the social science partners seem to have a stronger influence not only on the research methods but also on the definition of promising empirical hypotheses (see, for example, Dörner [10]).

Research Results

- For protocol studies, increasingly sophisticated and costly research methods have been developed.
- Researchers and sympathetic observers have developed a remarkable awareness of the questions, methods and findings of empirical research in the design field. The whole discussion now takes place on a much higher level than before. This may be the most important advance.
- Up to now the direct results have been less revolutionary and overwhelming than we had imagined twelve years ago. Why? As the interpretation of data from observation depends largely on the quality of the underlying hypotheses and assumptions, several researchers probably paid for the use of inappropriate and shallow underlying problem solving hypotheses. Thus we face many ambiguous and vague results. Nevertheless research has produced many more precise questions and stronger conjectures (see, for example, Günther & Ehrlenspiel [11]), mainly concerning the characteristics of 'successful designers'.

Application to Teaching and Practice

- Some results, such as the importance of the early design phases and the importance of graphical representations in the conception stage, immediately stimulated teaching in Engineering Design.
- A few former researchers are applying their advanced knowledge successfully on the job. Apart from that, practice is still waiting for new impulses.

2. Holistic Understanding of Design Processes

The current research focus is too narrow. With respect to recent developments in business management we should redefine the traditional role of design methodologies which consists of fostering the understanding of design and the competence and performance of fairly autonomously acting individual designers. The redefinition of this role may emerge from

- the view of designers in their organisational context;
- the modern understanding of design within the framework of Product Development;
- a more comprehensive and holistic understanding of Product Development as a business process subsystem;
- recognising stakeholder value as the final goal of every business process.

Designers in the Organisational Context

We need a more adequate recognition of the fields of work, and also the range of activities, work structures and professional profiles of designers as intellectual workers across diverse disciplines.

Design Within the Framework of Product Development

A new understanding of design as a specific aspect of Integrated Product Development has evolved in the last few years [12, 13]. The design process is the kernel of Product Development. Most attributes of a future product and therefore its success are determined within this process.

Product Development as a Business Process Subsystem

Design methodologies are embedded in a broad arsenal of methodologies concurrently used to plan and manage the diverse business processes [14]. Any applied design methodology is surrounded by and linked to a company-specific super-system of more general management and engineering methodologies. Thus the further development of design methodologies must be related to the development of their methodological super-systems.

Design methodologies belong to the most elaborate methodologies designed to support the planning and control of a large engineering process. But we have shown that their use in practice is not self-evident. One of the barriers to the standard application of design methodologies in practice may be that managers find them the too exclusive domain of engineers, and are not motivated enough to apply them themselves.

Recognising Stakeholder Value as the Final Goal of Every Business Process

The use of design methodologies within the framework of Product Development may ultimately be understood as one of many attempts to increase the stakeholder value of an enterprise, i.e. its aggregate value for all internal and external interest groups [15]. Management methodologies to raise the stakeholder value of business enterprises have mainly been proposed by Total Quality Management proponents such as Zink [16] and by representatives of Systemic Management such as Gomez [17]. Because of the stronger process focus of operational management, Product Development – as a decisive kernel process – has become of increasing importance [13].

Optimising running processes

Improving conditions for work performance

- Model, structure and organisation of enterprise
- Work design: work structures/ organisation, working conditions
- Human resources: competencies, qualifications, motivation
- Implemented methodologies of management, marketing, product develop ment, production, sales and service
- Direct support of brainwork: computer- and communication means, information systems

Integrating external impulses

- Development: markets, technology
- Societal targets (e.g. technology assessment)
- Standards and Regulations

Figure 2: Starting points for raising the stakeholder value of enterprises.

The decisive question now is in which respects and to what extent can the design research community contribute to raising the stakeholder value of business enterprises. Naturally by the development of better performing design methodologies, but there are other important starting points. Figure 2 shows some primary points ('developing areas') on the normative, strategic and operative management levels [18].

3. Orchestrated Research on the Design Process

Why Better Orchestrated Research is Needed?

The design field needs research that is consciously orchestrated because the fostering of design processes in the organisational context requires the combined efforts of all disciplines that are concerned with general aspects of design. More broadly orchestrated research on the design process seems to me a promising way of developing better-performing and better-integrated design methodologies. Above all we must know much more about the actual as well as the desirable integration of design processes into engineering and business processes as a whole. For that, the design research community needs partners from other disciplines who also care about the complex issues of business enterprises relevant to us, especially about work processes and human behaviour in organisations.

Involvement of Other Disciplines

We need stronger support to gain really new impulses (new paradigms, standpoints, views, concepts, validations) for both descriptively and prescriptively oriented design process research. How would you find out, for example, why design methodologies have not had as much influence on Product Development in industry as you would expect? Questions about the acceptance and fruitfulness of methodologies can hardly be researched without the cooperation of social scientists. And how would you define realistic objectives for innovative products if you were not generally aware of the manifold relations between acceptance and success of technologies and their social, cultural and economical contingencies? The history of technology is concerned about that!

The design science community should win over representative research partners from other disciplines to get a broader and stronger platform for the further development of design methodologies. The above envisaged research objects and objectives will involve several fields. In addition to **Cognitive Psychology**, which has been successfully involved since the early 1980s, it will primarily concern **Industrial Science/Sciences of Work** such as **Industrial/Work and Organisational Psychology** (including research on experts) and **Sociology of Work and Occupation**. We also need partners from **Business/Industrial Management** practice, training and research. Finally we should cooperate more closely with specialists in the **History of Technology and Industry** and the **Philosophy of Technology**.

Although scientists from these fields have sporadically cooperated with design process researchers or have contributed to design research following specific institutional or personal interests, until recently there has not been any well established collaboration.

As a second source of impulses, the design science community should also co-operate with authors, proponents and promoters of **other important methodological movements** that have similar development problems or have direct interests in and impacts on design methodologies. I am thinking of the general management schools and of a host of specific movements such as TQM, Value Management, Project Management, Innovation Management, Information Management, Environmental Management and others. Many of these concepts stem from industrial practice, are promoted by international standard organisations and have an increasing impact on real design processes.

Revised Research Strategy

The primary sources of methodological progress seem to me to be:

- the recognition and adaptation of fruitful concepts from other methodologies in engineering and management;
- the recognition of recent findings from whatever disciplines regarding conditions and performance of design processes in the industrial context;
- further interdisciplinary-based empirical investigations on the behaviour of designers and design teams;
- the testing of existing design methodologies in practice.

As proposed earlier [5] and revived recently [19, 20] we need an intensive open dialogue, a discourse that proceeds in cycles between, on the one hand, descriptive studies (those that describe and explain the behaviour of individual designers and design teams as well as those which test new and revised methods and instruments) and, on the other hand, prescriptive initiatives aimed at a revision and broadening of existing methodologies (Figure 3). After more than a decade of empirical research I find now, as before, that this type of research is critical for the development of better design methodologies. We need as many empirical studies as possible because of the ever-limited scope and means of the single project.

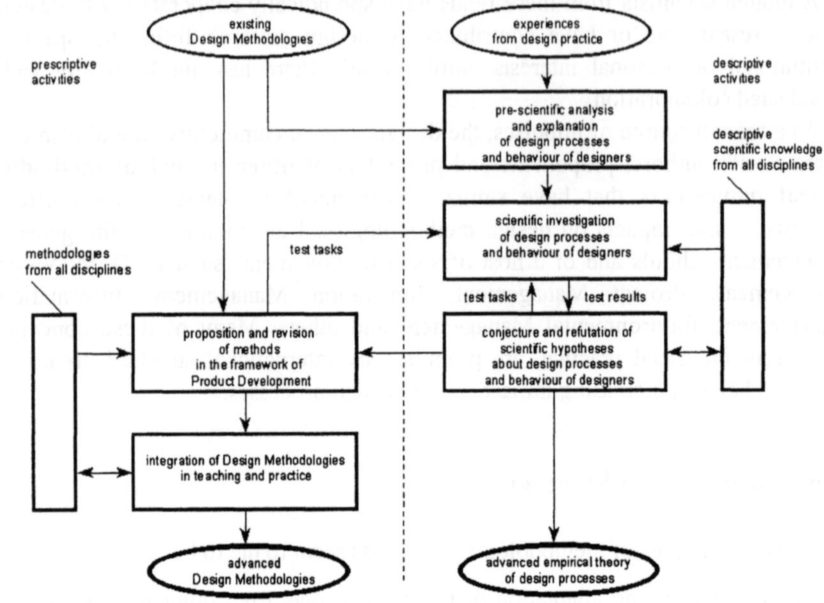

Figure 3: Further research activities towards better performing design methodologies.

Who Researches in Which Relevant Field?

Descriptive and prescriptive research [21] related to the design process (in the framework of Product Development) requires a broader effort. In the following I attribute a number of relevant publications from authors that I know in the formerly defined disciplines. Although the list is arbitrary, it may encourage design researchers to enter into discussions with one or other representative of these disciplines.

Cognitive Psychology
Badke-Schaub [22], Bromme [23], Dörner [24], Strohschneider & von der Weth [25]

Work and Organisational Psychology
Beitz *et al.* [26], Frieling & Klein [27], Hacker [28, 29], Kötter & Volpert [30], Ulich [31]

Sociology of Work and Sociology of Occupation
Brödner [32], Ekardt *et al.* [33], Helten [34], Kahsnitz *et al.* [35], Lange [36], Müller & Frech [37], Senghaas-Knobloch *et al.* [38], Wolf *et al.* [39]

Business/Industrial Management
Bleicher [18], Gomez [17], Warnecke [40], Watson [41], Zink [16]

History of Technology and History of Industry
Gugerli [42], Hellige [43], König [44, 45], Peters [46], Pfammatter [47]

Philosophy of Technology
Ropohl [48, 49]

General Problem Solving and Engineering Methodologies
Daenzer & Huber [50], Flood [51], Gomez & Probst [52], Hall [53], VDI [54], Züst [55]

Research Practice

How could cooperation with other disciplines be initiated and coordinated? An easy way to start discussion would be to look at recent publications from other disciplines. We must of course realise the different focus of interest of other research traditions.

We should first address researchers who are personally interested in some aspects of design methodologies. Joint and well balanced interests are crucial for joint projects. Also very important is an intensive dialogue with those in practice about their inclinations for and against design methodologies as well as their support, experience and criticisms of it.

Coordinated research requires regular meetings where all partners contribute. A steady exchange of ideas and results between all active research groups would surely foster research. Internet-based engineering research networks may become an attractive platform for this dialogue.

References

1. von Cranach M., Harré R. (eds) 1982: *The Analysis of Action. Recent Theoretical and Empirical Advances.* University Press, Cambridge (UK).
2. Schregenberger J.W. 1983: Probleme als Konstrukte. *Proceedings of ICED 83 Copenhagen.* Edition Heurista, Zürich.
3. Schregenberger J.W. 1985: Neue Impulse für die Konstruktionsmethodik. *Proceedings of ICED 85 Hamburg.* Edition Heurista, Zürich.
4. Schregenberger J.W. 1984: *Neue Impulse für die Konstruktionsmethodik des Maschinenbaus.* Arbeitsbericht Institut für Bauplanung und Baubetrieb, ETH Zürich.
5. Schregenberger J.W. 1986: Erfolgreicher Konstruieren – aber wie? *Schweizer Maschinenmarkt* 23, 46–49.
6. Cross N., Dorst K., Roozenburg N. (eds) 1991: *Research in Design Thinking.* University Press, Delft.
7. Cross N., Christiaans H., Dorst K. 1996: *Analysing Design Activity.* John Wiley, New York.

8. Blessing L.T.M. 1994: *A Process-Based Approach to Computer-Supported Engineering Design.* Thesis, University of Twente.
9. Frankenberger E., Badke-Schaub P. 1998: *This volume*, pp. 149–164.
10. Dörner D. 1998: *This volume*, pp. 3–11.
11. Günther J., Ehrlenspiel K. 1998: *This volume*, pp. 85–97.
12. Ehrlenspiel K. 1995: *Integrierte Produktentwicklung.* München: Hanser.
13. Fricke G., Lohse G. 1997: *Entwicklungsmanagement. Mit methodischer Produktentwicklung zum Unternehmenserfolg.* Springer, Berlin.
14. Schregenberger J. W. 1996: *The Position of Design Methodology in a System of Methodologies.* Memorandum (revised) for WDK-Workshop Integrating Design Methodology, Pilsen 1996. Institut für Bauplanung und Baubetrieb, ETH Zürich.
15. Janisch M. 1993: *Das strategische Anspruchsgruppenmanagement: Vom Shareholder Value zum Stakeholder Value.* Haupt, Bern.
16. Zink H. J. 1995: *TQM als integratives Managementkonzept.* Hanser, München.
17. Gomez P. 1993: *Wertmanagement. Vernetzte Strategien für Unternehmen im Wandel.* Econ, Düsseldorf.
18. Bleicher K. 1996: *Das Konzept Integriertes Management.* Campus, Frankfurt.
19. Blessing L.T.M., Chakrabarti A., Wallace K. 1995: A Design Research Methodology. *Proceedings of ICED 95 Praha.* Edition Heurista, Zürich.
20. Blessing L.T.M., Chakrabarti A., Wallace K. 1998: *This volume*, pp. 42–56.
21. Hubka V., Schregenberger J.W. 1989: Eine Ordnung konstruktionswissenschaftlicher Aussagen. *VDI-Zeitschrift* 3, 33–36.
22. Badke-Schaub P. 1994: *Gruppen und komplexe Probleme.* Peter Lang, Frankfurt.
23. Bromme R. 1992: *Der Lehrer als Experte. Zur Psychologie des professionellen Wissens.* Huber, Bern.
24. Dörner D. 1995: Konstruktion und Intuition. In: VDI EKV (ed.) *Effizienter Entwickeln und Konstruieren.* VDI-Verlag, Düsseldorf (VDI-Berichte 1169).
25. Strohschneider S., von der Weth R. 1993: *Ja mach nur einen Plan. Pannen und Fehlschläge – Ursachen, Beispiele, Lösungen.* Huber, Bern.
26. Beitz W., Langner T., Luczak H., Mueller T. 1990: Evaluation of a CAD-course in laboratory experiments. *Int. Journal of Human Computer Interaction* 2, pp. 111–135.
27. Frieling E,. Klein H. (eds) 1988: *Rechnerunterstütztes Konstruieren. Bedingungen und Auswirkungen von CAD.* Huber, Bern.
28. Hacker W. 1992: *Expertenkönnen. Erkennen und Vermitteln.* Verlag für angewandte Psychologie, Stuttgart.
29. Hacker W. 1994: Zur Struktur von Handlungswissen bei Entwurfstätigkeiten. In: Pahl G. (ed.) *Psychologische und pädagogische Fragen beim methodischen Konstruieren.* Verlag TÜV Rheinland, Köln.
30. Kötter W., Volpert W. 1993: Arbeitsgestaltung als Arbeitsaufgabe. Ein arbeitspsychologischer Beitrag zu einer Theorie der Gestaltung von Arbeit und Technik. *Zeitschrift für Arbeitswissenschaft* 3, 129–140.
31. Ulich E. 1994: *Arbeitspsychologie* (3rd edn). Hochschulverlag ETH, Zürich.
32. Brödner P. 1997: *Der überlistete Odysseus. Über das zerrüttete Verhältnis von Menschen und Maschinen.* Edition Sigma, Berlin.
33. Ekardt H.-P., Löffler R., Hengstenberg H. 1992: *Arbeitssituation von Firmenbauleitern.* Campus, Frankfurt.
34. Helten F. 1993: *Technik und Konstruktionshandeln.* Peter Lang, Frankfurt.
35. Kahsnitz D., Ropohl G., Schmid A. (eds) 1997: *Handbuch zur Arbeitslehre.* Oldenburg, München.

36. Lange H. 1995: Qualifikation und Rolle. Führungskräfte und wissenschaftlich-technische Fachkräfte im Übergang. In: H. Lange (ed.): *Man konnte und man muss sich verändern.* Lit, Münster.
37. Müller W., Frech U. 1996: *Die Grenzen der "Irgendwie"-Kommunikation. Vernetzte CAD-Systeme und Kooperation in Klein- und Mittelunternehmen.* Artec paper, Univ. Bremen, Bremen.
38. Senghaas-Knobloch E., Nagler B., Dohms A. 1995: *Industrielle Gruppenarbeit aus der Erlebnisperspektive. Wandel beruflicher Selbstbilder.* Artec paper, Univ. Bremen, Bremen.
39. Wolf H., Mickler O., Manske F. 1992: *Eingriffe in Kopfarbeit. Die Computerisierung technischer Büros im Maschinenbau.* Sigma, Berlin.
40. Warnecke H.-J. 1992: *Die fraktale Fabrik. Revolution der Unternehmenskultur.* Springer, Berlin.
41. Watson G.H. 1994: *Business Systems Engineering. Managing Breakthrough Changes for Productivity and Profit.* Wiley & Sons, New York.
42. Gugerli D. 1996: *Redeströme. Zur Elektrifizierung der Schweiz 1880–1914.* Chronos, Zürich.
43. Hellige H.D. 1995: Hierarchische Ablaufsteuerung oder kooperative Bewältigung von Problemzusammenhängen? Zur Geschichte von Modellen des Konstruktionsprozesses. In: H. Lange & W. Mueller (eds) *Kooperation in der Arbeits- und Technikgestaltung.* Lit, Münster.
44. König W. 1994: Konstruktionslehre zwischen Theorie und Praxis. In: Pahl G. (ed.) *Psychologische und pädagogische Fragen beim methodischen Konstruieren.* Verlag TÜV Rheinland, Köln.
45. König W. 1995: *Technikwissenschaften. Die Entstehung der Elektrotechnik aus Industrie und Wissenschaft zwischen 1880 und 1914.* Fakultas, Chur.
46. Peters T. F. 1996: *Building the Nineteenth Century.* MIT Press, Cambridge, Mass.
47. Pfammatter U. 1997: *Die Erfindung des modernen Architekten. Ursprung und Entwicklung seiner wissenschaftlich-industriellen Ausbildung.* Birkhäuser, Basel.
48. Ropohl G. 1995: *Die unvollkommene Technik.* Suhrkamp, Frankfurt.
49. Ropohl G. 1996: *Ethik und Technikbewertung.* Suhrkamp, Frankfurt.
50. Daenzer W.F., Huber F. (eds) 1997: *Systems Engineering. Methodik und Praxis.* Autorenkollektiv: Haberfellner R., Nagel P., Becker M., Büchel A., von Massow H.. Industrielle Organisation, Zürich (9th edn).
51. Flood R.L. 1995: *Solving Problem Solving.* Wiley, Chichester.
52. Gomez P., Probst G.J.B. 1995: *Die Praxis des ganzheitlichen Problemlösens.* Haupt, Bern.
53. Hall A.D. III 1989: *Metasystems Methodology. A new Synthesis and Unification.* Pergamon Press, Oxford.
54. VDI-GSP 1995: *Wertanalyse. Idee – Methode – System.* VDI-Verlag, Düsseldorf.
55. Züst R. 1997: *Einstieg ins Systems Engineering.* Industrielle Organisation, Zürich.

Résumé of Section I

The contributions in this section illustrate that there are several techniques and thus various design research methods that can obtain greater insight into design processes. On the one hand there are prescriptive approaches that are primarily product oriented, on the other hand some prescriptive approaches reflect the process by experience; descriptive approaches analyse empirically investigated processes and sometimes prescriptions also arise from these observations.

In attempting to obtain detailed information on design processes, emphasis has been put on the importance of detailed observation-based analysis of single cases of design work, if possible via multi-disciplinary research teams. There are two principal methods of observation, participant and non-participant, the latter perhaps leading to more valid data but being the more restricted approach. In any case, clear criteria and categories are needed in order to observe and describe design processes, especially if we want to compare different approaches or even to make prognoses about the future in order to evaluate the various models and methods.

Considering that the aim of design research is not just an understanding of design processes but also improving the design process and thus the product, we therefore need criteria that relate the process to the result. Eventually, we must know what results of which criteria are of importance, e.g. the resolution levels of cognitive processes, the time steps of the design process or the overall process, the cost, the quality, etc. Evidently, on the one hand there are *research criteria*, on the other hand *product criteria*, and both kinds of criteria may lead to completely different ratings. This dilemma illustrates that the rating of success depends on the value given to the results of the criteria, which itself may change with time as a process is always a dynamic structure.

Thus, it becomes obvious that design research has to evolve. The criteria described should be used to investigate design processes to make assumptions, and to prove them. Of course, the criteria may be totally different from case to case, because the aim and the subject of the investigations will extend from reasoning processes to organisational, industrial, philosophical and other aspects. In the sense of Schregenbergers 'orchestrated research of the design process' (in this volume) we need comparable criteria if we want to bring different perspectives and different research aspects together.

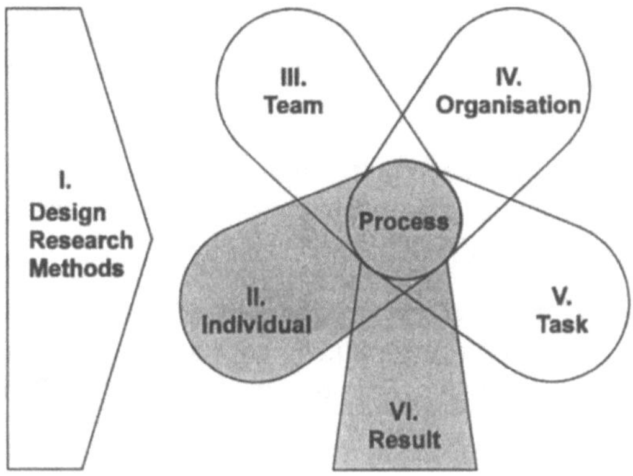

Section II:

Individual Influences

In engineering design, product characteristics are predominantly of major interest. However, questions like 'how can the product be optimised? are necessarily dependent on questions like 'what does the design process of a good designer look like?' and 'how can the designer be supported in an attempt to find the best route to the best product?'

To date, there have been many investigations and results about the characteristics of the individual designer, see for example Pahl in this volume. Among many individual prerequisites the *experience* of designers seems to be the most interesting criterion to look at, especially because it seems to be easy to compile; and groups can be compared using this obvious meaningful characteristic. In this section the following questions are of major interest:

How do expert designers design?

How does a design procedure influenced by experience differ from that influenced by methodical education?

What are the possible internal processes that explain these differences and what processes explain the process of an 'unmethodical' procedure?

The contributions of this section move forward from the *observation* of the activity to that of *modelling,* and close with the *explanatory effects* of individual influences:

- In the first contribution *Cross & Clayburn Cross* describe the successful strategies of 'outstanding' experts. Their approach is descriptive and stays close to the designers' activities.
- *Günther & Ehrlenspiel* analyse the processes of designers with and without education in systematic design. Their approach is structured and formalised in terms of categories derived from design methodology.
- Partly based on the same investigation as Günther & Ehrlenspiel and on investigations with marketing experts, *von der Weth* discusses the question of which internal mechanisms lead to non-methodological procedures (e.g. with very reduced task clarification).

Expert Designers

Nigel Cross & Anita Clayburn Cross

Abstract

Progress in design methodology will be restricted, or proceed in inappropriate directions, if studies of designer behaviour are limited to studies of novice or average-ability designers. We have studied two outstanding expert designers and are able to draw some parallels between their design strategies. We note that they both take a systemic view of the design situation, choose to frame their view of the problem in a challenging way, and draw upon first principles to guide both their overall concept and detailed design. Our observations are in line with those of studies of expert performance in other creative fields.

1. Introduction

Although teamwork is increasingly the dominant mode of design activity, many instances still occur of design practised almost solely by one individual. Even within teams, one individual may be nominated as the 'chief designer', or may spontaneously take on such a role through an apparently natural level of design ability. It is also the case that many highly creative or talented individuals become successful and highly-regarded designers, with international reputations both within and beyond their professional peer groups. The reality of design practice seems to be that some individuals have outstanding design abilities.

Nevertheless, most studies of designer behaviour have been based on novices (e.g. students) or, at best, designers of relatively modest talents. This is because it is easier to obtain such people as subjects for study. However, if studies of designer behaviour are limited to studies of rather inexpert designers, then our understanding of design ability will also be limited. This will in turn limit our potential for developing models of design activity, and restrict progress in the whole field of design methodology. Studying expert and even 'outstanding' designers may give us different, and more appropriate, insights and understanding of design activity, on which to develop models and methods of design.

There have been only a few studies of outstanding designers, such as Lawson's [1] studies of successful architects. In this paper, we report briefly on our own study of an outstanding designer – the racing car designer, Gordon Murray – and we make a new analysis of the strategies of the expert designer 'Dan', who was a subject of the 1994 Delft Design Protocols Workshop. In this analysis, we develop parallels between Dan's design strategies and those of Gordon Murray. We find several striking similarities, which suggest that general lessons may be constructed about the nature of expertise in design. Such lessons could be important steps towards raising general standards of design across the professions and towards improving design education and the nurture of design expertise.

2. Expert Designer 1: Gordon Murray

Our first study is of an engineering designer who has established a long record as a highly successful and highly innovative designer in a highly competitive environment; that of Formula One racing car design. Gordon Murray became chief designer for the Brabham racing car team as a young man, in 1973. Brabham cars designed by him were driven by Nelson Picquet to win World Championships in 1981 and 1983. In 1987, Gordon Murray moved to the McLaren Formula One team as technical director, and again his car designs, driven by Alain Prost and Ayrton Senna, won World Championships in 1989 and 1990. In some 20 years as a Formula One designer, he established an outstanding reputation not only as a successful designer (over 50 race wins) but also as a consistently radical innovator.

Gordon Murray then became technical director of an off-shoot company from the racing team, McLaren Cars Limited, and became responsible for the design and development of a completely new, road-going 'super car' – the McLaren F1, which attracted immense attention for its radical design features. Although not originally designed for racing (but designed to racing principles), GTR versions of the new McLaren F1 were entered in the 1995 Le Mans 24-hour race and came first, third, fourth and fifth. They continue to have considerable success in GT racing.

Gordon Murray is an example of an 'outstanding' designer – a designer who has achieved extraordinary success throughout a long career. In his case the measures of his success as a designer are absolute – his achievements have been in a competition field where absolute performance standards are the criteria. We have been able to gain some insight into Gordon's design strategies and approaches through discussion and informal interviews with him. We have reported this in more depth elsewhere [2]. Here, we will refer to the three aspects of his design strategy that we have suggested may offer keys to his design successes: systemic design, framing the problem, and designing from first principles.

2.1 Systemic Design

An example of Gordon Murray's radical design approach was the Brabham team's introduction of planned pit stops during a race, in the 1982 season. This was not a radical innovation in car design *per se*, but reflected a total systems approach to the overall goal of winning each race. At that time it was not normal to have pit stops as regular, planned parts of the race routine. Pit stops were for emergencies such as changing a punctured or badly-worn tyre.

In 1982, as part of his thinking about how to make a car lighter in weight, Gordon conceived of running the car with only half the normal, full-race fuel load, and including a pit stop for re-fuelling. Nowadays, Formula One pit stops have been refined down to an incredibly quick norm of about seven seconds actual stopped time, in which time all four wheels are changed and some 100 litres of fuel taken on. The total racing time lost is about 20 seconds. In 1982, a quick pit stop for tyre changes took about fifteen seconds of actual stopped time. Gordon calculated that to preserve the advantage gained from the lighter car he had to get the stopped time in the pit down to about ten seconds and to reduce the lost racing time to under 26 seconds.

To achieve this, an extraordinary development programme had to be undertaken, under the pressures of Formula One racing. For example, within three weeks, Gordon and his team had thought of, designed, made and tested a high-pressure re-fuelling system. To improve pit-stop procedures, he hired a film crew to film the team practising pit stops, and then played back the film, stopping it to identify difficulties and errors, and devising ways to improve the procedures. Such improvements included details such as re-designing the wheel-nut gun to improve its engagement with the nut. The new systems, the improvements and the training of the pit team reduced the actual stopped-time to under the target of ten seconds.

However, one 'big killer' remained, Gordon told us: 'When you put new tyres on they were cold, and it always took two laps to get back up to speed, and the time you lost in those two laps killed the whole thing. So then I thought, well I know the tyres start working at 70 degrees temperature ... so we designed an oven, a wooden oven with a gas-fired heater, and we heated the tyres up – and ten seconds before the car was coming in we opened the oven door, whipped the tyres out, put them on, and the guy was instantly quick. Now every Grand Prix team has tyre heating; that's where it started.'

The example of the introduction of planned pit stops illustrates how Gordon Murray adopted a total systems approach to the design task: the goal was winning a race over some 60 or 70 laps, which could include bringing the car into the pits during the race if that could be used to convey sufficient advantage. The systemic view of the design task led to a long series of detailed developments and innovations.

2.2 Framing the Problem

At the start of the 1981 season, the Formula One governing body, FISA, had introduced new regulations intended to reduce the 'ground effect' on racing cars. This effect had been pioneered on Lotus cars some three seasons earlier; smooth underbodies, flexible side-skirts and careful aerodynamic design provided a ground-effect downforce which increased the car's grip on the track surface. This meant much higher cornering speeds were possible, and by the 1980 season people were worried about safety and the *g*-force effects that were being imposed on the drivers. In 1981 FISA set a minimum ground clearance of 6 cm. For Gordon this change in the regulations was simply another stimulus to innovation.

Gordon's thinking on this – which he says came as a sudden illumination after a long period of worrying at the problem – was that the authorities had to accept that at some points during a race, any car's ground clearance is going to be less than the 6 cm minimum, simply because of the effects of braking, or roll on corners, etc. Knowing that any driver-operated, mechanical device to alter the adjust space ground clearance was illegal, he focused on the physical forces that act on a car in motion. The braking and cornering forces he felt unable to work with because of their asymmetrical effects on the car, but the downforce from air pressure on a moving car will, if the car is correctly designed aerodynamically, push the car down equally over its whole length and width. The design challenge, therefore, was to let the natural downforce push the car down at speed, and then somehow to keep it down when it slowed for corners, but allow the car to return to 6 cm ground clearance at standstill.

The ingenious solution that Gordon developed incorporated hydro-pneumatic suspension struts at each wheel, connected to hydraulic fluid reservoirs. As the car went faster, the aerodynamic downforce pushed the body lower on its suspension and the hydraulic fluid in each suspension strut was pushed out into the reservoirs. The trick then was to find a way of letting the fluid return to the suspension struts only very slowly when the car slowed down. At cornering speeds, the suspension would stay low, but on slowing down and stopping at the end of the race, the fluid would return from the reservoirs to the suspension struts, giving the required 6 cm ground clearance. Murray and his team developed such a system, using devices such as organic micro-filters borrowed from medical technology.

The hydro-pneumatic suspension system is an example of radical innovation arising through framing the problem in a new and productive way: how to lower the car and keep it there by natural forces. Like the systemic design approach, it also requires the motivation to follow through a basic idea into a finely-detailed implementation.

2.3 Designing from First Principles

Gordon Murray insists on keeping experience 'at the back of your mind, not the front' and to work from first principles when designing. For instance, in designing a component such as a suspension wishbone, he says 'it's all too easy, and the longer you're in design the easier it is, to say I know all about wishbones, this is how it's going to look because that's what wishbones look like.' But to make a step forward in design, to look for ways of making it much better and much lighter, then you have to go right back to first principles such as load-path analysis. It reflects an attitude of constantly assuming you are engaged in innovative design, rather than routine design.

As one example of his approach to designing from first principles, Gordon referred to a small, and perhaps seemingly insignificant part of the McLaren F1 – the steering column. 'Conventionally, it would have been, right, steering columns are typically three-quarter-inch solid steel bars.' He explained how this conventional solution arises because the column not only has to carry torsional forces from the resistance to the turning wheels but also bending loads from the driver leaning on it whilst getting in and out of the car. It also has conventional points of support, is mounted in rubber bushes to reduce noise, and it ends up being encased in a plastic housing for reasons of appearance and convenience. But it does not provide the sort of direct steering feel that a racing car has.

So Gordon decided to apply racing design principles to the steering column, starting by separating the needs to carry both torque and bending loads. However you design the steering column itself, you still need a cover to house electrical cables and to mount switches. So he reasoned that 'if you've got to have that anyway, why not use the insect principle where the skeleton's on the outside, and make that the structure that takes all the bending forces?' This thinking led to the design of an 'exo-skeleton' structural cover, with the steering column itself as an aluminium tube of just 1 mm wall thickness; 'it's only taking torque and it weighs nothing.' With other detail design changes, the steering system is now lighter but stronger than a conventional solution, and also has the right racing feel.

The steering column design process therefore stemmed from considering first principles – separating the torque and bending loads – and from an imaginative breakthrough – using the housing cover for structural purposes as well as appearance.

3. Expert Designer 2: 'Dan'

We propose to compare the design strategies of Gordon Murray with those of a designer who, though not such an internationally famous designer, is nonetheless another highly successful, outstanding designer. This second designer is 'Dan', the subject used in the 1994 Delft Design Protocols Workshop. Dan is an engineering designer with over twenty years' experience of designing both mechanical and electro-mechanical machines, and robotic systems and devices. He was one of the earliest designers of robotic devices, and he has won several design awards from the American Society of Mechanical Engineers.

Dan was video-recorded whilst he 'thought aloud' during a 2-hour experimental session in which he was asked to design 'a carrying/fastening device that would enable you to fasten and carry a backpack on a mountain bicycle'. Our observations of Dan's design strategy are therefore based on the artificial situation of a controlled, protocol analysis experiment. Full details of the experiment (and several other analyses of Dan's design activity) are reported in the Workshop proceedings [3]. Although he is highly experienced, the design task set in the experiment was, nevertheless, a novel task for Dan.

In the following analysis of Dan's strategy, quotations are taken from the transcript of his 'think aloud' comments, preceded by the timestamp for the quotation. The substantive experimental session began at timestamp 00.15minutes.

3.1 Rapidly Gaining Experience

A distinctive feature of the strategy followed by Dan was the time and effort he put into gaining an overview of the potential solution space; his aim appeared to be to determine the most feasible section of that solution space in which to begin work. This was a deliberate strategy, which he explained in 'think-aloud' comments such as:

> (00.21)
> there's no sense in starting from scratch if you can start at square two instead of square one

> (00.27)
> my general philosophy is don't try to reinvent the state of the art if it already exists

The early part of Dan's strategy is concerned with getting 'up to speed' – with informing himself of the nature of similar, rival products to that which he is designing, with identifying a section of the solution space where he is likely to generate an acceptable solution, and therefore with avoiding starting to work in areas that are likely to be unfruitful.

This strategy of 'starting from square two' – i.e. building on the experience of others – might appear to be borrowing from old ideas rather than initiating a new design. However, what Dan was doing seemed to be more like immersing himself as rapidly as possible in the domain of expertise relevant to the (for him) novel design task that he had been set. He was prepared to devote rather a lot of time to this – some 30 minutes were spent on gathering this kind of information, and a further 15 minutes were spent confirming and coming to the conclusion that the best location for the carrying device would be over the rear wheel of the bicycle. So at least 45 minutes of the 2-hour design session were spent getting up to speed, so that he would hit the ground running when he did start actual design work.

Most of this early information was gathered from a potential rival company already making bicycle luggage carriers. Dan gathered information from the company's catalogue and from a telephone call to the company. The catalogue information was scanned for general principles – the weight and cost limits of the rival products, and their mounting positions on the bicycle – as indicated in these comments:

(00.28)
aha, so he has a series of front mounting racks and rear racks, and he sells these things at between . . thirty dollars and fifty dollars

(00.29)
his frames weigh between . . 480 grams and . . 650 grams

(00.30)
he has more rear racks than front racks . . they are all frames that mount over the wheel

The telephone call was also used to gather general principles and expert advice about the preferred location on the bicycle for carrying a backpack. Dan commented as follows about what he learned from the telephone call:

(00.44)
I learned a few things; I learned about the fact that people originally thought that it was bad to have it on the front but if you keep the backpack pretty low on the front it's OK, high up is bad; on the rear the issues are related to heel clearance and thigh clearance, and he feels that keeping it as low as possible is good

So Dan appeared to be searching not for prior design examples *per se*, but for the criteria (such as weight and cost) that his design would have to match, and for experience that would guide the major design decision of whereabouts on the bicycle to locate the carrying device. In this way, Dan rapidly developed valuable, surrogate expertise in this novel design domain.

3.2 Systemic Design

As an expert designer, Dan also displayed aspects of his design approach that can be compared with those of an outstanding designer such as Gordon Murray. Firstly, Dan developed a systemic view of the problem he had been set. Very early in the session, reading the design brief, he made a comment that suggested he saw something special about the design problem:

(00.19)
it is to attach to a bicycle, a mountain bike and to me that makes it different

Dan was also able to draw on personal experience that helped him to formulate some of the implicit requirements for a good design solution:

(00.26)
having used a backpack on a bike in the past and having ridden over many mountains, unfortunately not on a mountain bike but I can imagine that the situation is similar, I learned very early on that you want to keep it as low as possible

He also drew upon personal experience to confirm that the preferred location for the backpack would be on the rear wheel rather than the front wheel:

> (00.51)
> my first thought is hey the place to put it is back here; there's another advantage by the way of having it in the back I can see immediately, and that is it's off the side in the front, and you're on a mountain bike trail and you hit something you're out of control in the front wheel

> (00.52)
> downhill work on mountain bikes, I know you want to keep your weight back rather than forwards

Dan's personal experience of biking with a backpack led him to identify an issue that only someone who has had such experience might be aware of:

> (00.55)
> when I biked around Hawaii as a kid that's how I mounted my backpack . . . and I have to admit if there's any weight up here this thing does a bit of wobbling, and I remember that as an issue

So the systemic view that Dan formed of the problem was that of the total task that encompasses the dynamic system of the rider plus bicycle plus backpack, and the issues of control of the bicycle that arise in the situation of riding over rough terrain with a heavy backpack attached to the bicycle. This is a different situation from that of everyday, smooth-surface, level-grade riding, and it accentuates the need to position the backpack low and to the rear. The view that Dan had of the design task was significantly different from a view that might be formed from considering the bicycle and backpack in a static situation, or without considering the effects on the rider's ability to control the bicycle with a mounted backpack. Dan's understanding of the dynamic situation therefore enabled him to formulate a systemic view of the design task.

3.3 Framing the Problem

The second aspect of Dan's approach that can be compared with that of Gordon Murray is how he 'framed' the problem. From a systemic overview of the total dynamic situation of rider + bicycle + backpack, Dan identified stability as a key issue. Quite early in the session, commenting on the prototype design that had been developed earlier by other designers, he surmised about the user-evaluation report on this prototype that:

> (00.22)
> it probably . . says the backpack's too high or something like that, and that bicycle stability's an issue

Dan therefore seemed to frame the problem as 'how to maintain stability', given that a heavy backpack had to be carried over the rear wheel of the bicycle, and given his experience of the 'wobbling' that can occur in the riding situation. This problem-framing and his prior experience led him to conclude that he must design a rigid carrying device:

> (00.59)
> the biggest thing that I remember in backpack mounting is that it's got to be rigid, very rigid

He then developed this viewpoint into the requirement that the structural members of any carrying device must be stiff:

> (01.06)
> making the carrier stiff enough for holding the backpack, that seems to be a big issue

So, at about halfway through the session, Dan had arrived at a framing of the problem which directed him to design a stiff, rigid carrier, mounted as low as possible over the rear wheel. Soon after, a secondary framing viewpoint emerged, which seemed to arise from considering the client's needs as well as those of the user (which had dominated Dan's thinking so far). The client for the design task was a manufacturer who wanted to sell the carrying device in conjunction with an already-existing backpack. The device therefore needed to have unique selling points that differentiated it from other, similar products. During the development of his design concept, Dan kept in mind that he needed the product to have a 'proprietary feature', as emerged from some of his comments, discussed below.

3.4 Designing from First Principles

The third aspect of Gordon Murray's approach as an outstanding designer that we identified was his concern with designing 'from first principles'. Dan also showed this aspect, as he developed his concept design for the carrying device. A 'first principle' that Dan identified and followed was that a triangulated structure is inherently rigid. This led him to avoid designing a rectangular, parallelogram form of structure, which was the form that rather naturally seemed to arise from considering the basic shape of the carrier and the location of its supporting structure on the bicycle. Whilst drawing Figure 1(a), Dan commented:

> (01.07)
> one of the problems with a bicycle carrier where the frame is mounted out here and it goes to that, is that you end up with a parallelogram; bad thing, bad thing!

He expanded on this comment, identifying his concern with stability as a key requirement:

> (01.08)
> if I were to make a frame that looked like this, that would be a very poor design because basically what I've got is, I've got a parallelogram which has very little lateral stability

He then introduced the 'first principle' of triangularity, whilst drawing the triangular form on to Figure 1(a):

> (01.09)
> it would be nice if I could, for instance, run these rods up here to some point and therefore create a triangle, this would give me great stiffness – good idea!

The 'first principle' of triangularity subsequently guided Dan's generation of the basic form and the detailed design features of his carrier. As he drew his design in more detail (Figure 2), he commented:

> (01.16)
> we're going to have this as a triangular structure here to provide the lateral stability

And as he developed the design he constantly referred to structural principles, seeking to avoid 'bad' configurations and to generate 'good' ones, making comments such as:

> (01.42)
> my detail here is going to have to be something like this because my forces along this tube are this way . . good, this is good; and then this detail is going to be, er, let's see . . alright that's bad . . that's bad . . that's bad, so I'm going to need something like that

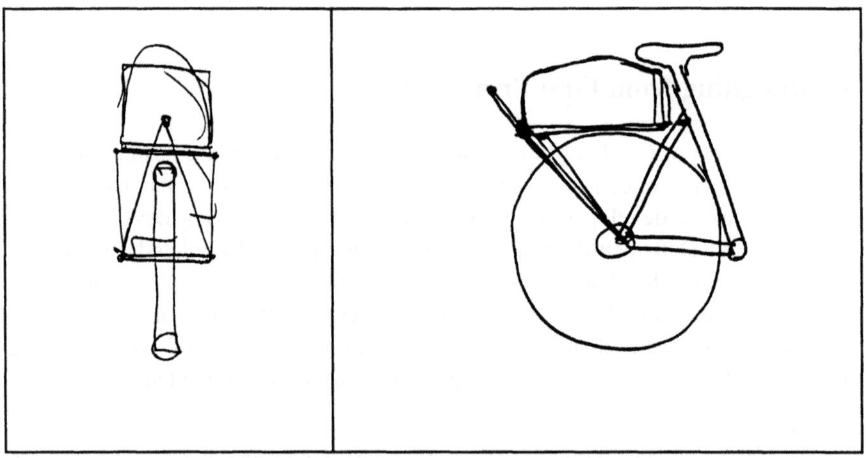

Figure 1: Dan's sketches of rear and side views of the bicycle and carrying device.

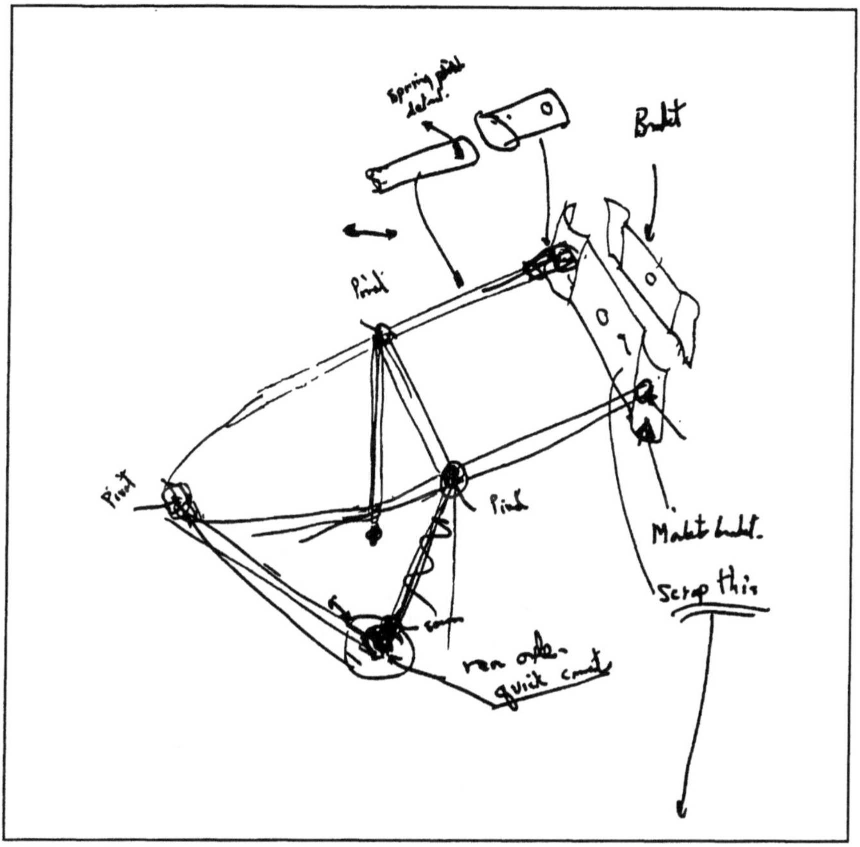

Figure 2: Dan's detailed development of his design.

In the meanwhile, as we noted above, Dan also used the client's requirement of a unique selling proposition to help guide and to reinforce his decision to seek a design based on triangular structures:

> (01.10)
> that is going to be our proprietary feature, a triangular, rigid structure with no bends in it; these rods are then going to be in tension and compression, no bending
>
> (01.41)
> I want to make sure that this rod here comes to a point, not stop right there . . that's to a point; that's going to be my feature

In these comments, Dan demonstrated that he regarded the pronounced triangular form at the rear of the product as something to be maintained as a feature that would help give the product an attractive, unique selling point.

3.5 Summary of Dan's Design Approach

Dan's design for the carrying device is an integrated design in which user requirements are addressed through the problem frame of stability, leading to the use of triangularity as the guiding first principle, which also addresses the client's goal of having a proprietary, unique selling feature to the product, as summarised in Figure 3.

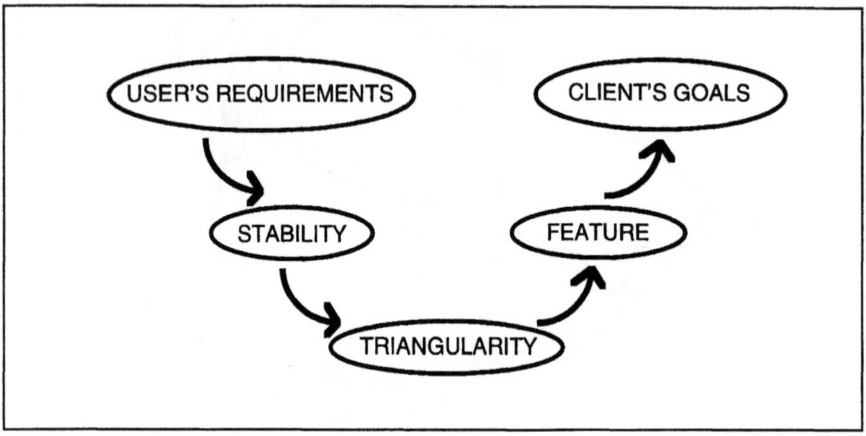

Figure 3: Dan's approach summarised as hierarchical levels of systemic design, problem framing and designing from first principles.

4. Discussion: Expertise in Design

We believe that we have been able to demonstrate that there are similar aspects in the design approaches of two outstanding expert designers. Both take a systemic approach to the overall design task; both frame the design problem in a way that challenges them to innovate; and both use first principles to guide their conceptual and detail designing. It is perhaps surprising that we see commonalities between these two designers, considering the great disparity between the design projects in which they were engaged – between Formula One racing cars and a simple bicycle luggage rack!

Therefore it seems that some common aspects of expertise are shared by these two outstanding designers. Outstanding designers may not have developed, or been gifted with, particular, unique strategies and approaches to design, but may be using similar strategies and approaches as other designers. In that case, we may hope that these successful strategies and approaches not only can be identified in outstanding designers but also can be coached and developed in less expert and novice designers.

Expertise has been studied in other fields (see, for example, Ericsson and Smith [4]), but so far there are few studies of expertise in design. What we can learn from these studies in other fields is also limited, because of the particular, distinctive characteristics of design activity. Many of the classic studies of expertise have been based on examples of game-playing (e.g. chess), or on comparisons of experts versus novices in solving routine problems (e.g. physics). These are all well-defined problems, whereas designers characteristically deal with ill-defined problems.

However, some studies of expertise in fields such as creative writing and computer programming, where problems are more ill-defined, do suggest some parallels with our observations of expert designers. Holyoak [5] has reviewed studies which suggest that some of the 'standard' results from studies of expertise do not match with results from studies of expertise in creative domains. For example, creative experts will define the given task so that it is problematic – i.e. deliberately treat it as ill-defined – which is contrary to the assumption that experts will generally solve a problem in the 'easiest' way, or certainly with more ease than novices. In some ways, therefore, creative experts treat problems as 'harder' problems than novices do. We have seen that both Gordon and Dan are not content to adopt an 'easy' view of the design problem that they are given; both of them choose to take harder, more innovative routes to finding a solution concept. In Gordon's case, it seems clear that deliberately challenging the given design 'rules', such as the technical constraints set by the governing body of his sport, has been a feature of his successful design approach.

Experts in creative fields are also reported to devote considerable time to constructing a problem representation, exploring constraints, etc. Again, we have seen this with both Gordon and Dan. In Dan's case, we saw that he devoted a major amount of his available time to reaching an understanding of the problem, and the potential solution space, before beginning to create a concept within the appropriate segment of that solution space. Gordon will allow himself 'incubation' time, which helps him to reach a new, creative insight.

In comparisons of experts and novices, experts are reported to make a breadth-first search for a global solution concept, whereas novices in the field will often get lost in depth-first searches in more detail. Creative experts are also reported as solving similar tasks from first principles each time, rather than recalling previous solutions. In these instances again, we see similarities with our observations of expertise in design, in our studies of Gordon and Dan. Neither of them gets lost in details, but always seems to have an overall, global concept that guides detail decisions. This is the importance of the 'problem frame' that they adopt. They also both have an approach in which experience of previous solutions is 'at the back of the mind, not at the front', as Gordon expressed it, and both refer, either directly or indirectly, to 'first principles' as the stimuli for creative design.

5. Conclusions

We have been able to explore the strategies employed by two outstanding expert designers, and have found some parallels in their ways of working. In order to create innovative designs, they adopt a systemic view of the design situation, frame their view of the problem in a challenging way, and then use 'first principles' of engineering to guide the generation of the design concept and its detailed development.

Our observations also tend to fit with other findings about the cognitive strategies employed by experts working in other creative domains where problems are ill-defined. It is important to recognise the distinctions between strategies employed by creative experts and those employed by experts working in well-defined problem domains and in routine problem solving. Any general theories of expertise, and any applications in education and practice drawn from more general studies of expertise, need to recognise these distinctions between expertise in creative and routine problem solving.

It is important that we learn more about expertise in design. Too many studies have been based on novices or, at best, mediocre designers. Studying novice and average designers may well limit our understanding of design, holding back progress in design methodology and leading to weak or even inappropriate models of design activity. Studying expert designers might enable us to identify the seeds of 'best practice', and then to transfer these insights more widely across the professions. This should also be useful in education, for guiding the development of better-than-average designers.

References

1. Lawson B. 1994: *Design in Mind.* Butterworth Architecture Press, Oxford.
2. Cross N., Clayburn Cross A. 1996: Winning by Design: The Methods of Gordon Murray, Racing Car Designer. *Design Studies* 17, 91–107.
3. Cross N., Christiaans H., Dorst K. (eds) 1996: *Analysing Design Activity.* John Wiley & Sons Ltd., Chichester.
4. Ericsson K A., Smith J. (eds) 1991: *Toward a General Theory of Expertise: Prospects and Limits*. Cambridge University Press, Cambridge.
5. Holyoak K. J. 1991: Symbolic Connectionism: Toward Third-generation Theories of Expertise. In: K. A. Ericsson & J. Smith (eds): *Toward a General Theory of Expertise: Prospects and Limits*. Cambridge University Press, Cambridge.

How Do Designers from Practice Design?

Joachim Günther & Klaus Ehrlenspiel

Abstract

In individual case studies the design processes of so-called p-designers (experienced designers from practice who have neither education at a university nor education in design methodology) were investigated and compared with those of m-designers (designers with education in design methodology at a university). The results are detailed descriptions of the design processes of eight individual cases and a set of hypotheses about the special design behaviour of p-designers.[1] Possible causes for their procedures are discussed. A minimum design guideline for p-designers is suggested to support their design processes. At the end of the paper, the limitations of the study and conclusions for research and design methodology are presented.

1. Introduction and Objectives

This contribution focuses on designers from practice who have had neither education at a university nor education in design methodology. These experienced designers, named **p-designers** in the following paper, have mainly had a practical education as draftsman, technician or master craftsman. p-Designers often do not use procedures of conventional design methodology. The support of design methodology is often looked upon sceptically [1]. Nevertheless p-designers are successful designers with regard to the quality and duration of the design processes.

In comparison with p-designers, we observed designers who had education in design methodology at a university and gained design experience from projects in cooperation with industry. To simplify the terms used, we named these **m-designers**.

Questions to be investigated by detailed analysis of the design processes are:

[1] The authors would like to thank Lucienne Blessing for suggesting the terms 'p-designers' and 'm-designers' when reading an abstract of this paper.

- How do p-designers without methodical training tackle the process of design?
- Do they use their own methods or systematic steps?
- How do p-designers and m-designers differ in their procedures?

Our objectives were to identify findings for further development and application in design methodology:

- What can we learn from p-designers?
- What hints for design procedure – with regard to quality and time – can be drawn from their processes?
- How can design methodology support p-designers?

2. Methods

In a set of experiments, 20 subjects of varying background (p-designers, m-designers, students) solved a design problem under the same external conditions in a laboratory situation. They had to design a mechanical device without being under time constraint. The design task and working situation, which were near to design work in practice, are documented in [2].

The results of the design process are video-recordings of several hours, sketching and notes, and layout drawings with parts lists. The investigation of the design process was done using detailed case studies with the help of protocol-analysis – see [3]. The resulting designs documented in the layout were evaluated by two groups of experts using a point-rating method in respect of quality. The evaluation method of the designs is documented in [4].

The results in this paper refer to six analysed individual cases of p-designers out of the set of 20 designers. Two additional cases are two experienced m-designers as a basis for comparison with real processes driven by design methodology. Statements on the design processes of the p-designers are always made in comparison with the processes of the m-designers.

3. Results

Following the guidelines of design methodology [5,6,7] the design phases of task clarification, conceptual design, rough embodiment design and final embodiment design should follow in consecutive periods. The phase of detail design was not included in the experiments. The orientation of the methodical procedure is shown as an example in the design process of the m-designer Hans (Figure 1). His process differs only slightly from the ideal procedure.

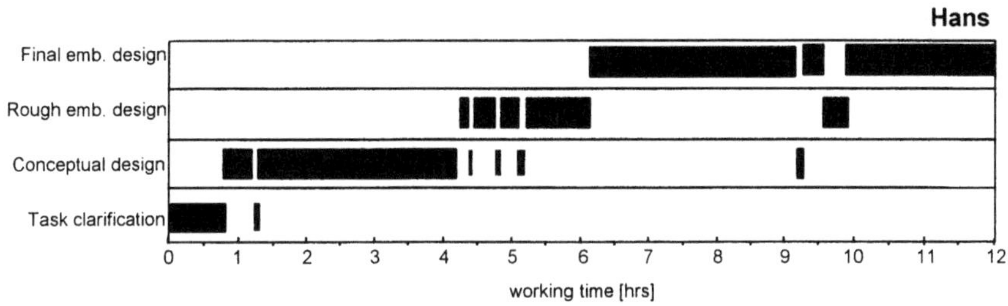

Figure 1: Course of the design phases of the m-designer Hans.

The division of the design process into four phases is the basis for analysis of the process and the character of the design problem. Because all the designers had no experience regarding the problem – it was new to them – they had to go through all the phases.

The following section presents the main hypotheses that can be derived from the results of the investigation and gives examples of results. A more extensive account is in preparation [8]. The individual cases presented here are from one completed part of a series of case studies conducted in cooperation with researchers from Bamberg, Darmstadt and Munich [2, 4, 9–11].

3.1 Procedure for Clarifying the Task

The time needed for clarification of the task does not differ (in absolute value) between p-designers and m-designers, when all steps throughout the process are taken into account. The **p-designers** tend not to clarify the task in the outset, but during the process when the initial conceptual solutions are elaborated. They often take the first conceptual solution as a means of clarifying the task. At the beginning of the design process they immediately start the conceptual design, after the task has been roughly clarified. The **m-designers** begin the process by clarifying the task extensively before attempting the conceptual design.

Hypothesis 1: p-Designers do not clarify the task in detail at the start of the design process, but during the process, when the first conceptual ideas are elaborated.

There is an interesting sub-result referring to clarification of the task: the time taken for clarification does not correlate with the quality of the design (this statement applies to all 20 persons). It was shown that the design experience of each individual was related to the quality of the design result. The following hypothesis is therefore put forward:

Hypothesis 2: The experienced designer is able to clarify a task in a short time, while the novice has to invest much more time into clarification.

3.2 Procedure in Conceptual Design

p-Designers differ from m-designers because of their shorter conceptual design phase which is not confined to one time block. They proceed by concentrating on subproblems (see section 3.4). They elaborate only a few of the principal variants, and often only one particular solution for a subproblem is considered.

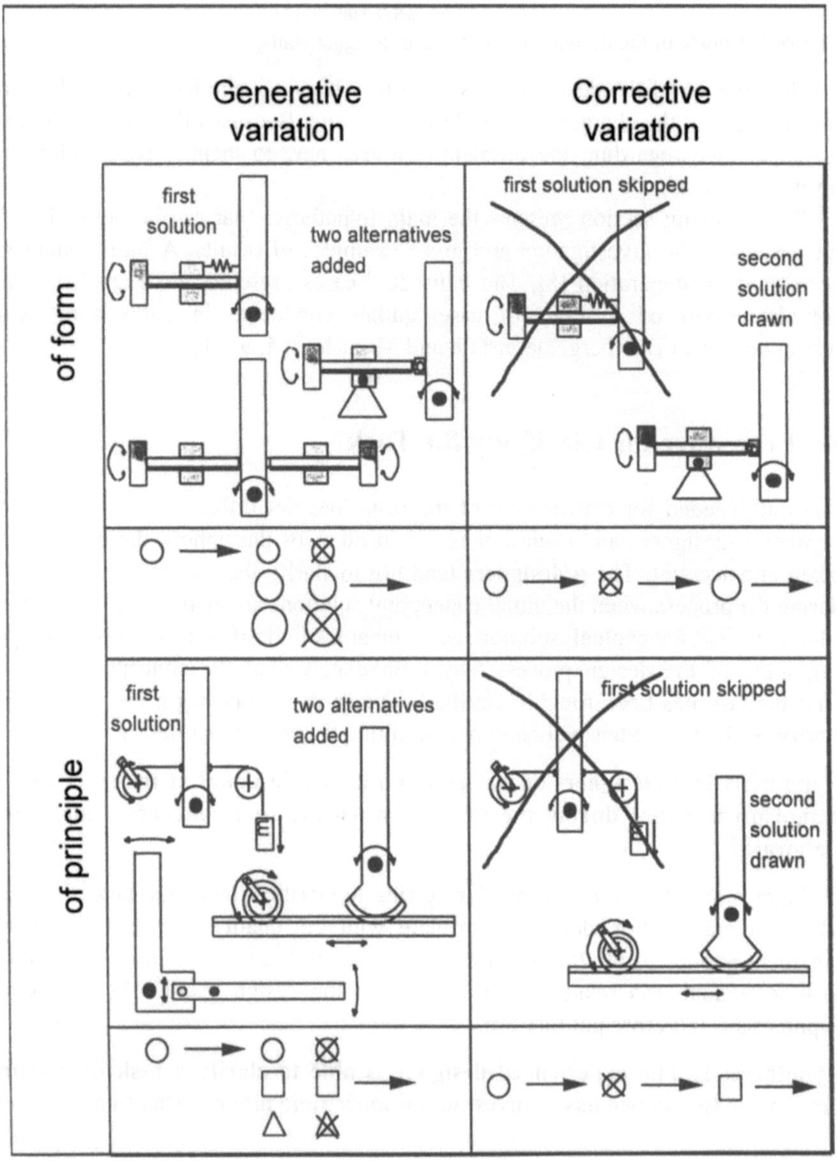

Figure 2: Generative and corrective variation of principle and of form.

p-Designers also document their conceptual work less intensively in sketches and notes. Only m-designers note down the advantages and disadvantages of the conceptual solutions or do a written evaluation. p-Designers elaborate concepts mainly in the head and these results are then used to draw up the rough embodiment design.

In elaborating variants, two kinds of variation are possible – see [2]:

- **Generative variation** means that several variants are elaborated with the aim of choosing the optimum solution via a selection method. This means that several variants exist in parallel.
- **Corrective variation** means the several variants do not exit in parallel, but in series. This means that elaborating a new variant automatically cancels the previous one.

A further differentiation of variation is possible on the levels of *principle* and *form*:

- The generation of a new variant is not based on the **principle** of an earlier solution (e.g. first variant ball bearing, second variant slide bearing for a bearing problem) is called variation of principle.
- In contrast, the elaboration of a new variant that changes **form** (shape, size, orientation) means variation of form.

Figure 2 gives an overview of the four possible levels of variation. p-Designers lay stress on corrective variants and usually on variation of form. Generative variation appears very infrequently. m-Designers lay stress on generative variation which leads to work with several alternatives.

Hypothesis 3: p-Designers document conceptual design less intensively and elaborate fewer variants than m-designers.

Hypothesis 4: p-Designers prefer corrective variation and variation of form.

3.3 Procedure in Embodiment Design

Observation of the course of rough embodiment design and final embodiment design shows different methods of tackling these design phases. In the m-designers' processes, the phases of rough and final embodiment design appear in a sequence (as shown in Figure 1). In practice, slight differences from this ideal process are exhibited, e.g. several minor changes between the rough and the embodiment design.

Interesting examples in the p-designers' group are: a) the designer who skips the rough embodiment design and goes immediately from the conceptual design to the final embodiment design (see Rolf, Figure 3); b) the designer who exhibits changes between the conceptual and the embodiment design but a non-interrupted phase of final embodiment design (see Franz, Figure 3).

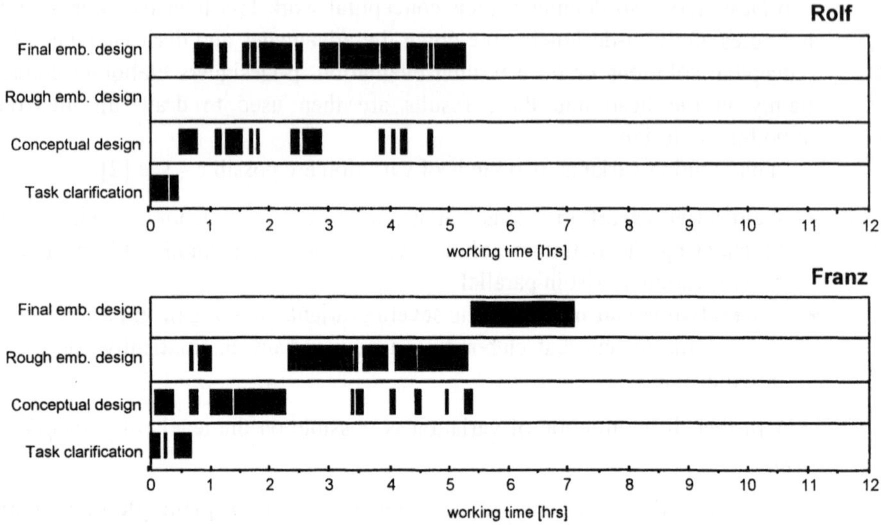

Figure 3: Course of the design phases of p-designers Rolf and Franz.

An important difference between m-designers and p-designers is that the m-designers mainly concentrate on the phases while the p-designers mainly concentrate on the subproblems. The latter work on one subproblem in the conceptual, rough and final embodiment design, and then move on to the next subproblem. This procedure is called the **subproblem-oriented procedure** in contrast to the **phase-oriented procedure** (see Figure 1 and Figure 4 below).[2]

3.4 Subproblem-oriented Procedure

From the processes used by p-designers, various alternatives of the subproblem-oriented procedure can be derived (see Figure 4).

3.5 Main Characteristics of the Procedure of p-Designers

The processes of p-designers differs from those of m-designers in several aspects.

The time of the design process is shorter – more than 30% below the average of that of m-designers (see Figure 5). This results partly from the shorter conceptual design phase having fewer variants, and partly from the skipping of steps in the process (see procedure C, Table 1).

---

2 Fricke [4] used the terms 'ablauforientiertes Vorgehen' for orientation at phases and 'bereichsorientiertes Vorgehen' for orientation at subproblems.

Table 1: Variants of the subproblem-oriented procedure.

Procedure	Explanations
A: Strict subproblem-oriented procedure	For all subproblems the designer goes through the phases of conceptual, rough and final embodiment design.
B: Subproblem-oriented procedure in conceptual and rough embodiment design.	At the end of the process there is an uninterrupted phase of final embodiment design.
C: Subproblem-oriented procedure between conceptual and final embodiment design.	The same as procedure A, but the phases of rough embodiment design are omitted.
Notice that for procedures A, B and C the guideline of the design phases (conceptual, rough and final embodiment design) follows from the orientation of subproblems.	
D: Subproblem-oriented procedure without phase-oriented micro-structure	The procedure follows the work on the different subproblems, but the design phases do not follow in a definite sequence.

The design process shows many changes in the design phases because p-designers prefer the subproblem-oriented procedure. The main kind of variation is the corrective variation of form.

p-Designers do not make notes (e.g. no list of requirements, no notes why they favoured one solution over another). They utilise only a few sketches of the solutions and a limited number of rough embodiment design drawings. Solutions are worked out on a very concrete level and very rapidly. p-Designers lay stress on documenting the final result rather than the process.

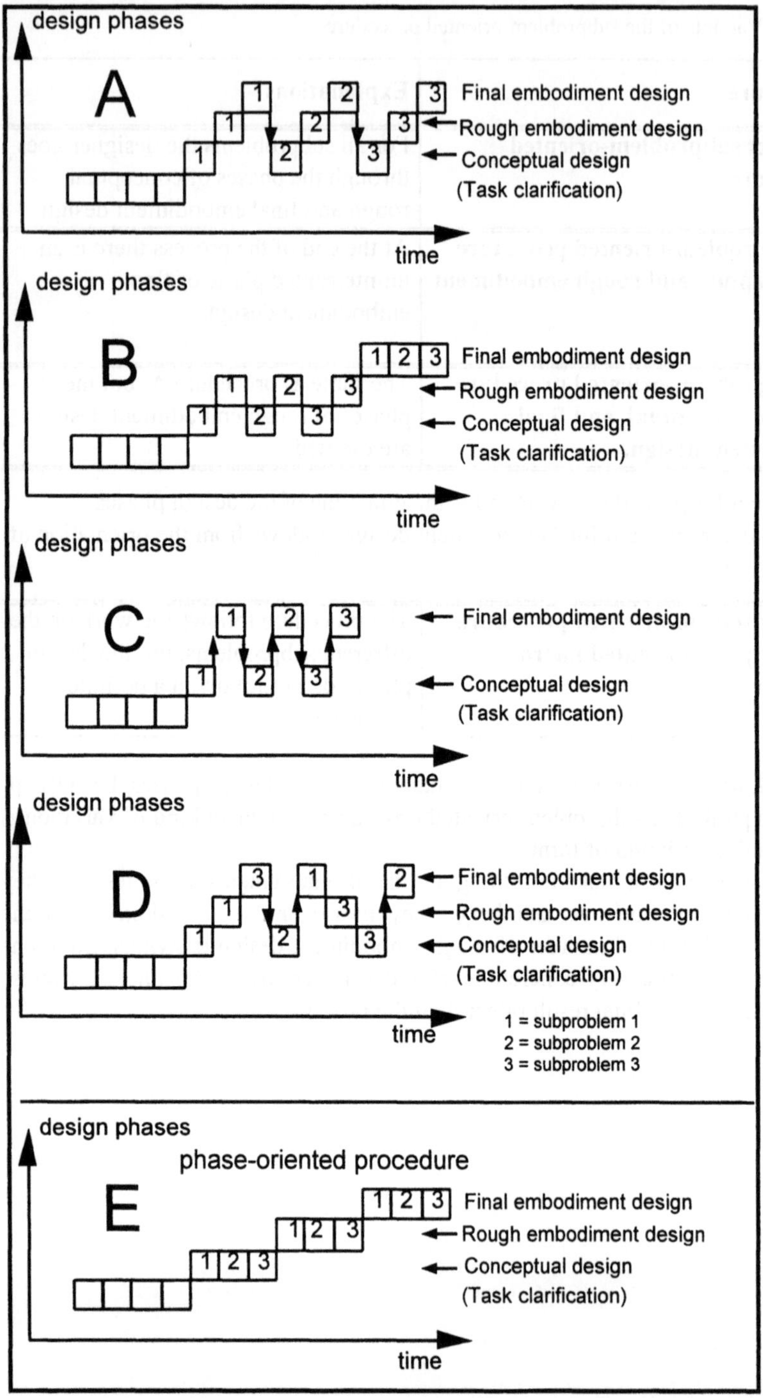

Figure 4: A, B, C and D are alternatives of the subproblem-oriented procedure. E shows the phase-oriented procedure for comparison.

4. Reasons for the Procedure Used by p-Designers

At the outset it must be stated, that some p-designers (e.g. Rolf and Franz) reach a design quality that approaches the design quality of successful m-designers (e.g. Hans). Nevertheless, p-designers in general use a procedure that differs from the recommended methodical design guidelines. What brings this about?

- The educational process given to the p-designer as draftsman, technician or master. According to this process the design process is not paramount. The **product**, its manufacture and the technical drawings are the important topics, **not the design process** itself. For this reason, p-designers do not receive much training in conceptual design.
- p-Designers have a very **concrete idea of their solutions** from their practical experience. They have a store of solutions in their memories, which can quickly be recalled. Therefore conceptual design is often done in the mind without the use of sketches, and hence is not visible to an external observer.
- **Time constraints exist in design practice**. Because of the limited time available for a project, there is often no possibility of considering several solutions in the conceptual design stage. Often **one satisfactory solution** in two days is more important than **one optimal solution** selected from three concepts in a week. Under this time constraint – which is almost standard in design practice – designers have worked the same way under experimental conditions, although time was not limited in this experiment.
- From a psychological view, **minimisation of cognitive effort** – a typical human behaviour – may also play an important role [12]. To got from the first conceptual solution to the embodiment design directly is less mental effort than to elaborate several conceptual alternatives. The quick concrete elaboration of one solution gives a feeling of security – that part of the task has already been solved.

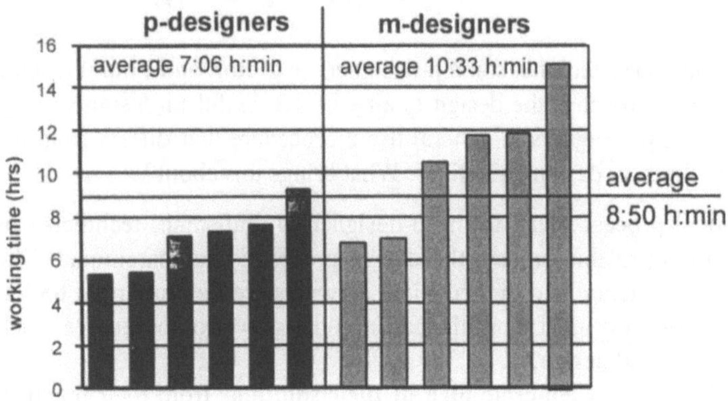

Figure 5: Working time of p-designers and m-designers.

5. Disadvantages and Advantages

From observation of individual cases with low and high design quality, we are able to distinguish some disadvantages and advantages of the procedure of p-designers. **Disadvantages** are:

- Requirements are either not recognised or recognised too late in the process.
- Subfunctions are not recognised in their context or are assessed wrongly.
- No justification of a concept is drawn up, or it is elaborated after the embodiment design has already been decided. Therefore the combination of subsolutions can only be judged in the embodiment design at the end of the process.
- Extensive changes – even to the embodiment design – are necessary when new requirements are recognised.
- Extensive changes to the embodiment design are necessary when it is recognised that subsolutions cannot be combined.
- If redesign is necessary, there is no possibility of reverting to alternative solutions, because they have not been documented.

An **advantage** of the procedure of the p-designers is the shorter design process. Their procedure is optimised with reference to working time and documentation of sub-results. Some characteristics of the process connected to time are:

- Design work starts on a concrete level at a very early stage. Therefore it is not necessary to work with conceptual representations, which are often difficult to evaluate.
- The main working time is used for documentation of the final embodiment design.

- Sub-results (e.g. documentation of the conceptual phase) are not worked out or quickly done in the head, but not recorded on paper.
- The search area for solutions is kept small. Often the first acceptable solution found is used for the embodiment design. A pre-selection of solutions is normally done quickly in the head.
- p-Designers are not aiming at optimal solutions, but at simple solutions that fulfil the needed function adequately.

6. A Minimum Design Guideline

If we want to suggest a minimum methodological guideline for p-designers, then we would propose the following steps. These steps are basic and can be implemented after a short training period without intensive education in design methodology. Such steps help to reduce the risks of the p-designers` procedure. A designer need not work through all six steps, he could choose to consider only the first four steps for example.

Table 2: Steps of the minimum design guideline.

Step of the procedure	**Explanations**
1 List of requirements A written description of the requirements for the design task should be made.	This step minimises the risk of recognising requirements too late.
2 List of subproblems Subproblems are written down and sorted into main problems and secondary problems.	This step helps the structure of the design problem to be recognised and helps to identify the main problem.
3 Principal solutions For each main subproblem there are two or more principal solutions to develop. One of these should be chosen.	This step ensures that, for the main problems, two or more alternatives are considered. Thus the method of redesign is made easier.
4 Concept The principal solutions should be combined in an overall sketch which can be evaluated against the requirements. When the concept fulfils the list of requirements then the embodiment design can be started.	With this step the likelihood that no conceptual changes will occur during embodiment design is increased. Documentation of the overall concept is achieved.
5 Rough embodiment design A rough embodiment design layout should be elaborated before the final embodiment design is considered.	The rough embodiment design provides the opportunity of analysing and judging the design before working on the final embodiment design.
6 Advantages and disadvantages of solutions Advantages and disadvantages of the principal solutions and concepts should be noted down.	This documentation assists later understanding of interim results and represents the history of the design work. It acts as an aid to understanding for customers or co-designers.

7. Limitations of the Results

The hypotheses above and the suggested minimum design guideline were derived from detailed analyses of individual cases. The cases studied are based on design **experiments** with only **one** design task.

Individual case analysis of this kind should highlight the main influencing factors on the design process. Further investigations with a greater number of designers are needed to support the results and give them statistical significance.

If we want to transfer the results found here to design in general, we must keep these limitations in mind. It is not useful to consider '*the* designer' and '*the* p-designer' on the whole, because individual characteristics, individual processes of education and the practical experience of designers are very different. Design tasks also differ greatly in practice. In design practice, cooperation between designers is very important, and this was not considered here.

The results have to be regarded as hypotheses and suggestions, which need to be proven by further research and implementation in practical design work.

8. Conclusions and Outlook

What conclusions can be drawn for design methodology from these results?

1. It is important to know the **basic goal** of a design process. This could be one of two different aims: a) a short design process under time constraint with minimum documentation – resulting design need not be an optimal solution, but should be a fairly good one; b) the resulting design must be optimal, i.e. innovative, safe and of low cost – the design process necessary for these goals may be longer than average. The authors gained the impression that most practical design work falls into category a). p-Designers are used to producing a single solution quickly, which must be satisfactory and able to be manufactured easily.
2. It is important that research in design methodology widens its objectives. To date, the main goal has been to develop products that meet an optimal quality. Time and effort were investments to reach this goal. A further and new goal must be to develop methods and techniques that support **design in the minimum time with minimum effort**. Also teamwork design must be considered under these aims. These aspects are especially important for adaptive design and variant design in practice [7].

Up to the present, the empirical research on individual influences for single designers [2, 4, 8] and teams [11] has made some progress. This line of research is worth continuing. Influences of serious importance that have been recognised are

design experience and **emotional behaviour** (e.g. tendencies to give up, ways of coping with failure).

A still open question is how the personal characteristics of a designer are generated and developed and how they their development can be supported. The **development of design experience in education and practical work** is one of the challenges for design research.

The authors hope that this contribution will encourage other researchers to further investigate the design progress and that designers from practice will use the suggestions made for self-observation or practical design work. Good luck!

References

1. Birkhofer H. 1991: Methodik in der Konstruktionspraxis – Erfolge, Grenzen und Perspektiven. In: V. Hubka (ed.) *Proceedings of ICED 91. Schriftenreihe WDK 20.* Heurista, Zürich.
2. Dylla N. 1991: *Denk- und Handlungsabläufe beim Konstruieren.* Hanser, München.
3. Günther J., Frankenberger E., Auer P. 1996: Investigation of Individual and Team Design Processes. In: N. Cross, H. Christiaans & K. Dorst (eds) *Analysing Design Activity.* John Wiley and Sons, New York, pp. 117–132.
4. Fricke G. 1993: *Konstruieren als flexibler Problemlöseprozeß – Empirische Untersuchung über erfolgreiche Strategien und methodische Vorgehensweisen beim Konstruieren.* VDI-Verlag, Düsseldorf (Fortschrittberichte VDI, Reihe 1 Nr. 227).
5. VDI-Richtlinie 2221, 1993: *Methodik zum Entwickeln und Konstruieren technischer Systeme und Produkte.* VDI-Verlag, Düsseldorf.
6. Pahl G., Beitz W. 1993: *Konstruktionslehre. 3rd ed.* Springer, Berlin.
7. Ehrlenspiel K. 1995: *Integrierte Produktentwicklung.* Hanser, München.
8. Günther J. 1998: *Individuelle Einflüsse auf den Konstruktionsprozeß.* PhD thesis in preparation.
9. Rutz A. 1985: *Konstruieren als gedanklicher Prozeß.* PhD thesis, Technical University, München.
10. Weth R. 1994: Konstruieren: Heuristische Kompetenz: Erfahrung und individuelles Vorgehen. *Zeitschrift für Arbeits- und Organisationspsychologie* 38, 102–111.
11. Frankenberger E. 1997: *Arbeitsteilige Produktentwicklung. Empirische Untersuchung und Empfehlungen zur Gruppenarbeit in der Konstruktion.* VDI-Verlag, Düsseldorf (Fortschritts-Berichte. VDI Reihe 1 Nr. 291).
12. Müller J. 1990: *Arbeitsmethoden der Technikwissenschaften.* Springer, Berlin.

Having a Nose for Good Solutions – The Development of Individual Strategies for the Design Process

Rüdiger von der Weth

Abstract

In several studies, successful individual design strategies were found to deviate from the path of 'classical' design methodology. Important decisions were made early in the design process without careful analysis of the task and possible ideas and solutions. Very similar behaviour could be observed in marketing experts. From these results we have derived a model for the development of individual action strategies, which also explains designers' behaviour as described above. Starting from a short check of the situation's complexity subjects activate knowledge about useful features of action in situations of this type. These features limit the number of possible action strategies to only one or a few alternatives. The details are planned later in the problem–solving process.

1. Objectives

The following contribution will concentrate on the earliest stages of the design process. In this way, the author hopes to answer some questions in connection with research on the design process. These questions arise from the following descriptions of what experienced designers do.

Dylla has provided a very detailed study on the behaviour of six designers with different designing experience [1,2]. He has identified typical approaches in seeking solutions. Dylla differentiates between 'generating variation' and 'correcting variation'. 'Generating variation' implies searching for several solution variants according to the requirements at hand. Thus 'correcting variation' deals with the permanent modification of an established overall solution. 'Correcting variation' means taking one working solution principle for the whole design task, and correcting and adjusting the chosen idea while elaborating its environment. According to [1], this approach is appropriate for meeting the demands of experienced designers: they are able to choose suitable solutions, to grasp even complex interrelations between the subproblems, and to work out good solutions (although the correcting variation approach is not applicable to every design task). In spite of the fact that in the cases studied the correcting variation approach provides solutions in a shorter period of time, it involves a greater number of risks. The study makes it clear that managing ideas for a solution is a vital part of any strategy of procedure in design.

Using the same investigation methods, Fricke [3] compared the actual behaviour of thirteen designers, each with a different educational background, with the suggested ideal way of proceeding in systematic design, and then interpreted his findings in relation to the quality of the solution (see [3]). Thus, the strategies and tactics of design have been identified on different levels.

Concerning design strategies Fricke distinguishes 'step-oriented' and 'function-oriented' modes of working. The 'step-oriented' strategy refers to the design steps proposed by Pahl & Beitz [4, 5]. Fricke [3] considers it very successful if engineering designers make flexible use of the methods proposed to support the design steps. This implies adjusting the emphasis on the various steps in accordance with the task at hand. It is not necessary, for instance, for a formal evaluation to be made according to technical and economic criteria in order merely to fulfil the plan of procedure if the result is obvious.

The 'function-oriented' strategy differs from the 'step-oriented' strategy in so far as the subproblems are treated entirely in order of importance, from the clarification of the task through to the layout drawing. As a result of this sequential work on subfunctions, the solution of the subfunction in question has to be adjusted to other subfunctions already elaborated. By fixing the sub-areas immediately, a function-oriented approach in design is very time-saving. It may, however, complicate the optimisation of a solution. Another similarity to the correcting variation approach is that during this study, experienced designers without systematic design training preferred the function-oriented strategy.

The function-oriented designer and the designer using the correcting variation approach do things that offend against principles of design methodology and rational decision making in general. Important decisions are made early in the design process without careful analysis of the task, solution principles and concepts. Surprisingly, experienced designers have been found to cope with complex design problems and to produce good solutions in this way [1, 3, 6]. In many cases function orientation and correction variation are connected ('p-designer' [7]). It was found that they often needed less time than designers using methodological principles, but that their approach was more risky and was thus inappropriate for novices and for individual styles of designing [3, 6, 8].

Yet there are still open questions in connection with those experts who do use design methodology. Even a designer 'making flexible use of design methods' does not analyse the problem space completely. The process of what Fricke [3] terms 'ausgewogene Suchraumerweiterung' (which means a methodological search for solutions, which abstains from careful consideration of *all* possible solution principles) leads to an analysis of only a few seemingly promising ideas. The criteria according to which these potential solutions are selected are not yet known, however.

These deliberations raise the following questions:

- What information do the function-oriented designers use when making important decisions at the outset of the design process? How do they organise their problem-solving processes? Under which circumstances are their decisions successful?
- What makes the behaviour of a successful step-oriented designer 'flexible'? What information do these designers use to decide which methodological steps are important and which are not?

2. Findings from Studies Investigating the Problem-Solving Processes Employed by Experts in Other Fields

During the past years I have conducted research concerning another group of professionals [9], namely managers in the field of marketing and sales. There are, of course, some differences in comparison with the work demands of designers. Managers in this field are not required to design technical objects, but they do have to organise and control business processes. Despite these differences in their work demands, very similar phenomena were evident as regards the individual problem-solving processes employed, and also with regard to the explanations given for this kind of behaviour in both groups.

How do managers develop individual action strategies in complex, dynamic and novel situations? In two studies, their problem-solving processes were analysed by means of two different tasks. In the first task the subjects had to deal with a complex and dynamic computer simulation of a marketing problem, in which they had to plan public relations campaigns, market research and other activities, and to connect them to a marketing strategy that then had to be translated into action over the course of several simulated months. In the second task they had to propose solutions for a series of short scenarios describing typical problems in the fields of marketing and sales.

Both studies revealed that most of the participants began their activities without planning and without the use of project management or marketing strategies in arriving at decisions. Important and far-reaching decisions were taken early on in the problem-solving process, yet the subjects did not gather and analyse the available information in an adequate manner. This is illustrated by the following results.

Hasty establishment of important features of proceeding ...

Strong support for this feature of strategic acting can be found when we examine the correlation between the measure for success across successive simulation cycles of the marketing game (Figure 1).

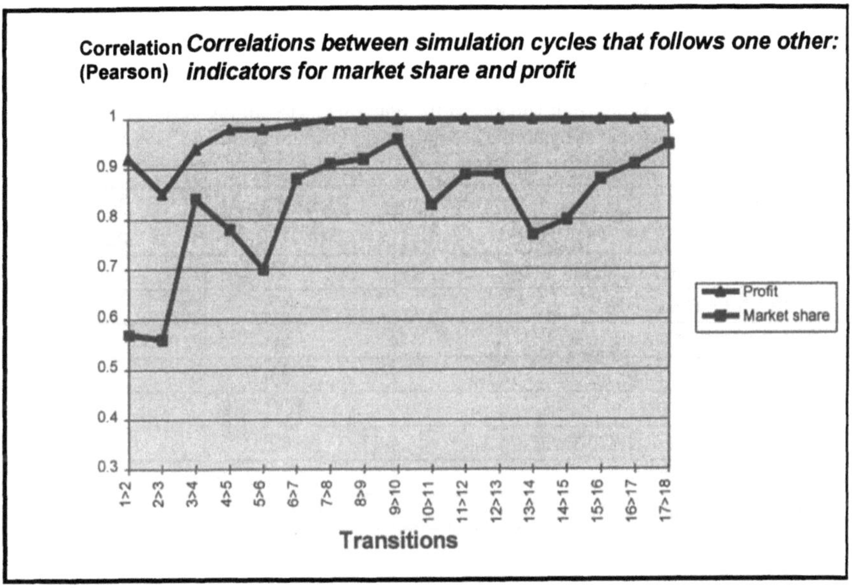

Figure 1: Success in a marketing simulation game (profit, market share): correlation between simulation cycles that follows one other. All correlations are highly significant *($p<0.01$)*. Very similar results could be found in both studies [9].

The correlation coefficients are soon very high. Individuals who are successful in the initial stages of the game are more likely to achieve good results in the end. Indeed, relevant decisions seem to be made early on in the problem-solving process.

The scenarios also provide evidence to support this assumption. In each 'story' (a typical individual work problem in the field of marketing), the subjects were provided with very little information, which from the point of view of normative methods does not allow the making of well-founded decisions, even when the experimenter's additional information is taken into account. One only has to look at the frequencies of proposed measures to see that their actual behaviour is more similar to the strategies of experienced designers described above than to descriptions derived from normative decision-making methods. In the first study we analysed our subjects' responses to 528 scenarios. In 91.2% of the cases, measures to change the situation were taken into consideration, and in 58.3% of the cases subjects even decided upon definitive measures.

... without careful analysis of the task at hand

We have already reported that most of the subjects made definitive decisions without gathering sufficient information. According to the rules concerning planning methods, the situation and the requirements of the task must be analysed carefully before decisions are made. This includes the development of alternatives, the consideration of conditions, etc. Both studies show that these decisions were not based on such a careful analysis of the material that the subjects received. They did not even *attempt* to analyse the situation carefully (Table 1).

Table 1: Percentage of different categories of 'planning steps' in the scenarios. In the first study we analysed 528 scenarios, in the second study the number was 431 [9].

Variables	First study	Second study
Proposed measures in total	91.2 %	98.1 %
Definitive measures	62.7 %	66.4 %
No conditions	69.1 %	69.4 %
No ramifications	91.9 %	86.1 %
No alternatives	100.0 %	93.5 %
No additional information	42.0 %	22.5 %
No assumptions	74.2 %	33.4 %
No action rules		87.2 %
No examples		91.0 %
No consideration of own capacities		71.2 %
No consideration of proceeding		100.0 %

When the results of research concerning problem-solving in marketing are compared with the behaviour of engineering designers a great many similarities can be discerned. Most of the marketing managers and many experienced designers establish important features of strategies and solutions early in the problem-solving process. In many cases a careful analysis of the task cannot be observed.

3. A Psychological Model to Explain these Phenomena

The development or use of strategies is not necessary in routine activities. The individual regulation of action in such situations is described by Hacker [10], Rasmussen *et al.* [11] and Anderson [12]. Individual action can be seen as a sequence of actions ruled by a hierarchical structure of goals, an ***action plan*** (Figure 2). The more novel the situation, the more gaps we find in these internal plans for action. In such situations ***strategies*** can be used. In a former publication we have defined strategies as follows ([8], p. 361):

- Strategies are oriented towards an ultimate goal of an action process. In attaining subgoals or intermediary goals we cannot use the term strategy; e.g., strategies in chess playing do not aim at taking a chess-piece but at placing the other player into checkmate.
- Strategies have a methodological character – they contain information about ***how*** to proceed in order to reach the final goal under certain conditions. Strategies define subgoals, distinctive features of proceeding (e.g., offensive, defensive) and in that way limit the possible operations.

- Suitable strategies structure and simplify the action. They help to subdivide a problem into clear units of subproblems so that it becomes superfluous to reconsider over and over again the whole way of proceeding, unless a modification in strategy becomes necessary. By following a strategy, not every single step of the continuation of action has to be planned. It is rather the preconditions of the strategy which allow control of the problem-solving process based on a limited number of alternatives. Thus one cannot speak of a strategy if action is controlled by routine or automatically and/or a complete plan has been made up for the further course of action.

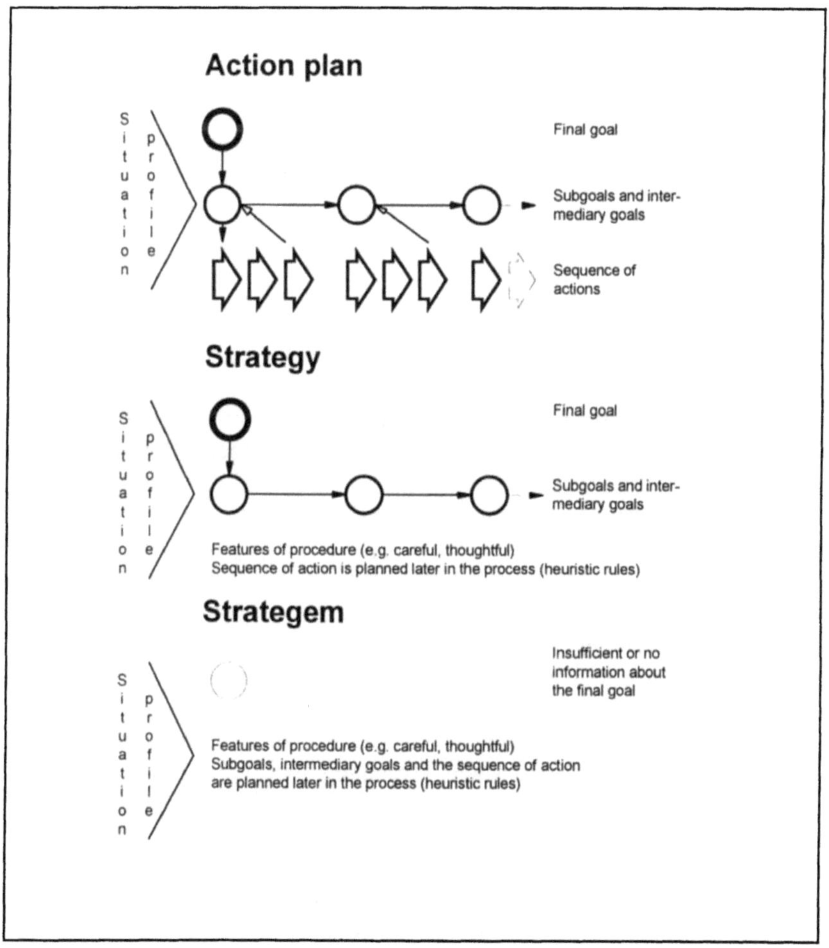

Figure 2: Different types of strategic knowledge.

But even in complex and novel situations humans are able to start activities, long before action plans and strategies are developed. In such a case heuristic knowledge is necessary to build up new strategies and plans. Heuristic knowledge can be described as a system of rules, governing the process of the construction of new strategies and plans. Von der Weth & Frankenberger described ***heuristic rules*** in the following way ([8], p. 368).

- Heuristic rules are 'rules for making rules'. They are used to generate action plans for situations for which no useful routine behaviour exists.
- Therefore, these rules must be very abstract. As concrete features of new problems are very different, similarities between different novel situations can be matched only on a very abstract level.
- Moreover, heuristic rules must be applied in a very differentiated way. For example, new situations which, at first glance, appear to be similar to situations already experienced, often require completely different ways of acting [13]. Hence very differentiated patterns of matching processes must trigger the right rule at the right moment. The quality of these pattern-matching processes probably decides how heuristic knowledge can be usefully applied in certain situations (in a heuristically competent way).
- Heuristic rules have to refer to certain data in order to cope successfully with novel situations. These situations have to be judged as to their complexity and dynamism and as to which of the aspects are really new. This is only possible with a good knowledge of one's own individual prerequisites (e.g., experience, typical emotional reactions, habits).

However, in very complex and novel situations a range of possible heuristics and strategies often exists. To select the best one on the basis of the principles of rational decision making would be very time-consuming or would not be possible with the available information. Our model envisages the activation of a further set of rules for cases such as these, which we will term 'strategems' following Senger [14].

Deviating somewhat from Senger and the English term 'stratagem', we define strategems as rules by which the initial approach to a problem-solving task is regulated by specific features (i.e. that this approach should be careful, thoughtful, etc.) in accordance with perceived aspects for a given situation. We believe that individuals use certain indicators to assess the complexity, novelty and dynamism of a situation. This does not necessarily happen consciously. For instance, an atmosphere of hectic rush indicates dynamism, whereas a great amount of information that is not immediately comprehensible indicates novelty. These indicators activate knowledge about potentially useful features of procedure, based on the individual's experience concerning typical indicators for specific aspects of complexity. These features limit the number of possible strategies or heuristics that might be useful in a situation of this particular type, until only a few, or even just one, need be taken into consideration. Thus, the individual is able to start acting without a conscious decision in favour of a certain plan or strategy. In most cases, a change of this initial direction of procedure is very improbable. According to Gollwitzer & Heckhausen [15] and others, a change in the course of action is improbable because once a course of action has been initiated any subsequent search for information is directed towards the realisation of the chosen plan, rather than towards its falsification. In most cases, therefore, the direction of action initiated by the strategems is preserved.

The strategem concept allows us to explain how individuals start acting with only a rough idea of a possible strategy and of the further development of a novel and complex situation. Nevertheless, many questions still remain unanswered. Under what conditions are certain strategies implemented? Which internal and external signals allow the relative importance of various features of complexity in a specific situation to be assessed appropriately – i.e., roughly, but more or less correctly – by experts?

The most important *internal signals* are emotions [16, 17]. They trigger the process of action by adapting certain parameters (like activation or vigilance) according to the demands of the moment. Thus, according to Dörner *et al.* [16], anxiety arises when an individual fails to build up a sufficiently well-defined 'expectation horizon' concerning the further development of a problem-solving activity in relation to the major goals of that activity.

According to these theorists, emotions are based on information about individual resources, and about the goals and features of a given situation. Hence, they are important indicators for the assessment of a situation's complexity. Anxiety, as defined by [16], indicates a high level of intransparency in the further development of the activity, for example. In order to obtain an adequate assessment of complexity using this method, an appropriate selection of internal and external information must be taken into account. The most important internal prerequisite is the individual's knowledge about his/her own problem-solving resources. This knowledge enables the designer to perceive which elements of his/her own knowledge (e.g. heuristics, strategies, strategems, qualifications etc.) and other individual prerequisites (e.g. physical fitness) will be conducive to success in a specific situation. Experts know what they know and what they do not know.

The appropriateness of an estimation of complexity also depends on the appropriate perception of *external information.* How is this possible without laborious and time-consuming analysis of the demands of the task and the possible strategies to be employed in finding an effective solution. An answer might possibly be supplied by the following example.

Let us consider the situation faced by a designer who returns from holiday to find a pile of papers on his desk: an order for a new design, along with a 150-page file of supplementary information. This information is structured and formulated in a way that is familiar to the designer, and nothing of relevance comes to mind that is not listed in the index. When initially leafing through the pages he finds familiar pictures and headings, and when reading a short paragraph on a randomly selected page he has no difficulty in understanding the text. Thus, the designer will look upon the problem as fairly transparent from the very outset of the problem-solving process. This initial assessment depends on the familiarity of the format in which the information is presented, the recognisability of particular keywords, and the fluency with which the designer is able to process the information whilst reading the text. These variables (along with others) are combined with experience from similar situations in the past.

In creating a complexity profile, it is not only this type of external information concerning the task that is used. Other aspects of the situation also play a role in assessing the different features of complexity. A third type of information used for developing a 'rough and ready' profile consists of social signals. The level of intransparency might be reduced if, for instance, a colleague one regards as more competent than oneself seems to be very calm and sure about what needs to be done.

Hence, intransparency and all other features of complex and novel situations can be assessed very quickly (Figure 3), and the resulting complexity profile causes the designer to draw upon various features of his knowledge – heuristics, strategies and potential solutions – in order to find an appropriate response to the specific problem.

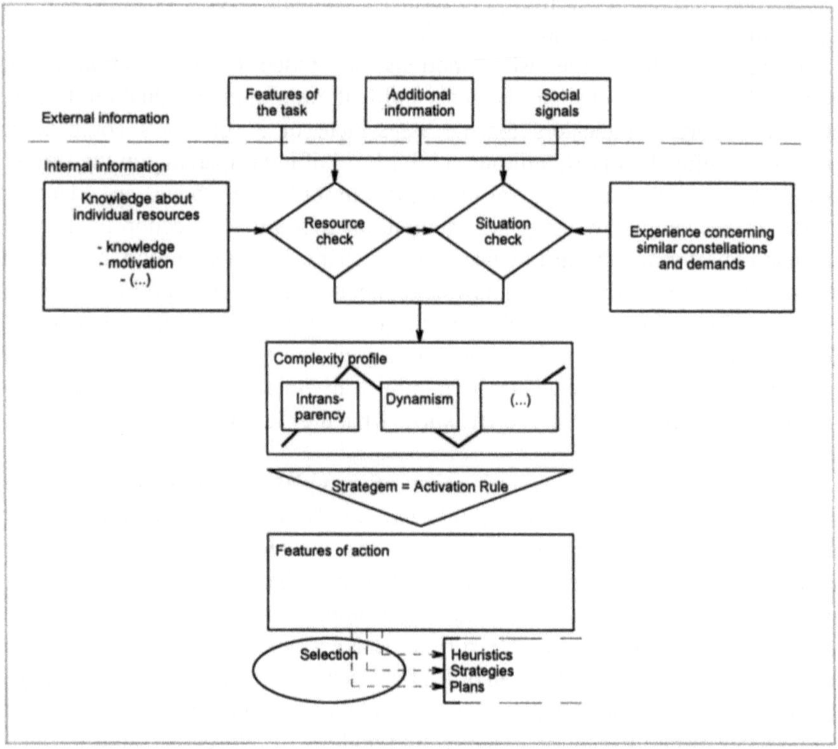

Figure 3: Development of a complexity profile and activation of strategems in complex and new situations.

How can a problem-solving process that begins in this way lead to good solutions? There are certain specific prerequisites. The first of these is 'knowledge about one's own knowledge'. A further condition is knowledge about significant external signals that provide a more or less accurate indication of the complexity of the situation.

How is it possible, for instance, that a superficial procedure like that described above can, under normal circumstances, permit intransparency to be estimated to a sufficiently accurate extent? It would seem probable that those 'rough and ready' predictors develop in an evolutionary process of life-long learning that is specific for each individual and dependent on the tasks and conditions with which the individual has been confronted over the course of his/her life.

4. Conclusions

This model describes how some experienced designers are able to develop good solutions to design tasks early on in the design process without conducting systematic analysis of important information and employing methodological strategies. It permits conclusions to be drawn for further research and has consequences for the use of design methods. There follow some brief remarks on these themes.

The validity of the predictors depends strongly on the type of experience that the individual designer has at his disposal. A designer who has had the opportunity to test them under numerous different conditions (various tasks, different companies and colleagues, different materials and tools) has a wider variety of predictors than someone who has always worked under the same or very similar conditions. The strategy of a designer of the second type can be seriously upset, for instance, when his normal set of predictors is removed and his 'habitual environment', in which he normally estimates the complexity of a new task, is altered. Empirical support for this model could be obtained by means of careful observation of designers' successful strategies and the set of predictors used. In a psychological laboratory, conditions can be constructed under which these signals do not work in an appropriate way. According to this model, it is to be expected that under such conditions the designers would adjust their strategy and the quality of the solutions would suffer.

These deliberations also have practical consequences. Especially when the designer's perception of complexity is unable to function in the normal way, design methodology can be usefully employed. Organisational and technical changes can render many experienced designers' complexity predictors obsolete, even when these changes are not directly related to the work of the particular designer. Therefore in such cases the individual designer's usual strategy when faced with complex and new design problems is often inadequate. In these situations design methodology can assist the designer in finding new ways of coping with complex design problems, because maybe then experienced designers are more open-minded and adaptable about methodological proposals.

Acknowledgements

The following institutions were engaged in this research: Lehrstuhl für Psychologie, Universität Bamberg (Prof. Dr. Dietrich Dörner), Bundesinstitut für Berufsbildung Berlin (Guido Franke) and Lehrstuhl für Absatzwirtschaft, Universität Bamberg (Prof. Dr. Frank Wimmer). The project was supported by a grant of the Bundesinsitut für Berufsbildung (Federal Institute of Professional

Qualification).

References

1. Dylla N. 1991: *Denk- und Handlungsabläufe beim Konstruieren* [The Process of Thinking and Acting in Engineering Design]. Hanser, München.
2. Ehrlenspiel K., Dylla N. 1993: Experimental Investigation of Designer's Thinking Methods and Design Procedures. *Journal of Engineering Design* 4, 734–748.
3. Fricke G. 1993: *Konstruieren als flexibler Problemlöseprozeß – Empirische Untersuchung über erfolgreiche Strategien und methodische Vorgehensweisen beim Konstruieren* [Engineering Design as a Flexible Process of Problem-solving – Empirical Studies on Successful Strategies and Systematic Design]. VDI-Verlag, Düsseldorf.
4. Pahl G., Beitz W. 1984: *Engineering Design*. The Design Council, London.
5. Pahl G., Beitz W. 1993: *Konstruktionslehre* (3rd edn) [Engineering Design]. Springer, Berlin.
6. von der Weth R. 1994: Konstruieren – Heuristische Kompetenz, Vorerfahrung und individuelles Vorgehen [Engineering design – Heuristic competence, Experience and Individual Proceeding]. *Zeitschrift für Arbeits- und Organisationspsychologie* 38 (N.F.12), 3, 102–111.
7. Günther J., Ehrlenspiel K. 1998: *This volume*, pp. 85–97.
8. von der Weth R., Frankenberger E. 1995: Strategies, Competence and Style – Problem-solving in Engineering Design. *Learning and Instruction* 5, 357–383.
9. von der Weth R., Endres K. submitted: The Development of Individual Problem-solving Strategies for Complex and Dynamic Management Problems. *Journal of Occupational and Organizational Psychology.*
10. Hacker W. 1986: *Arbeitspsychologie*. [Occupational Psychology]. Huber, Bern.
11. Rasmussen J., Duncan K., Leplat J. 1987: *New Technology and Human Error*. Wiley, Chichester.
12. Anderson J. R. 1983: *The Architecture of Cognition*. Harvard University Press, Cambridge, Mass.
13. Dörner D., Pfeifer E. 1993: Strategic Thinking, Stress and Intelligence. *Ergonomics* 36, 1345–1361.
14. Senger H. 1988: *Strategeme* [Strategems] (3rd ed). Scherz, Bern.
15. Gollwitzer P.M., Heckhausen H. 1987: *Breadth of Attention and the Counter-plea Heuristic: Further Evidence on the Motivational vs. Volitional Mind Set*. Preprint 11/87, Max Planck Institut für Psychologische Forschung.
16. Dörner D., Schaub H., Stäudel T., Strohschneider S. 1988: Ein System zur Handlungsregulation oder: Die Interaktion von Emotion, Kognition und Motivation [A System of Action Regulation or: The Interaction Between Emotion, Cognition and Motivation]. *Sprache & Kognition* 7, 217–232.
17. Auer P., Frankenberger E. 1994: *Vorgehensstile beim Konstruieren – Flexibilität und Invarianz beim Lösen unterschiedlicher Probleme* [Action Styles in Engineering Design – Flexibility and Stability in Problem-solving]. Memorandum 8 des Lehrstuhls Psychologie II der Universität Bamberg Neue Serie, University of Bamberg.

Résumé of Section II

The investigation of individual influences on individual design processes can be ascribed to many different personality traits or individual characteristics such as intelligence, social competence, action strategies, experience, etc.

One important aspect concerning investigation research relates to the question of whether extreme cases are useful for design research. For example, it seems very difficult to compile the strategies of outstanding experts in a systematic manner. In general, the design process could be recorded only *post hoc*. In consequence, the data are necessarily self-reports (e.g. sometimes of work done years before), which are difficult to verify.

However, what is the origin of the remarkable success of designers like Gordon Murray? One may argue that this designer has a great deal of experience, but we must not forget that there are many designers with years of experience who are less successful. Therefore it seems necessary to analyse these outstanding designers and their design processes in great detail. By doing so we may perhaps learn unusual but significant aspects that might be important for success, such as obsessive motivation or neglecting rules and traditions. In any case, it is not enough to reduce success only to the experience of outstanding designers.

Günther & Ehrlenspiel illustrate another way of examining individual differences by comparing designers who have received education in systematic design with designers who are in practice without this education. Although the authors consider there are several differences with regard to the steps of procedure' between these groups, it is necessary to emphasise the fact that there is no such person as '*the* experienced designer from practice', just as there is no such person as '*the* designer with education in design methodology'.

In general, it seems to be have been established that methodological education is useful – even for experienced designers – especially in situations in which the designer is faced with a new and complex problem, and what serve as his usual strategies will not deliver a successful result. It would seem easy to distinguish designers using the criterion 'methodical education' vs. 'no methodical education', or 'experience' vs. 'no experience'. However, some designers without a methodical education may use a combination of different strategies according to fixed rules as their routine behaviour.

The same problem occurs with the 'experience' criterion: evidently, experience is only a verbal declaration that contains a complex pattern of several aspects determining the design process and thus the result. Furthermore, experience is a continuum, not a two-dimensional category, and differentiation between experienced and non-experienced designers has to be complemented by a detailed observation and analysis of the procedure of the designer. It would be interesting to know the detailed methods applied by the designers in practical situations, and how they gain advantage by using such methods.

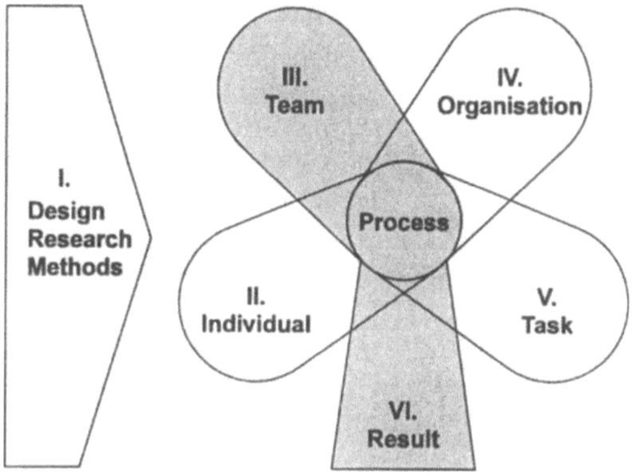

Section III:

Team Influences

In industrial design processes the division of work has turned out to be a technique that should ensure the reduction of time and cost. Therefore organisations change traditional structures into project organisation thus teamwork becomes an important constituent. However, practice shows that teamwork is a very global 'something' that may represent for different arrangements of individual and group processes. Usually we notice that organised project meetings alternate with conferences, group discussions, informal discussion and individual work in design. Therefore, the particular group process as well as the group structure are created by a mixture of individual characteristics and organisational demands, and influence thus the design process and the result. The questions addressed in this section are:

How does 'shared understanding' between designers evolve?

How can tools and team concepts be used to improve teamwork in *education* and *how* can we *measure their performance*?

What are the *sources of mistakes* in design with regard to individual, external, management, team and information processing influences?

> *How do* the influences of the group, the individual and external conditions interact in the design process and how can we *integrate* and *explain* these influences?

The course of the 'team section' proceeds from the *descriptive* investigation of teamwork to the *implementation of changes* in education and to the broader perspective of a *post hoc* and *in situ* compilation of the causal interrelations in the socio-technical system of engineering design practice:

- *Lloyd* follows an ethnographic approach in observing design teams. He differentiates between the use of storytelling and metaphors in the search for shared understanding in a design team.
- *Leifer & Mabogunje* describe their approach of design education in teams by describing tasks from industry using modern communication tools (e.g. www), and ask how these tools change design work.
- *Hales* investigates the source of catastrophic failures in design processes. This analysis illustrates the complexity of various influences from different fields.
- *Frankenberger & Badke-Schaub* integrate the influences of the group, the individual and of external conditions in empirical investigations in practice. Critical situations are modelled by factors and mechanisms in order to explain positive and negative outcomes.

Storytelling and Metaphor in the Engineering Design Process

Peter Lloyd

Abstract

'There seem to be two ways of understanding things; either by way of a metaphor or by way of a story ... either [a person] can call on a model – a metaphor – of rationality ... or [they] can tell a story' [1]. The two conceptions are not mutually exclusive, but the distinction seems sound. I argue that, although the popular way of representing and understanding the engineering design process is through metaphor – 'models of rationality', there is also another way in which *designers* understand designing, and that is in terms of 'storytelling'.

I draw on empirical data from a two week study in a small engineering design company to show: how metaphors relate to designing behaviour, how social experience is encapsulated in stories, and how stories from different organisational departments lead to different understandings of common objects – product specifications for example. I conclude that storytelling is the key mechanism by which a well defined 'common language' evolves, and thus a highly economical way of interacting.

1. Introduction

In a fascinating paper entitled 'Storytelling in Economics' [1] Donald McClosky describes the way in which economics, in both theory and practice, operates just as much by what he terms 'storytelling' as it does by the perhaps more conventionally understood mechanism of 'rationality' which he refers to as 'metaphor'. McClosky sees these as two interacting modes of explanation for economic reality and the distinction is strikingly illustrated in making an analogy with the understanding of academic literature. Here the distinction is between aesthetic reading and efferent reading:

> In efferent reading (*effero*, carry off) the reader focuses on what she will carry off from the reading. Efferent reading is supposed to characterise model-building and science. In aesthetic reading the reader focuses on her experience at the time of the reading, which is

> supposed to characterise storytelling and art. Yet an aesthetic reading of a scientific text commonly carries the argument. The feeling 'yes: this is right' ...(p.16)

The quote seems to suggest that rationally understanding how something works, in effect building a causal model of the something (be it a scientific paper, economic reality, whatever), is not the mechanism that we use for that something to become part of our experience, i.e. to become valuable to us. I have argued elsewhere that a designer's experience, in a cognitive sense, is the single largest factor in the solution of design problems [2–5] and there is indeed good evidence to support this view [6, 7]. A mechanism that could describe how 'social' design experience is acquired and used cognitively might, then, be useful in explaining design activity.

A common understanding of what the design process *is* would probably run along the lines of: something that proceeds from specification to concept to detail to manufacture; ergo the field of design methodology. This suggests that the design process, and by extension designing, is commonly understood as a rational metaphor. This is not, however, how designers qua designers understand the design process. Design here is understood as a particular: *this* drawing, on *this* project, in *this* organisation. In this respect the distinction between storytelling and metaphor in explaining a phenomena is one I'd like to pursue in terms of engineering design, a subject that, prima facie, consists of rational, and mainly technical models (metaphors) of the design process.

There is, it should be pointed out, an assumption that I will make throughout this paper. Although often talked of as an individual cognitive process – and undoubtedly processes such as sketching and thinking are individual – there can be no doubt that designing is primarily something that is social in nature. As Bryan Lawson illustrates in his description of the design process [8] commercial constraints *necessarily* motivate the need to communicate with clients, suppliers, users, legislators, team members, mentors, etc. A purely cognitive description is not in itself sufficient, there has also to be a social cognitive description.

2. Metaphor and Rationality

McClosky [1] defines rationality as a logically systematic metaphor, a normative model that shows how the various parts of a system or process interrelate. In terms of engineering design something like British Standard 5750, explaining the procedures necessary to maintain a certain level of product quality, might be considered as a rational metaphor. A sample quote from BS5750:

> the specification and design function should provide for the translation of customer needs from the product brief into technical specifications ... the [design organisation] shall draw up plans that identify the responsibility for each design and development activity. [9]

The metaphorical content, which hinges on the meaning of the word 'translation' in the quote, comes when we consider that what is being said is not so much that designing *is* this activity, but rather that designing is, or should be, *like* this activity. And similarly with other normative models of the design process [10] (to reference the paradigm example). Again there is no suggestion that design *is* this, merely an implication that design should be *like* this.

There are also more traditional metaphors that describe designing, though still rational in the sense that they have an internal, systematic, logic to them:

> If you look at the whole process of design, and how designers think, think of it [...] as if you've seen an explosion on film run backwards very slowly. What you see is a whole load of dust and rubbish and bits and pieces all floating about in the air, and as you wind the film back it goes uuuuuwwwwwvvvvvvvvvump! very quickly, and the design is finished. [11]

Again the crucial point here is that we would not wish to say designing *is* this – we would not expect to see, for example, bits of dust and rubbish floating around in the air if we walked into a design organisation, we would merely wish to make the point that designing *is like* this. The appeal is metaphorical, and poetically logical.

A metaphor, however, goes almost no way in explaining designing behaviour in terms of how the design organisation is experienced or constructs its own particular culture. When a metaphor prescribes action, by saying what the process, for example, *should* be like, there is an awkward ring to any interaction, and we almost arrive at a 'no explanation'. In the following excerpt[1] a software designer, phoned by his customer, is asked to outline how far through the design he is. He tries to explain this by making direct reference to the customer's metaphor of the design process:

---

[1] *A note on method:*
This paper is structured with an analysis of the terms 'metaphor' and 'story' and uses as its data several vignettes culled from an observational study of a design and manufacturing organisation. The organisation in question employed approximately 100 people, producing bespoke test equipment systems for the automotive industry, for example airbag testers, automatic braking systems (ABS), rolling roads, automatic wheel alignment systems, etc.
The fieldwork for the study took place over a two week period and consisted of two researchers observing different organisational departments, tape recording and taking notes of situations, activity and conversations, and conducting informal interviews. The approach adopted was ethnographic [12], although the period of study was far shorter than the proper sense of 'ethnographic' might suggest.
The company was treated in many senses as a distinct culture, and practices and procedures were looked at in holistic terms. As much data as possible were tape recorded and subsequently transcribed. This led to a 'text' being produced of mainly qualitative data, consisting of all the transcriptions of interviews and activity, corporate documentation, and our own field notes. Our method of analysis was to draw systematic and triangulated interpretation from the original text and this method is outlined in a sister paper [13].

	[phone rings]
Designer	Hello Blair
Customer	[How are you?]
Designer	Fine thanks, and you?
Customer	[How are you getting on with the design?]
Designer	All I've said to various people ... only time to finish it off if you like, but that's all ... I've got to the, how do you put it, 'the requirements from other people for information' at the moment ...

Now clearly the reference to the customer's metaphor – 'the requirements from other people for information' – suggests a further question. Both people understand what stage the design process is at in the terms of the metaphor, but in terms of the meaning of 'the requirements from other people for information' what has actually been done, i.e. the evidence for or experience of, having reached that stage, is not spoken of. One might imagine the conversation continuing along the lines of:

Customer	[Ah right, but what does that mean?]
Designer	Well I did *x*, then I did *y*, then I did *z* ...

And here the reference to the metaphor would be explained with a story, a time-based narrative of events. In fact the conversation continues with:

Customer	[...]
Designer	No, I've, well going from a previous conversation where you said 'it's all done on the same machine ...

So the common narrative thread is picked up with a reference to a shared experience and the conversation continues more comfortably.

Similarly a meeting between a customer and a designer to discuss progress on the installation of a new piece of equipment goes through a list of requirements that *should* have been met. Stopping at the requirement stating that the supplier must use a software design methodology the customer says: 'you *have* used a software methodology haven't you?', 'I don't know ...' comes the reply, the designer not having written the actual program code himself. The customer goes on: 'you use Yourdon [14] don't you?', the designer answers: 'yes we do'.

Again the interaction rings false. There is an ambiguity to the discussion – it is not clear whether Yourdon *was* actually used on *this* particular design. The only satisfactory way of resolving this issue is with a story that recounts the experience of using Yourdon on this particular design. As it is, a story is not forthcoming, the checkbox is ticked, and the next requirement is 'discussed'.

A metaphor, or model of rationality, is an appeal to the general, not the particular. As such it is a normative appeal; a characterisation, not a personification. In terms of design this provides tension between what a designer thinks he *should* do, and what, in actual fact, he does. Storytelling provides a way in which designers can explain what they actually do, *specifically*. There is no doubt that metaphors of the design process are pervasive, and I am not in any way suggesting they are divisive. However as an explanation for design or, to be more accurate, how the designer explains the design process, they are on their own insufficient. The metaphor and the story interact; the metaphor provides a rhetorical structure, the story provides the personal experience.

3. Constructing a Story

Storytelling introduces a narrative element into designing, a description of related events that link people over time. Storytelling in this technical sense comes equipped with added terminology, a story becomes a 'text', which can be interpreted or 'read', different narrative 'viewpoints' might be included, there is a sense of 'closure' in a story, a definite ending, and a 'name' might be invented that references the complex of action.

We explain why John the chief designer did not turn up for work on time by explaining that he had to travel to see a client the previous day, and missed a train connection on the way home. That is why the supplier, who'd arranged to see him the next morning to discuss his latest line of products, had to wait around for two hours before being told that everyone was too busy to see him. The 'supplier waiting in reception' story is born, it enters the experience of the chief designer and the people around him, *those* words refer to *that* story which then becomes an instrument by which to add to, or measure, other story experiences.

The next excerpt is taken from a meeting that discussed a new rolling road being commissioned at the motor manufacturer's site. The problem was that the rollers rotating the front wheels of the car, and the rollers rotating the rear wheels of the car were rising at different rates. This had caused the customer to complain to the designers that the rolling road was not working properly. The service engineer Phil outlines the problem to the designers Ian, John and Tony:

Phil	The next problem is the balancing of the lift edge, which I had a look at ... the problem is that the front lift is a lot longer than the rear lift. If both valves are actuated at the same time, the same valves obviously respond a lot slower, so what happens is you get the rear of the car coming up [before the front of the car]... the only thing I would suggest we could do is possibly to put a timer on it ...
Ian	In other words delay the rear coming up?
Phil	There's no other way, John would probably back me up on that ...
Ian	[to John] is this reasonable?
Tony	Does it make any difference if the back of the car comes up first?
Phil	No, but it's something they're going to pick up on ...
Ian	Unless there's a mechanical way that we can get over that then, it's down to me to do it electrically ...

Initially Phil explains the problem of the rolling road by making his experience of it explicit. This is, in effect, his *story*, his version of events. Ian tries to make sense of this story, firstly by summarising: 'in other words...', and secondly by asking John, who has more experience of this problem, if he thinks the story makes sense: 'is this reasonable?' Tony queries the wider implications of the story by asking if it makes any difference, Phil counters by adding to his initial account, saying that, regardless of whether it makes any difference, it is something that the customer is 'going to pick up on'. Ian, having understood Phil's story, suggests that he, an electrical designer, try to solve the problem.

What is constructed over the course of this discussion is a common understanding of a situation, and the way that that situation is initially described is by telling a story. Later in the meeting the problem was referred to as 'the lift edge problem', these words now referring to this commonly agreed story.

There is an economy of language here. Complex social and technical scenarios are deconstructed, understood socially, and then reconstructed into a word or phrase. Experience is encapsulated, the story no longer needs to be retold. The implications of this become apparent when Phil generalises the 'lift edge problem':

Phil You know that's a problem that we have on every rig we build

In any future meetings, the 'lift edge problem' can be used as a 'story index', a pointer to a common discussion. Any new design can take account of the 'lift edge problem'.

Conversation in the engineering design organisation is peppered with such words and phrases: the 'Tudor fiasco' refers to a design project for a company called Tudor Webasto that went seriously wrong. When a project manager visits a designer and says only: 'mini airbags' the designer knows exactly that he is referring to the current state of the corpus of information built up in a project that has lasted several months. Similarly three letter abbreviations would seem to have this economising and indexical function. Reference to 'ABS systems' or 'ECUs' need not be explained, they are the agreed units that form part of the ongoing and larger stories within the design project and design organisation.

To the designer within the organisation, specialist 'story' words and phrases have *particular* meaning. They encapsulate personal experience and contribute to a larger narrative at either team, project, departmental, or organisational level. For the researcher entering the organisation the same words leave 'narrative gaps'. It is simply impossible to understand what is being talked about without further explanation. In effect a new language has to be learned. This dual function of story words – that they are simultaneously inclusive and exclusive – has a direct relationship with team, group, departmental and organisational culture. The more specialist one gets the more inclusive the culture becomes. The limiting case is, of course, individuals and how they understand and encode their own cognitive experience before it becomes social. In this respect the work of Donald Schön is important, having successfully analysed this type of vocabulary [15] (see also [16]).

4. Deconstructing a Story

The previous examples have shown that particular words – 'Spikes', 'CB40', 'Floflex', ECU, ABS system – have a commonly agreed meaning which is negotiated through a particular narrative discourse. There are, however, more familiar words –'product specification', 'callout', 'shopfloor', 'quality', 'engineering drawings', etc. – which intersect everyone's experience. There is a certain ambiguity to these types of words which allows appropriation into different, and often conflicting, stories. In terms of *individual* experience these appropriated meanings might be explained by Bucciarelli's term 'object world' [17]. In this view the designer literally constructs a 'world' of meaning in designing. What is of interest in social terms, however, is the way in which different organisational functions appropriate a commonly existing 'object' to build a narrative that reflects its own sub-culture.

As an example of this social appropriation from our study the 'product specification' was an object of some disagreement, particularly between salesman and designers. The disagreement was not about the existence of the product specification, but about what it actually meant. Designers claimed that specifications were 'vague', 'underspecified', 'one-pitch' and 'open'. Salesman on the other hand explained the significance of the design specification in a different way:

> it's a balance between giving the customer sufficient information to give him confidence ... and not dotting i's and crossing t's.

The specification for the salesman is something to inspire confidence in the customer, not something to detail a working design, which is how the design engineer saw it:

> we [should] provide a sufficiently detailed specification so that if the customer places the order on that basis, we can say 'look this is what you've paid for.

The different functions have incorporated the product specification into different explanations – stories – of their work. The word 'sufficiently' in both quotes suggests that the common object, the product specification, has (at least) two necessary purposes. To the salesman a 'sufficient' product specification is one that sells the design project with minimum effort: 'if you do ten quotations and get two orders, you've probably not done too bad', why 'overspecify' when you might not get the order anyway? To the salesman there is a reason for leaving out information. The 'story' of the salesman is of efficient selling, and this explanation fits this narrative. The designer, on the other hand, understands the specification as providing information: '[you need] to find out what is *really* required'. To the designer the specification should be a succinct summary of 'what the customer wants' – a well-defined starting point for the design.

The point here is not the validity of either claim, but that there *are* two different claims, two different understandings, and hence two different narratives that use the same object. And similarly with quality procedures, with engineering drawings, and with the design object itself. In a sense these are competing narratives trying to appropriate the object as their own, to tell the definitive story, to locate a permanent meaning to a set of words. The design vocabulary might, then, be defined as much by common disagreement as by common agreement.

'What is *really* required' brings us back to where we started from – the commonly understood metaphor of design: specification, to concept, to detail, to manufacture. By suggesting that it *is* possible that a complete product specification can be detailed prior to work starting on design, 'what is *really* required' invokes a metaphor to challenge the explanation of the salesman's story.

5. Discussion

The first objection that needs to be met is the underlying assumption behind this paper, which was stated in the introduction, that design *is* essentially a social process. This may seem a strange claim to make given the primarily graphical nature of communicating design information in schematics, sketches, drawings, etc. [18] but there are two remarks that can be made here. The first is that graphical design communication is metaphorical in the sense that a schematic, sketch or drawing presents a logically coherent system which in itself does not express personal experience, i.e. does not contain narrative elements. The second is that schematics, sketches and drawings have their own indexes in language[2] and really help to perpetuate the narrative of any particular design project by themselves being objects of discussion, i.e. they are there to communicate. One striking observation in our study was that designers spend very little time actually designing, or what one would commonly think of as designing: quietly sitting in front of a desk, CAD station or drawing board (or alternatively specifying, conceptualising, detailing, etc.). There can be very few parts of a design process that are not socially explored [20].

As the foregoing sections illustrate there are many linkages between stories and metaphors; they rely on one another, they interact. There are also not always clear-cut differences and the process of how a story might turn into a metaphor is maybe a point of future consideration. The purpose of the paper has been primarily to redress the 'models of rationality' – the metaphorical – balance that seems to have dominated theorising about the engineering design process. In doing this it was first necessary to make the story/metaphor distinction, and then justify it with examples. Above all the purpose of a story is to explain individual experience. The difference between storytelling and metaphor in describing aspects of engineering design might then be summarised thus: by telling a story one is saying 'this is how this something came to be this something'; by using a metaphor one is saying 'this is how this something works'. The story particularises, the metaphor generalises.

---

[2] Bryan Lawson [19] claims that the ambiguity of words is fundamental to designing, and suggests that future technology might exploit this insight.

An awareness of this distinction has two obvious implications. The first is methodological. If designing 'reality' might be thought of as the collection of particulars (in Wittgenstein's famous phrase 'the world is everything that is the case'), what are the prospects for so-called 'reductive' descriptions of engineering design – the compilation of empirical evidence into a 'system'? As a description of designing behaviour the prognosis does not look good. As a means of producing new metaphors the prospects look better. The key point to note is that the production of metaphors invariably leads to a rhetorical use rather than, I suspect what is hoped for, a qualitative change in behaviour.

The second implication concerns the instrumental benefits of understanding 'storytelling' as a mode of relating experience, and more generally the development of a common language surrounding a particular product. If one is to agree that the quality of different products varies then might there be a corresponding variation in the quality of a related common language? This is tantamount to saying that, in terms of designing, some experiences are better expressed and re-expressed through words than others. Frankenberger and Badke-Schaub [21] describe 'critical situations' in the design process, and it just might be that these 'experiences' are worth holding on to in words. In our own studies we have noted that language evenly balanced between expressing an aspect of the product – in effect 'the what' – and expressing part of a process – 'the how' – tends to be more effective than only product focused conversation. The 'lift edge problem', a noun-phrase relating to the product, might equally have been expressed as a verb-phrase relating to a process: 'customer communicating' for example, and would tend to underscore the time-based nature of events. This situation is similar in some respects to *helping* a designer to 'reflect in action'. There seems an intervention that it is possible to make that would improve communication in the design process by developing and making explicit the opportunity to converse about experiences that were previously private. The theory remains speculative however.

The philosophical problems of relating language to experience are complex and unfortunately beyond the scope of this paper. Suffice to say that a psychologist, Vygotsky, studying how the developing child acquires and uses language, comes close to the arguments advocated in this paper in saying that: 'a word is a microcosm of human consciousness' [22]. The phrase might be appropriated to say that 'a story is a microcosm of human experience'.

6. Conclusions

The argument of the paper might be outlined as follows: the distinction is made between storytelling and metaphor as separate, though interacting, ways of explaining. Designing is described as an activity that depends largely on experience, and as storytelling is a way of explaining experience, it seems a particularly apposite means of explaining designing, particularly designing as a primarily social process. In the detailed analysis of several textual segments from an ethnographic study of a small design and manufacturing organisation, the idea of 'design as metaphor' and 'design as storytelling' is explored. We conclude that storytelling is a central mechanism in the development of a 'common language', something that is often noted in design organisations.

Acknowledgements

Funding for the study reported here comes from an EPSRC grant 'Pragmatic Models of Concurrent Product/Process Design' (references GR/K39417 and GR/K36348), together with contributions from British Aerospace PLC and Lucas Industries PLC. I would like to thank Tim Ellis for both his part in the fieldwork stages of this project and his helpful suggestions for this paper. Thanks are also due to Peter Deasley, Ken Swift and Mike Dooner, managers of the project.

References

1. McClosky D.N. 1990: Storytelling in Economics. In: C. Nash (ed.) *Narrative in Culture: The Use of Storytelling in the Sciences, Philosophy, and Literature.* Routledge, London, pp. 5–22.
2. .Lloyd P.A., Scott P.J. 1994: Difference in Similarity: Interpreting the Architectural Design Process. *Environment and Planning B: Planning and Design* 22, 383–406.
3. Lloyd P.A. 1994: *Psychological Investigations of the Conceptual Design Process*, PhD Thesis, Sheffield University, Sheffield, UK.
4. Lloyd P.A. 1994: Is There a Case for Generic Design? *Languages of Design* 2, 243–262.
5. Lloyd P.A., Scott P.J. 1994: Discovering the Design Problem. *Design Studies* 15, 125–140.
6. Holt J.E. 1997: The Designer's Judgement. *Design Studies* 18, 113–123.
7. Lawson B. 1994: *Design in Mind.* Butterworth Architecture, Oxford, UK.
8. Lawson B. 1990: *How Designers Think.* Butterworth, London.
9. British Standards Institution 1987: *BS5750.* London.
10. Pahl G., Beitz W. 1984: *Engineering Design.* The Design Council, London.
11. Channel 4 1994: *Designing Dream Machines.* Channel 4 Television, http://www.cityscape.co.uk/channel4/big_bytes/equinox95/.
12. Lloyd P., Deasley P. in press: Ethnographic Description of Design Networks. *Journal of Automation in Construction.*
13. Lloyd P., Deasley P. 1997: What's the Story? Text Based Analysis of Engineering Design. In: A. Riitahuhta (ed.) *International Conference of Engineering Design (ICED).* University of Tampere, Tampere, Finland, pp. 371–376.
14. Yourdon E. 1988: *Managing the System Life Cycle: A Software Development Methodology Overview, 2nd edn.* Yourdan Press, Prentice Hall, NJ.
15. Schön D.A. 1988: Designing: Rules, Types and Worlds. *Design Studies* 9, 181–190.
16. Roozenburg N.F.M., Doorst C.H. 1998: *This volume,* pp. 29-41.
17. Bucciarelli, L.L. 1994: *Designing Engineers.* MIT Press, Cambridge, MA.
18. Dörner D. 1998: *This volume,* pp. 3–11.

19. Lawson B. and Loke S.M. 1997: Computers, Words and Pictures. *Design Studies* 18, 171–183.
20. Leifer L., Mabogunje A.A. 1998: *This volume,* pp.124-136.
21. Frankenberger E. and Badke-Schaub P. 1998: *This volume,* pp. 149-164.
22. Vygotsky L.S. 1962: *Thought and Language.* MIT Press, Cambridge, MA.

A Framework for Instrumenting Design Teams

Larry Leifer & Ade Mabogunje

Abstract

A research programme based on the pedagogic concept of Product Based Learning (PBL) is described. The similarity of this kind of learning to design learning provides an opportunity to observe design teams at work and to develop an instrumentation model to monitor their performance. Examples of some of the metrics developed are presented and discussed.

1. Introduction

Over the last ten years we have been observing design teams and intervening in their work in order to better understand the effect of novel design tools such as electronic notebooks and internet-mediated technologies on design performance. While our initial focus was on the tool technology, it has evolved into a socio-technical one encompassing a social network of people working with, and generating a repertoire of interlinked technical knowledge, as shown in Figure 1. We have been applying this socio-technical framework, in one sense a paradigm, to the education of designers, with several interesting results and implications for improving product development performance.

To put the findings into context, we must first introduce the pedagogic framework that guides our value judgements. Hence, we introduce the concept and practice of 'Product Based Learning' (PBL) and the feedback instrumentation model we use for assessing PBL performance. Within this framework, the discussion will focus on technology specific issues, the rationale for technical intervention and a sampling of performance assessment results.

2. Context

Economic and technologic forces are changing the landscape of engineering. There is an increasing need for organisations to form joint design development teams that collaborate for the life of a project and then disperse. These teams need to quickly locate, evaluate and make effective use of the best resources available (people, facilities, tools, information), wherever they may be found. We will concentrate on a subset of this problem that is critical in the early stages of product development when new ideas, processes, components and materials are being explored and prototype solutions are developed. We seek to enhance the ability of teams to exploit their own shared knowledge, novel tools and manufacturing processes for improved product performance, reduced cost and documentation of their exploration for those who will inevitably follow the next generation re-design team. In this way we believe that the slow cycle of exploration, maturation and widespread adoption of advances in information handling, materials and processing capabilities can be compressed from decades to years.

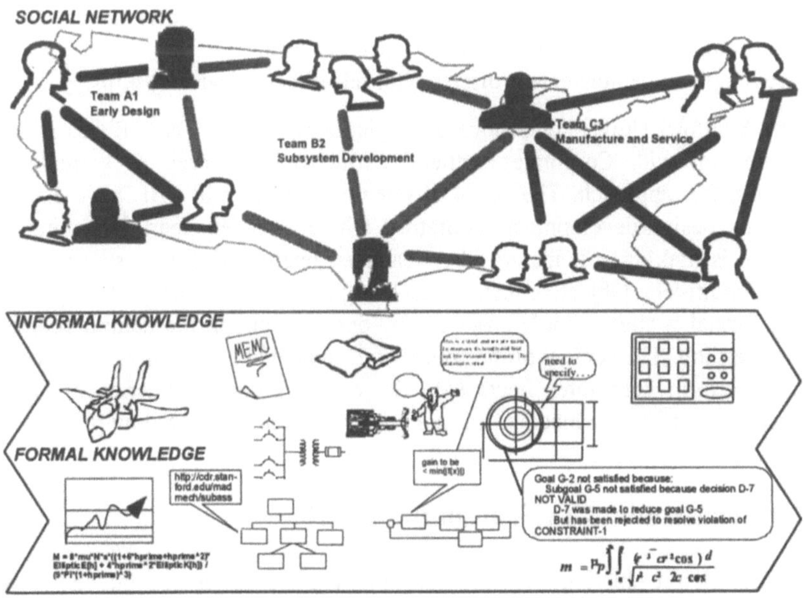

Figure 1: Engineering design as a socio-technical system encompassing a social network of people generating and working with a repertoire of interlinked technical knowledge. This knowledge consists of an informal part, examples of which are back-of-the-envelope calculations and jottings in the designer's notebook, and a formal part such as component specifications, mathematical formulae and technical drawings.

Our work is ongoing in the form of an annual series of experiments involving 12 to 15 design teams in a product design-development laboratory at Stanford University. Ten years of careful observation and analysis of teams, at work in their graduate studies and in industry, are beginning to yield objective predictors of design quality outcome and management strategies for ensuring uniformly high team performance. The pedagogic framework of this test-bed course and associated research programme is underpinned by work in the social and cognitive sciences [2, 3].

2.1 Product Based Learning and ME210

PBL is defined as problem oriented, project organised learning activity that produces a product for an outsider. Unlike other versions of PBL that focus only on the project or problem, the product focuses our attention on delivering something of value beyond a 'training exercise', something suitable for evaluation by outsiders. The objectives of PBL curricula may be stated briefly as follows [4]:

1. Familiarise students with problems inherent in their future profession;
2. Ensure content and process knowledge are relevant to these problems;
3. Ensure competence in applying this knowledge;
4. Develop problem formulation and problem solving skills;
5. Develop implementation (how to) skills;
6. Develop skills to lead and facilitate collaborative problem solving;
7. Develop skills to manage emotional aspects of leadership;
8. Develop and demonstrate proficiency in self-directed learning skills.

The Stanford University School of Engineering course 'Team-Based Design-Development with Corporate Partners' (ME210) has served as a model for exploring this approach. The seeds of our understanding of PBL have emerged from this design-development laboratory where eighteen years of cumulative evidence is making soft factors like 'team personality' ring true statistically as well as intuitively. Objective metrics are being derived from detailed examination of the formal written records produced by design teams in the natural course of doing their job. The content of the course is documented in a brief video recording used to inform corporate clients [5]. The design process (experience) of four representative teams is documented in a longer (52 minute) recording [6].

2.2 Mechatronics Curriculum in ME210

Our curriculum technology focus is Mechatronic systems (smart products) design-development. Mechatronics is frequently defined as the integration of real-time software, electronics and mechanical elements for imbedded systems. In our definition, we include human–computer interaction and materials selection. Typical products in this domain include: disk drives, camcorders, flexible automation-systems, process-controllers, avionics, engine controls, appliances and smart weapons. Mechatronics is a particularly good medium for introducing PBL because it is dependent on interdisciplinary collaboration.

2.3 ME210 as a Product Based Learning Model

Corporate clients of the Stanford Center for Professional Development (SCPD) who sponsor ME210 projects are increasingly adamant about the need to give their employees a continuing, life-long education opportunity without losing them to full-time study on campus. Encouraged by this demand, ME210 has been offered since 1994 to SCPD students across the US and since autumn 1995 to students and corporations around the world. The course is intensely experiential, i.e. hands-on. Distributed design teams composed of 2 to 4 members work on different industry funded projects. They are supported by a four-person teaching team, a practising design-team coach and a corporate liaison. The 1994–95 academic year included Ford, GM, FMC, Lockheed, Pfizer, 3M, Raychem, NASA-JPL, Redwood MicroSystems, Quantum, Toshiba, Seiko and HP Medical Products. The deliverable is a functional product prototype and detailed electronic documentation of the product and development process. These teams won 11 of 12 awards in the Lincoln Foundation Graduate Design Competition in 1995. The competition is based on a formal external peer review of the design documentation.

3. Instrumented Team Learning

ME210 is supported by a World-Wide-Web based information system created to meet the dual challenges of design knowledge capture and interactive distance education. ME210 events are distributed by video and internet channels. The focus is on capturing and re-using both informal and formal design knowledge in support of 'design for re-design'. Video (preferably a digital video server) is used for high-bandwidth, real-time transmission of lectures, design reviews and demonstrations. The internet is used for low-bandwidth data transmission, email, knowledge capture and retrieval. Electronic mail is used extensively for communication between the teaching staff, students, coaches and industry liaisons. Communication is automatically archived and organised by a HyperMail utility on the web. The ME210-Web also includes the class schedule, syllabus, assignments, five years of past student work, design-team background information and a HyperMail archive of working notes. The web-archive functions as a cumulative team and curriculum memory. It facilitates informal knowledge sharing within the class, and shares with subsequent generations the legacy knowledge of prior experience.

In addition to distance education benefits, the dynamic nature of the ME210-Web fundamentally changes the model of interaction between student teams, teaching staff, coaches and industry liaisons. Before the ME210-Web, product development team progress was only observable at formal meetings and quarterly reviews. With the ME210-Web, work-in-progress is available for review by all members of the community, at any time or place. This option augments and strengthens traditional briefings. It facilitates feedback from the teaching team, coaches and corporate partners to the design team. Importantly, teams can share their 'lessons-learned' in real time. All of this gives our corporate partners a basis for judging the validity of our curriculum and the value of their financial investment in engineering education.

The course and the web become a model for high-tech 'instrumented' learning. This extensive use of high-tech delivery systems has led to the formulation of a

PBL assessment instrumentation model which is shown in Figure 2. The model contains three distinct levels of feedback – formative, summative and validative – which we have found to be common to all design teams. The formative assessment loop is primarily concerned with incremental explorations, like prototyping and back-of-the-envelope calculations. It is carried out directly by the team members. A great deal of informal knowledge is used and generated at this level. The summative assessment loop concerns more formal knowledge and generally involves people external to the team but within the organisation. This level of assessment will include measures of efficiency and effectiveness such as described by Duffy and O'Donnell [7] elsewhere in this volume. Finally, the validative assessment loop involves the concrete experience of the customer. As is shown, the feedback loops are related to specific nodes on the input side.

A later addition to our model was technology, and more specifically, the computer. We believe that as a tool, it has crossed an important threshold in multi-functionality so that it can now play a major role as a design team member. We then reformulated the concept of 'noise' to model the disturbance we observe each time we introduced a new technology tool into the design situation. As it turns out, the concept was equally applicable to other players in the design situation – the client/customer, the team members and the instructor. We believe this formulation is important in the following sense: if 'noise' can be better defined especially in relation to design-team performance, and its occurrence can be predicted, then we can focus on finding means of filtering it or exploiting it, in order to move it to a position of real value. The study by Frankenberger and Badke-Schaub [8] in this volume provides a very valuable way of defining various types of noise and understanding the symptoms and probable origins.

Through longitudinal analysis of the input to the teaching situation in the form of students' prior experience and personal learning style preferences we have seen that the structure, membership and self-management of product development teams can be addressed in new and productive ways. The union of an explicit feedback assessment model and advanced technology for instrumenting and facilitating team activity has begun to yield performance metrics that can be used by a team to monitor and accelerate its own productivity. This information is also available to aid project supervisors, including instructors, in the assessment and prediction of team outcome performance.

We report two examples of our experience with metrics that have proven valuable in both understanding and facilitating Product Based Learning within a collaborative, team environment. The first is a subjective, behaviour-based index for controlling the 'noise' of student learning preferences as an input to team performance (*Node 4* – see Figure 2). The other is an objective, content-based measurement of work in progress to predict the quality of a team's final product. Data from work-in-progress assessment are used by teams for self-assessment *(Node 3)* and by project supervisors as an in-course corrective mechanism *(Node 2)*. Both types of metrics depend on advanced technology in the workplace for their implementation and acceptance.

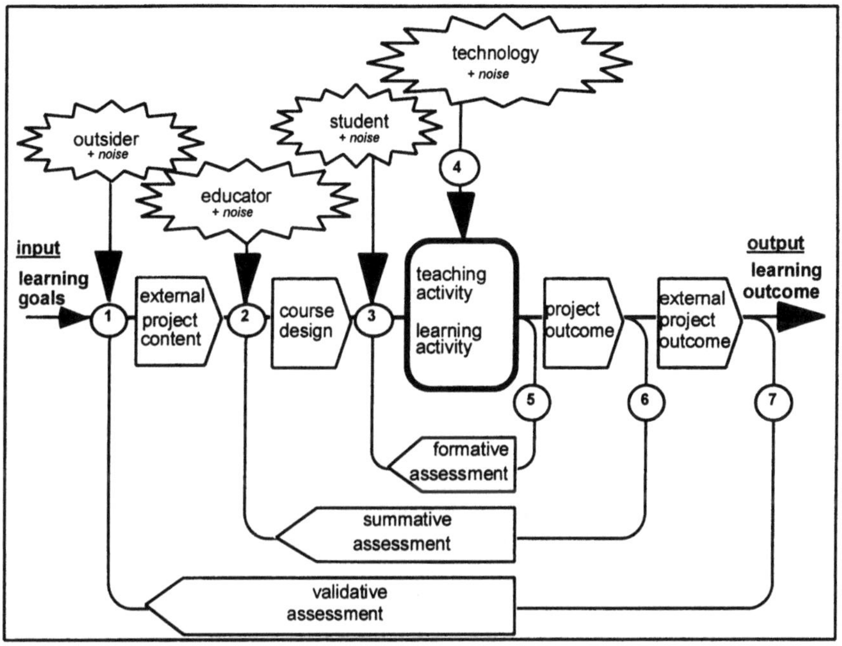

Figure 2: A 7-node design instrumentation model. The three levels of assessment feedback in general correspond to a team perspective (Nodes 3 & 5), a management perspective (Nodes 2 & 6) and a client perspective (Nodes 1 & 7). We later introduced technology (Node 4), and the concept of 'noise'. The latter was an attempt to model the disturbance we observed each time a new technology tool was introduced into a design situation. The concept was later found to be equally applicable to each of the preceding three nodes.

4. Team-Preference Profile Management

The idea of using questionnaires to guide the formation of design teams had its origins in 1989 in a Workshop on Creativity for design professors organised at Stanford by Bernard Roth, Rolf Faste and Douglas Wilde. To determine the effectiveness of the workshop, participants were surveyed both before and after the two-week intensive experience to determine their 'Gough Creativity Index' (GCI), a measure of creativity devised by the psychologist Harold Gough. The GCI is an empirically determined linear transformation on the four scores generated by the Myers–Briggs Type Indicator (MBTI), a questionnaire used widely for psychological, educational and vocational counselling [9]. As reported by Wilde [10], the workshop enhanced the creativity index of the participants and promised to be a useful guide for the design of effective teams.

In 1990 this idea was extended to ME210. After responding to the MBTI, the students were asked to voluntarily form teams composed of individuals whose MBTI scores differed in at least two of the four variables. Subsequent performance confirmed that teams having a high GCI member in balance with other preference profiles tended to win more and better design competition awards. Initial findings were refined in subsequent years and applied to teams of two, three and four members. Mathematical transforms were developed to use MBTI numerical scores and vector analysis to decompose the score vector into the 2, 3 or 4 components required to guide team formation [11]. It was also possible to compute Jungian cognitive modes, one of which corresponds closely to the GCI [12].

The MBTI has since been replaced by a much briefer Preference Questionnaire. This saves time and de-emphasises the counselling aspect of the survey. Students wishing to investigate the counselling side further are encouraged to do the MBTI under the supervision of a licensed counsellor. The Preference Questionnaire is viewed simply as an expression of personal preferences rather than an indicator of personality type. The data are used to place students into four broad preference groups named North (N), South (S), East (E) and West (W). While the basis for this categorisation is different from that of p-designers and m-designers described by Günther and Ehrlenspiel [13] in this volume, we see a common purpose in seeking a better understanding of the 'chemistry' of design teams based on the characteristics of their individual members.

Teams in the class are formed in a peer-to-peer protocol that rewards five team diversity factors:

1. include one person from each personal preference group;
2. balance gender representation;
3. maximise ethnic diversity;
4. maximise technical background diversity; and
5. maximise the members' geographic distribution.

Beginning in 1977, ME210 final reports have routinely been submitted for consideration in the Lincoln Foundation Graduate Design Competition. The reports are read and evaluated by designers and design professors, who award twelve prizes to those they consider best. Judges change from year to year and are blind to the authors and their university affiliations. Since 1991, balanced high GCI teams have clearly been more successful at winning Lincoln awards, not only more of them, but also of higher quality [14]. Table 1 summarises the awards won during the five-year period of our experiment and the preceding thirteen (control group) years.

The table shows that the percentage of prize-winning teams from ME210 doubled, from 29% to 60%, during the experimental period. High GCI teams had a higher winning frequency (63%) than did the other teams, whose performance (50%) had already been improved from 29% simply by diversification. The study shows that the quality of prizes won was also considerably better for high GCI teams, which won a high proportion of Silver, Gold and Best of Program medals (42%) triple that for the other teams (14%). In the best year to date, 1995, student teams won eleven out of the twelve prizes awarded nationally. Team preference diversification thus corresponds to the quantity of prizes won, an indication that distributing human talent diversity amongst the project teams raised overall quality especially at the top.

Table 1: Summary of Lincoln Prize Awards

		High-GCI Team			Other Teams			All Teams		
Period	*Yrs*	*N*	*Aw*	%	*N*	*Aw*	%	*N*	*Aw*	%
1977–90	13	–	–	–	–	–	–	138	40	29
1991–95	5	49	31	63	14	7	50	63	38	60

Yrs = number of years studied; N = number of teams; Aw = number of awards won; % = percentage of teams winning awards.

The personal preference questionnaire and group transform algorithm have been implemented as World-Wide-Web forms [http://me210.stanford.edu]. Given internet access, anyone, anywhere can complete the questionnaire to identify their preference profile. Another web form is used to characterise the team-preference diversity map. As documented in Hong and Leifer [15], WWW-mediated peer team formation, using voluntary personal preference data, is one of the most unique and useful applications of technology to team performance management in ME210.

5. Design Documentation Content Analysis

Engineers will, of course, be more at ease with objective, technical content factors that predict team performance. Accordingly, and in response to the increasing number of Engineering classes that require open-ended problem solving, for example, design assignments, we have been studying product design documentation in the interest of supporting design knowledge re-use and grading. Traditional tools for measuring student performance, such as written examinations and multiple choice questions, are not appropriate assessment instruments in project courses. Grading product documentation is uncomfortably subjective. As the instructional situation moves from one in which there is one right answer and method to one in which there are several alternative answers and methods, there is a need for assessment tools that can accommodate multiple points of view, and be context dependent and content specific.

Studies of redesign behaviour demonstrated that designers ask for information about prior designs in terms of noun-phrases [16]. It was also observed that an automatic content search engine that could use noun-phrases in simple associative relationships was far superior to key-word searches in terms of the designer's appreciation of search results. Hence, noun-phrases have become the object of attention in recent documentation content analysis studies.

The raw data for our study of noun-phrases [17] come from a set of three reports, named 'design-requirement documents', submitted by students in ME210. The projects analysed were taken from the 1992–93 academic year and dealt with a wide range of products including a catheter for gene therapy in the human body, a control mechanism for maintaining the focus of an infrared optical system, and a power locking device for an automobile door. In Figure 3, team grades have been superimposed on a graph of the number of distinct noun-phrases found in each of their quarterly design-requirement documents.

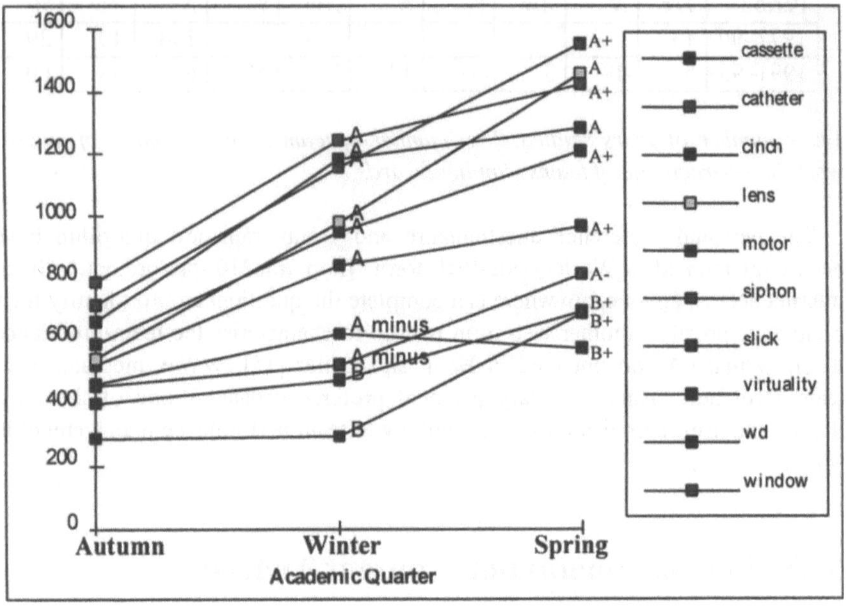

Figure 3: When the winter and spring quarter grades were superimposed on the plot of noun-phrase score versus academic quarter, it was found that the reports with low number of noun-phrases corresponded those projects with low grades, and correspondingly, those with a high number of noun-phrases had a higher grade.

To calculate the level of association between the incidence of unique noun-phrases and peer assessment of design quality (the grade), a gamma measure of association was calculated. For the ten project teams studied, gamma was equal to 1.0, a perfect ordinal prediction, for the winter quarter. For the spring quarter, gamma was 0.71 (strong but less than perfect), following intervention by the teaching team to stimulate the performance of slack teams. Projects assigned the letter grade B or B+ corresponded to reports with the lowest noun-phrase counts while teams assigned the letter grade A or A+ produced reports with the highest noun-phrase counts. It was not possible to make this sort of differentiation based on the readability of the documents for which the gamma association with the class grade was low (0.36 for the winter quarter and 0.31 for the spring quarter). The gamma association between the grade and the number of words in the document was higher than that for readability but lower than that for noun-phrases (0.93 in winter and 0.49 in spring). Hence the number of distinctive noun-phrases in formal project documentation appears to be an objective predictor of product quality ranking.

6. Discussion

While detailed examination of the learning implications of these results must be deferred, it is important to point out that the results are, in part, validated by the performance of these same teams in Lincoln Graduate Engineering Design competition where the top awards were given to projects with the highest number of distinctive noun-phrases.

Additional corroboration comes from recent empirical studies, which indicate a similar pattern of knowledge development in product development teams. Based on studies of several Japanese companies, Nonaka and Takeuchi [18] for example describe a process they labelled 'the spiral of organisational knowledge creation', a depiction of which is shown in Figure 4. Without going into the details of the mechanisms indicated on the diagram (externalisation, socialisation, etc), it is useful to observe the similarity between the increase in noun-phrases shown in Figure 3, and the growth in the amount of explicit knowledge, as shown in Figure 4. Elsewhere in this volume and based on psychological studies, Dörner [19] describes a number of mechanisms that have been observed for single subjects. One of these mechanisms is the 'Picture–Word Cycle', for which we think the evolution of noun-phrases is a satisfying marker. Dörner describes other mechanisms such as mental simulation and sketching, for which noun-phrases may not be appropriate, and these may be important sources of tacit knowledge as shown along the epistemological dimension of Figure 4.

The ontological dimension of Figure 4 shows several levels at which knowledge is created in an organisation – individual, group, which is the team level, organisation, which is the company level, and inter-organisation. The latter three levels coincide with the feedback loops of our design instrumentation model (Figure 2). Furthermore, as educators, the variables along the ontological dimension tie in very well with the work of some educational theorists such as Vygotsky, who, as represented by Moll [2], argues that knowledge is social before it becomes personal. The implication of this is that learning must be external and shared before it can be internalised and made personal. While our instinct as designers is to reverse that order, the overall notion reinforces the socio-technical framework of our work and provides an additional context for interpreting the findings of our study.

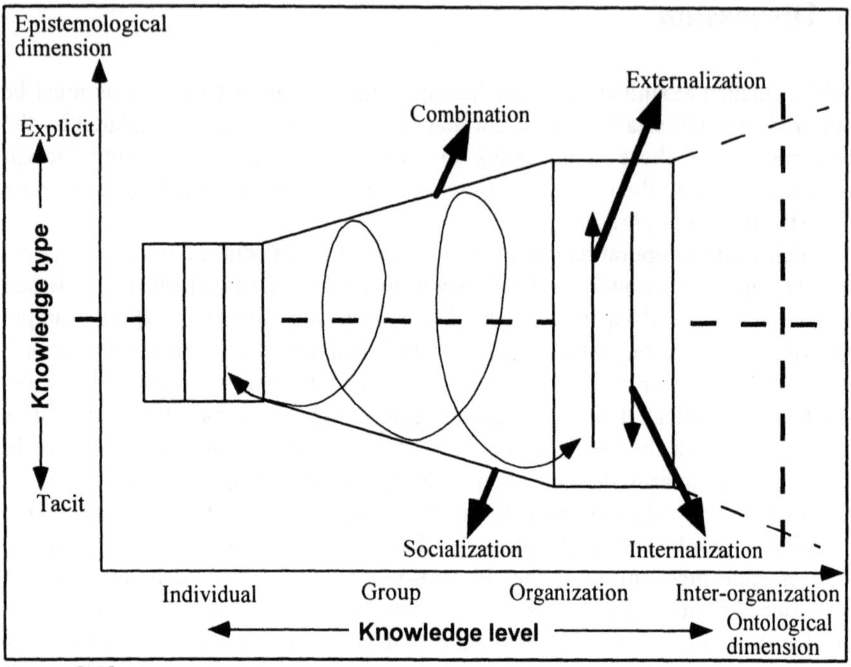

Source [15]

Figure 4: The spiral of organisational knowledge creation. In this framework two types of knowledge (tacit and explicit) and two loci of knowledge (individual and group (at multiple levels)) are identified.

7. Summary

We have described a socio-technical framework for thinking about engineering design. We have presented the concept and practice of 'Product Based Learning' (PBL) and the feedback instrumentation model we use for assessing PBL performance. A theme that pervades our work is the role of technology both in implementing learning environments and in assessing performance outcomes. Utilisation of internet capabilities to facilitate team-based, distance learning has changed the character of the learning environment in profound ways. Learning in PBL environments is applied within the context of the course. Work-in-progress data, again technology instrumented, are used for self-assessment as the work of the teams evolves. The metrics produced by these data have been shown to predict quality of the final products. The overall quality of designs produced by these diverse, collaborative teams stands up well against industry standards as reflected in the Lincoln Design Competition. Finally our objective metrics are corroborated by more descriptive studies in industry. We believe this is just the tip of the iceberg, there is still a lot of work to be done. However, we are much encouraged by our strategy of using university-based PBL classes to study the design process, and our focus on developing performance metrics. This we believe is an important key to design process improvement.

Acknowledgements

We would like to thank the following people whose ideas have helped in shaping the work presented here: David Cannon, Vinod Baya, Sheri Sheppard, Margot Brereton, Doug Wilde, John Tang, Scott Minneman, George Toye and Jack Hong.

References

1. Leifer L.J. 1997: Design Team Performance: Metrics and the impact of Technology, forthcoming. In: S.M. Brown & C. Seidner (eds) *Evaluating Organizational Training.* Kluwer Academic Publishers, Netherlands.
2. Moll L. (ed.) 1990: *Vygotsky and Education.* Cambridge Univ. Press, UK.
3. Harb J., Durrant S., Terry R. 1993: Use of the Kolb Learning Cycle and the 4MAT System in Engineering Education. *ASEE Journal of Engineering Education* 3, 70–77.
4. Bridges E.M., Hallinger P. 1995: *Problem Based Learning: In Leadership Development.* ERIC Clearinghouse on Education Management, Eugene, Oregon.
5. Leifer L.J. 1995: *Global Product Development Teams.* A 9 minute video focusing on the content deliverables of ME210. Anne Flatte (Producer), Stanford Center for Professional Development, Stanford, California.
6. Leifer L.J. 1996: *An Experience in Product Based Learning.* A 52 minutes video tracing the experience of four teams in ME210. Anne Flatte (Producer), Stanford Center for Professional Development, Stanford, California.
7. Duffy A.H.B., O'Donnell F.J. 1998: *This volume,* pp. 269-283.
8. Frankenberger E., Badke-Schaub P. 1998: *This volume,* pp.149-164.
9. Briggs-Myers I., McCaulley M.H. 1985: *Manual: A Guide to the Development and Use of the Myers–Briggs Type Indicator, 2nd edn.* Consulting Psychologists' Press, Palo Alto, California.
10. Wilde D.J. 1993: Changes among ASEE Creativity Workshop Participants. *Journal of Engineering Education* 82, 167–170.
11. Wilde D.J. 1993: Mathematical Resolution of MBTI Data into Personality Type Components. In: *Proceedings of the ASME Conference on Design Theory and Methodology,* pp. 37–43, Albuquerque, New Mexico.
12. Wilde D.J., Berberet J. 1995: A Jungian Theory for Constructing Creative Design Teams. In: *Proceedings of the ASME Conference on Design Theory and Methodology,* pp. 525–530, Boston, Massachusetts.
13. Günther J., Ehrlenspiel K. 1998: *This volume,* pp. 85-97.
14. Wilde D.J. 1997: Using Student Preferences to Guide Design Team Composition. In: *Proceedings of the ASME Conference on Design Theory and Methodology,* DTM-3890, Sacramento, California.
15. Hong J., Leifer L. 1995: Using the WWW to Support Project-Team Formation. In: *Proceedings of the Frontiers in Education Conference on Engineering Education for the 21st Century,* Atlanta, Georgia.

16. Baya V., Gevins J., Baudin C., Mabogunje A., Toye G., Leifer L. 1992: An Experimental Study of Design Information Reuse. In: *Proceedings of the ASME Conference on Design Theory and Methodology* pp. 141–147. Scottsdale, Arizona.
17. Mabogunje A. 1997: *Measuring Mechanical Design Process Performance: A Question Based Approach.* PhD thesis, Stanford University, California.
18. Nonaka I., Takeuchi H. 1995: *The Knowledge Creating Company.* Oxford University Press, New York.
19. Dörner D. 1998: *This volume,* pp. 3-11.

Forensic Analysis of the Engineering Design Process

Crispin Hales

Abstract

Forensic engineering, or the application of engineering knowledge to legal problems, is a well-established field with highly developed techniques available for gathering of evidence, analysing failures, investigating accidents, reviewing safety issues and presenting conclusions in the form of opinions. A particular area of concern within the general field is the engineering design process itself. When legal claims are made such as: the design didn't work; the performance did not meet the specifications; the designers made mistakes; the machine destroyed itself; the design was copied or the patent was infringed – then the process by which the design came about is called into question.

Some common issues involved in forensic analyses of the engineering design process are presented, together with illustrative examples. It is concluded that the substantial hidden costs of insurance and product liability built into the price of every product may be reduced by effective use of the symbiotic relationship between engineering design research and forensic engineering.

1. Introduction

Forensic engineering analyses cover a huge variety of unusual and peculiar circumstances where a number of factors happen to coincide in an unforeseen way to create a major problem or accident. Although the causes may at first sight seem straightforward, and indeed the claim may well be that if only the lever went up instead of down the person would not have died, investigation often reveals a complex interplay of engineering and human issues not easily unravelled.

As the engineering design process is a human activity it does not exhibit the same hard physical evidence usually associated with forensic engineering. This complicates the analysis. However, the results of research into understanding of the design process, together with the techniques and tools developed to carry it out, now provide a solid basis for investigating how and why a design came to be the way it was. By use of a structured approach, timeline analysis, logic diagrams, review of activity records and detailed inspection of documents such as design specifications and drawings, it is often possible to develop precise opinions, each with an adequate basis for defence through depositions and trial if necessary. 'Opinions' in the sense of an expert in a forensic engineering case are concise summary statements of a conclusion developed from review and analysis of all the available evidence at the time. They are likely to be challenged, word by word, not only during the course of the particular case but also every time the issue arises in future cases. A set of opinions therefore represents a personal position which needs substantial evidentiary support to survive.

In this paper the design process is first considered from an analysis point of view. Then from a selection of typical cases some recurring forensic issues have been identified and these are presented in the context of the case files. Conclusions are drawn regarding the relationship between design research and forensic engineering.

2. Analysis of the Engineering Design Process

It is possible to arrive at a design in many different ways, ranging from informal experimentation through to a highly organised formal procedure [1]. When it comes to analysing what was done in any particular case, however, it is essential to have a clear and structured approach to mapping whatever evidence is available, no matter what type of design process was used. The more it can be broken down into measurable components the less subjective will be the conclusions and the better the chance of developing defensible opinions. For the purposes of this paper, the schematic shown in Figure 1 represents the basic design and manufacturing process adequately and the discussion will be confined to the project and personal levels of resolution [2] within the wider context, as represented diagrammatically in the figure.

Although it may be argued that in practice the design process will not proceed through the apparently sequential set of project phases as shown in the diagram, the sequencing is not as important as the existence, nature and effectiveness of the actual design activities implied within each phase [3]. For example, there must be some kind of specification to which the design relates and there must be some kind of concept from which a final design evolved. The concept must be developed to a greater or lesser degree to result in a practicable overall design and the details of every component must be established to the point where the product or system can be manufactured. A combination of human activities is required to reach each of these end points or outputs, and these activities have now been the subject of much research as described, for example, in [2, 4, 5–7] and elsewhere in this volume. In practice, a wide variety of influencing factors acts on the design process at different levels of resolution and at any particular point in time, as focused on in five out of the six symposium sessions represented by the papers in this volume. The *activities of* the design team, the *influences on* the design team and the consequent *output from* the design team are key issues in any forensic analysis of the design process [2].

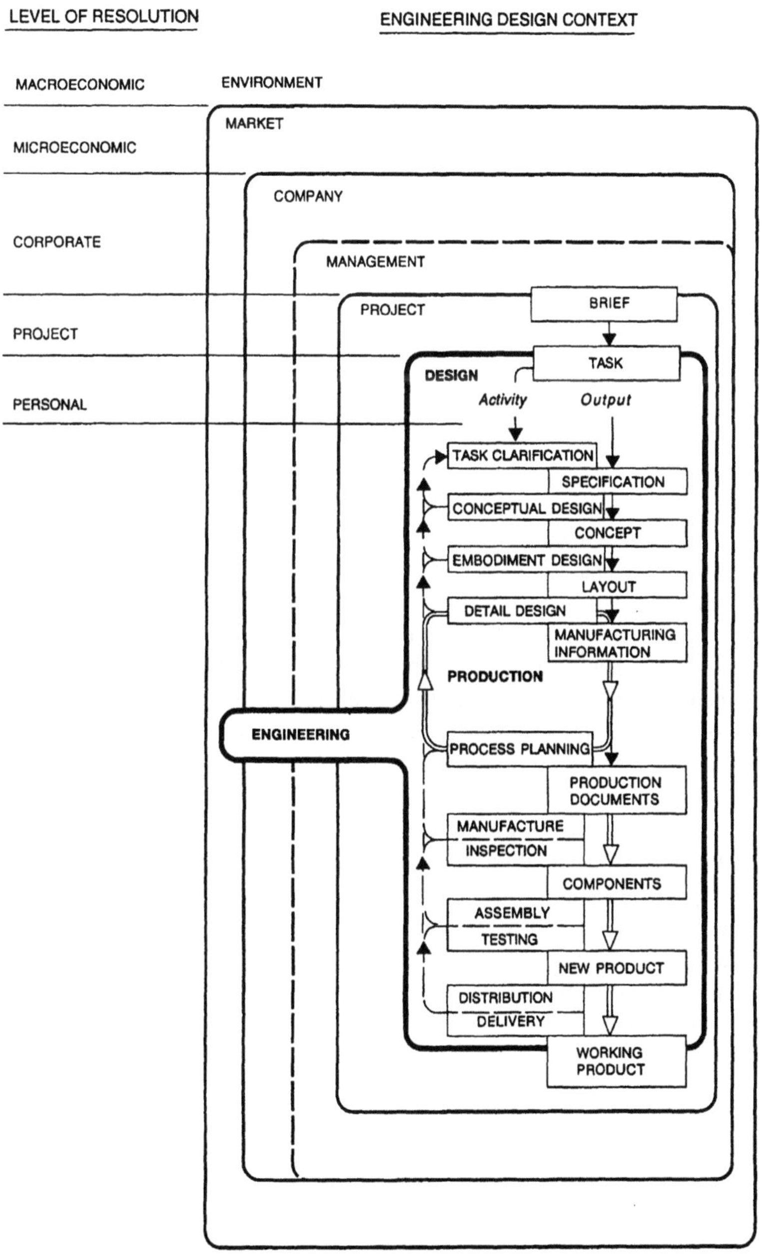

Figure 1: Levels of resolution related to the engineering design process [2].

3. Forensic Analysis Issues

When something goes wrong with a design, inevitably there is confrontation [8]. If the parties involved cannot come to an amicable agreement on solving the problem, lawsuits will be filed and the matter will become both more complex and far more costly. Lawyers for each party will look for legal issues to help them blame other parties for the problem and will hire technical experts who they feel will offer opinions which support their case. The issues as seen by one party may not even be at the same level of resolution as those seen by another party.

Technical experts must review and analyse the issues in the light of available evidence and remain as objective as they can while developing their opinions. They may have to combine analyses at different levels of resolution in order to present a complete argument. For example the ultimate claim by one party that certain parts failed and caused downtime may be countered by another party's claim that the original design specification was deficient. The expert would have to address both claims and respond with appropriate opinions 'to a reasonable degree of engineering certainty'.

It may well be that the issues seen by the expert to be important from an engineering point of view are not the same as those that the lawyer wants to concentrate on from a legal point of view. If the technical expert has been hired by the lawyer then the analysis must be carried out from the lawyer's point of view, even if some concerns which seem important to the engineer get pushed aside. However, the engineer must at all times be faithful to the professional code of ethics, and if the lawyer pushes to the point of compromising the engineer's position then of course the engineer should consider withdrawing from the case. It is the job of a lawyer to try and 'win the case', whether or not it helps resolve any technical issues. Depending on the size of the claims, the depth of antagonisms, the strength of the case, the nature of the lawyers and numerous other factors, the dispute may be settled by informal mediation, formal arbitration, negotiated settlement or trial. In general the cost rises dramatically through this spectrum of possibilities, and the clarity and certainty of the engineering expert's opinions often have a decisive effect on the stage at which the process concludes. Sometimes a case continues beyond trial through appeals and further trials.

The simpler the argument, the easier it is for a lawyer to 'present it to the jury'. However, the design process by its very nature is complex, so a key part of a forensic analysis of the design process is reducing the issues to the simplest terms, then presenting the results in a clear and understandable fashion.

4. Typical Forensic Issues in Design

4.1 The Designers Made Mistakes

In almost every case where a design fails to perform as expected, the finger points first to the designer. Many times it is true that there is a problem with the design, but usually broader issues such as communication breakdown and the effect of external influences need to be investigated [9].

Case Example: Dynamometer Bearing Failure

The lawyer representing a dynamometer company called up requesting an inspection to investigate the cause of a catastrophic bearing failure on a large remanufactured dynamometer. A request for drawings and other documentation prior to the inspection set off the following chain of events. Firstly, during the inspection the engineer in charge of the bearing selection and remanufacturing process volunteered in detail his version of what happened, together with all the documentation. It turned out that he had recently joined the company to take over in that position from a long-standing engineer who because of personality problems had been moved into more of a technical sales role by the company management. Shortly after this first meeting came an unsolicited and confidential visit by the reallocated engineer who presented a different but no less valid version of what happened, together with more drawings and documentation. The focus of the design problem itself concerned the bearing dams used to control the amount of oil fed to the bearings during operation. However, the underlying problem which caused the catastrophic failure was jealous withholding of essential operational feedback data by the one engineer, coupled with an overconfident assumption of personal technical knowledge by the other. The requested forensic analysis of a bearing failure metamorphosed into a design management seminar for the manufacturing company, followed by a jolly good dinner!

Case Example: Tri-axis Transfer Press

Two huge tri-axis transfer presses for making car body parts failed to perform adequately in service. During the analysis of what went wrong some critical features of the project were identified as follows:

- Custom-built, one-off machines
- Largest units ever built up to that time
- Minimal design specification cobbled together from previous projects
- Stringent, novel performance requirements compared with prior machines
- More complex and massive transfer mechanism than on previous machines
- High degree of technical and financial risk

- Total separation of design from manufacture
- Miscommunications from translation problems with language and units
- Unreasonable degree of urgency with penalty clauses for late delivery.

The contracts called for the machines to be manufactured by the supplier company in Europe, but designed by a subcontracting company located in North America. Language difficulties created misunderstandings throughout various memoranda and when the drawings were translated the intent of the designer was not fully communicated. Mixed imperial and metric units were used which had to be rationalised and created confusion. Material specifications were not exact equivalents from one country to the other and it was not clear that specified surface treatments and heat treatments were ever carried out according to the original drawings. Approximately 60,000 hours of engineering design work was involved and some 4,000 manufacturing drawings were shipped from the designing company to the manufacturing company. The time-scale was so short that the contracts permitted use of drawings from previous machine designs where practicable, and to reduce late delivery penalties the machines had to be built directly in the customer's plant without proper testing. They were then put into production without any systematic shakedown procedure. The customer ended up with machines that could work only at low speed, so the customer refused to pay the manufacturer and the manufacturing company wouldn't pay the designing company on the basis that the design was faulty.

The failure of the machines to perform to the customer's expectations caused such loss of revenue to the customer that the claim for damages against the supplier company and the subcontracting company far exceeded the cost of design and manufacture of the machines. Most of the problems could have been avoided had there been closer communications between the two companies during detail design. Both the supplier and subcontracting companies were long-established and well-respected designers and manufacturers of large metalworking presses with a history of working together on previous contracts.

By analysing the complete project in terms of a systematic approach to the engineering design process it was possible to identify a series of contributing factors leading to the failure of the machines. This forensic analysis played a major role in effecting an acceptable monetary settlement between the opposing parties prior to the planned international arbitration hearings. Nevertheless, both companies were bankrupted by the proceedings.

4.2 The Design Failed to Meet the Specification

If a machine or product fails in service, an obvious potential claim is that the design did not meet the design specification. From a selection of typical cases, it was found that about a third of them involved this issue. The development of a comprehensive design specification is a critical activity in the design process and often one that is poorly carried out in practice [10, 11]. In the case of products used by consumers, the 'specification' referred to is generally the 'product specification' [12] as printed on the box or package, rather than the original design specification.

Case Example: The Dog that Bit the Girl

A Rottweiler dog allegedly broke free of its outdoor tether and attacked a small girl, severely biting her hand. The dog owner claimed that the 'accident' happened because the vinyl coated 'Tie-out Cable' which he had recently purchased was incapable of restraining the 38.6 kg (85 lb) dog and failed at the swaged loop end as shown in Figure 2. The product package contained the following words: *Breaking Strength 1750 lbs. 1/8'' Galvanized, Vinyl Coated Aircraft Cable for Dogs Over 50 lbs.* The plaintiff's expert did tensile tests on exemplar Tie-out Cables in which the end pulled out of the swage ferrule at less than half the stated 'breaking strength', despite the cable itself meeting the specified tensile strength. The dog-owner claimed that the product was therefore defective as it did not meet the specification, and that this was the cause of the accident.

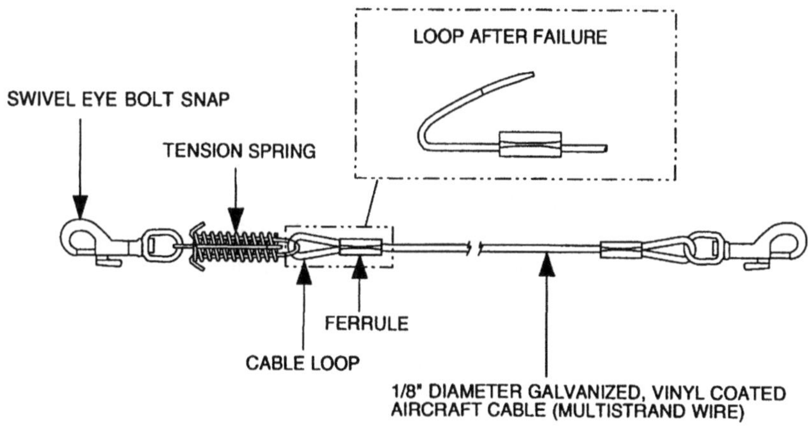

Figure 2: Alleged failure of Tie-out Cable.

An investigation into the design process for the product revealed that it was simply an assembly of commercially available parts and that there was no record of *any* design activities within the manufacturing company. However, a careful analysis of the complete product package, historical information from the company, information on dog pulling performance and a detailed inspection of the failed Tie-out revealed the following:

- The cable was far stronger than necessary to restrain a dog, but its oversized diameter made it more visible on the ground, less likely to twist up into knots, more tolerant of physical damage and less likely to cause injury to people or the animal.

- No dog wearing a normal loop collar could have exerted enough force to pull the loop apart, even with multiple impact loads.
- While the accused cable was galvanised for better weather resistance, the accident cable was not, and therefore it was a different manufacturer's product.

At trial the jury agreed that the accused product was different from the accident product and thus the case against the product manufacturer failed. From an engineering design point of view, however, and not related to the accident, there remains an issue over customer expectations. The wording of the product specification was technically correct, but it miscommunicated the intent of the design and led to false user expectations.

Case Example: Vertical Lift Conveyor

A large number of custom-designed vertical lift conveyors were required for use in a series of new automated US mail sorting facilities. These facilities comprise essentially a series of code-reading sorting units at ground level with a vast array of horizontal roller conveyors above. Vertical lift conveyors are used to transport plastic trays of sorted mail from ground level up to horizontal conveyor level and the reverse. The vertical lifting concept was developed to overcome the slippage problem encountered with inclined roller conveyors when plastic trays superseded cardboard trays in earlier facilities.

A massive standard design specification had been prepared for the Post Office by the general contractor. This was to apply to all facilities, but for each particular facility there was also a detailed set of special requirements. Within these documents were embedded the requirements for the conveying systems in general, and within those the requirements for the vertical lift units in particular. Prototype vertical lift units which used twin in-running belts to deliver the trays to three-fingered lift platforms already had been developed and tested off-line to the point of acceptable performance for this application. The Post Office design specification for the lifting units was compiled with the prototype units in mind, even to the point of *requiring* in-running entry and out-running exit conveyors having two or more belts, and lift platforms with three or more fingers. In such a way the specification unnecessarily constrained the design to a specific concept which was known to have inherent operational problems.

The many required units were designed, built, tested, delivered and installed at a cost of more than $500,000. While some design weaknesses and manufacturing problems were evident which detracted from the performance of the units, nevertheless they did pass the acceptance tests laid down in the design specifications, right up to the final production trials. At this point, however, multiple failures and jam-ups were encountered because the operators were using the vertical lift units on a start–stop basis instead of continuously as intended. The Post Office sided with the general contractor in claiming that the units did not meet the specification. They refused to pay, then proceeded to tear out and scrap all the machines, at the same time contracting with a different supplier for completely different replacement units. In the ensuing legal battle it was revealed that a modified design specification had been issued to alternative manufacturers who were invited to bid on supplying the replacement units. The ones finally selected were based on a different concept which could never have met the requirements of the original design specification. The modified design specification deleted the requirements for belt conveyors and for the lift platforms to have three or more fingers. This removed the fictitious constraints on the design and allowed the use of an articulated slat conveyor, far superior in concept to the suspended tray type of conveyor for this particular application.

In engineering design terms this was again a matter of poor communication and mismatched expectations, but also combined with a deficient design specification. It resulted in a huge waste of effort, money and materials as well as the bankruptcy of the original vertical lift unit manufacturer.

4.3 The Design Was Copied

The infringement of patents and the direct copying of consumer products have long been areas requiring forensic analysis to resolve disputes. More recently, however, new forms of copying have appeared which take advantage of computer technology and are concerned directly with the activities of the engineering design process. The essential information necessary to manufacture a product or system is usually encapsulated in a set of drawings, which today may be stored in a series of computer files. If someone is able to obtain a complete set of drawing prints or equivalent computer files, then they can short-circuit the design process shown in Figure 2 and are likely to be able to reproduce the product or system with very little design activity at all. This represents an enormous saving in time and cost.

When a claim is made by one party that a design has been copied in this fashion, with a response from the other party that 'we designed it all ourselves', there are a number of complementary avenues to investigate, such as the following:

- Review of the design specification with the evidence of how it was developed.
- Check on what alternative designs were generated and how they were evaluated.
- Check on all design activity records, including notebooks, calculations, letters, memos, telephone calls and e-mail, throughout the alleged design process.
- Analysis of the team composition, experience and time records in regard to the nature and extent of design work claimed to have been carried out.
- Detailed examination of all the drawings and their interrelationships, including notes, revisions, dimensions, tolerancing, drawing numbers, dates, materials and compatability of parts within assemblies.
- Interrogation of the computer files to establish the sequence of generating each feature on each drawing and the time taken to create the documents.
- Analysis of the quality of any actual engineering design activity in evidence.

Case Example: The Public Domain

A very small company claimed that it developed the design of a large and complex piece of equipment independently, simply based on what information was available in the 'public domain' and the experience of its design team. The large company that had originally developed the concept, and had a long-established market position to defend, thought otherwise. Volumes of project documents, drawings, computer files and historical documents were produced by the small company to support its claim and a lengthy investigation was carried out on behalf of the large company to come to conclusions on the copying issue. Although the nature and details of the investigation must remain confidential, all the above avenues and more were explored. From a design research point of view this was a case where a small company had to present its idea of what a 'clean sheet' engineering design process entails and evidence of how they claimed to carry it out right through to a working product with a minimal design team. It was less than convincing.

4.4 The Design Failed During Transportation

Some of the most bitterly fought disputes involving the design process arise out of accidents during transportation, before a product or system ever reaches its destination. Just as with any vehicle accident, the immediate focus is on the factors which caused the accident at the time, but further investigation usually reveals prior circumstances that made an accident situation almost inevitable. Often it is possible to trace the original problem to design specification deficiencies which either were never identified or which led to fatal compromises during later phases of the design process. One typical scenario is when a crated piece of new equipment, being transported by truck, tips over unexpectedly and smashes through the side of the vehicle. The transportation company treated the crate simply as a box to be moved from one place to another but, unbeknown to them, the piece of machinery inside was top-heavy with a displaced centre of gravity. The designer and manufacturer see in their mind's eye only the wonderful machine they created, now destroyed by the clumsy transporter. The transporter, on the other hand, sees only a box which turned out to have peculiar and unexpected characteristics. It is another case of mismatched expectations and inadequate interface communication.

Case Example: The Great Boiler Module

Several huge boilers were purposely designed in a modular fashion to facilitate transportation to a new plant site by rail. Each module was matched to a specific type of railroad car, and a delivery programme was planned with the railroads, using available train services. This was a complicated business because not all modules were made in the same location, and from region to region along the routes there were different regulations, different railroad companies and different railroad track conditions. The largest and heaviest modules were a problem from the start. They required the heaviest duty cars in the US of which only two or three existed, their centre of gravity when loaded was extremely high, and it was off to one side. To meet railroad regulations the cross-sectional envelope of the loaded car had to fit within fixed limits, and if the centre of gravity was higher than a defined limit then the cars had to go by special train which was both slower and more costly. The designers calculated the weight of ballast needed to be bolted to one side of the module in order to control the horizontal location of the centre of gravity, and did approximate calculations for the vertical location. It was worked out that because the car had a 'depressed centre' then extra ballast could be added below the module to lower the centre of gravity just enough to avoid having to use a special train service. However, this would bring the overall weight very close to the maximum capacity for the car. As the module weight could not be known accurately until the unit had been manufactured, extremely careful coordination was required to ensure that the geometry, weight and centre of gravity of the loaded car were within the specified limits. In fact this did not occur. Nobody took responsibility for ensuring that all conditions were met, each party assuming that the other had taken care of it. Although the horizontal location of the centre of gravity of the loaded car was checked (easy to do by measuring tilt), the additional bottom ballast was never added, the loaded car was never weighed and the final vertical location of the centre of gravity was never established. Normal train service was used.

The modules joined a train which had a variety of other cars for other customers, including some loaded with hazardous materials. Along the way a massive derailment took place, which not only destroyed the entire train but involved the evacuation of a nearby town. A large number of complex lawsuits followed and it took ten years before the matter was finally settled.

5. Conclusions

There is a strong symbiotic relationship between engineering design research and forensic engineering which has development potential yet to be fully realised. Design research has led to effective systematic approaches for design and continues to help our understanding of the design process. Both those results assist in the forensic analysis of design failures, accidents and unexpected phenomena. In turn the forensic analysis of a wide variety of cases in practice offers a rich source of hard won data and documented design experience to design researchers.

The more precisely, succinctly and confidently a forensic analysis of the design process can be carried out, the more likely a legal battle over engineering design issues can be settled by lower-cost alternative dispute-resolution methods. In this way design research has a direct bearing on reducing the enormous hidden costs such as insurance, product liability and legal representation currently built into the price of every product and system.

Furthermore, future design research could be targeted towards improvement in areas identified by forensic investigations as a prime cause of design failures, such as inadequate communication at each level of resolution.

References

1. Dörner D., Ehrlenspiel K., Eisentraut R., Günther J. 1995: Empirical Investigation of Representations in Conceptual and Embodiment Design. In: *International Conference on Engineering Design, ICED '95, WDK 23, Vol 2.* Heurista, Zürich, pp. 631–637.
2. Hales C. 1993: *Managing Engineering Design.* Longman Scientific & Technical, Harlow.
3. Raine J.K., Whybrew K. 1997: Speeding Up Product Development Through Systematic Management of Design. In: *Proceedings of 1997 World Manufacturing Congress*, Auckland, New Zealand (Nov. 1997).
4. Horton G.I., Radcliffe D.F. 1995: Character of Contextually Grounded Prototyping. In: *International Conference on Engineering Design, ICED '95, WDK 23, Vol 2.* Heurista, Zürich, pp. 638–643.
5. Dwarakanath S., Wallace K.M. 1995: Design Decision Making Process: Observations from Individual and Group Design Experience. In: *International Conference on Engineering Design, ICED '95, WDK 23, Vol 2.* Heurista, Zürich, pp. 555–560.
6. Frankenberger E., Birkhofer H. 1995: Teamwork in Engineering Design Practice. In: *International Conference on Engineering Design, ICED '95, WDK 23, Vol 2.* Heurista, Zürich, pp. 828–833.
7. Stetter R., Ullman D.G. 1996: Team-Roles in Mechanical Design. In: J. Cagan & K.L. Wood (eds) *8th International Conference on Design Theory and Methodology, 96–DETC/DTM–1508.* American Society of Mechanical Engineers, New York, pp. 1–8.
8. Hales C. 1995: Five Fatal Designs. In: *International Conference on Engineering Design, ICED '95, WDK 23, Vol 2.* Heurista, Zürich, pp. 662–667.
9. Nijhuis W. 1995: Management of Design Process: Management of Conflicts. In: *International Conference on Engineering Design, ICED '95, WDK 23, Vol 2.* Heurista, Zürich, pp. 755–760.
10. Hurst K., Hollins B. 1995: Improved Product Design Specification Compilation. In: *International Conference on Engineering Design, ICED '95, WDK 23, Vol 2.* Heurista, Zürich, pp. 584–589.
11. Otto K.N., Wood K.L. 1996: A Reverse Engineering and Redesign Methodology for Product Evolution. In: J. Cagan & K.L. Wood (eds) *8th International Conference on Design Theory and Methodology, 96–DETC/DTM–1523.* American Society of Mechanical Engineers, New York, pp. 1–15.
12. Hales C. 1990: Project Proposals, Project Briefs and Design Specifications. In: M. Oakley (ed.) *Design Management, A Handbook of Issues and Methods.* Basil Blackwell, Oxford, pp. 293–303.

Integration of Group, Individual and External Influences in the Design Process

Eckart Frankenberger & Petra Badke-Schaub

Abstract

The contradicting conditions of modern product development are well-known and often claimed: engineers have to collaborate in differently organised projects under time constraints, and they have to meet high quality requirements at the lowest possible price. But how can they be supported? Attempting to answer this question we need to know, *what* factors influence teamwork in engineering design practice, *when (in what situation)* they are of importance in the course of a project and, of cause, *how* they interact.

This paper presents an approach designed to answer these questions by empirical investigations in industry and highlights selected results on mechanisms that lead to success or failure in product development.

1. Introduction and General Aims

Engineering designer B designs a height-adjustable equalising-device for a particle-board production plant. While designing a lifting-frame he asks his colleague F from the planning department for the dimensions of the frame-structure of the plant, which is a prior spatial requirement for the component that he is working on. The planner F cannot help him at once, but he promises to look into the existing layout drawing of the framework and to inform B later on. In order to continue his work B tries to derive an average measurement of the dimensions by analysing the neighbouring components, but this attempt is not successful. Thus, B insists on definitive information three days later. Now F tells him that he can design a support for his lifting-frame independently, because the frame structure of the plant is yet not decided and too far away from the equalising-device anyway. This new information makes B angry: '... if I had known this earlier, I would have made the whole thing differently ... but now I'll not make any more changes'.

In further discussions with his colleagues F and J about the design of the lifting-frame, B justified his decisions.

This short episode is no fiction. It is an excerpt from an extended collaborative design project in a successful German enterprise. With regard to the difficulties occurring in this episode, it could have taken place almost everywhere in the world: obviously, product development is not just determined by the technical aspects. Moreover, we experience additional influences on design work from the prerequisites of *individual* designers, from designers as a *group*, and from the *external and organisational conditions*. In this initial example the non-technical aspects were of great importance for the later product: a sub-optimal solution was developed over a higher-than-necessary time. Additionally, this experience caused frustration and a loss of motivation by designer B, which also led to a tense group atmosphere in the following days.

Besides those short-term effects, each design decision has various consequences on the following stages of the project. This raises questions such as: How can we organise the collaboration in an effective way, what form of leadership and communication is adequate for the actual situation, and what are the individual prerequisites of a 'good' team-member?

Thus, the question of how to develop products effectively contains, in addition to technical and organisational factors, several psychological aspects.

2. Approach

In the research project 'Teamwork in Engineering Design Practice' [1], engineers and psychologists are investigating engineering design processes of teams in industry. Assuming that the effectiveness and efficacy of design processes in industry are determined by the characteristics of non-technical factors from the different inputs of the individual, the group, and external and organisational conditions (see Figure 1), we based our investigations on a general starting model of these central influences on design processes in practice. The overall aim of this empirical study was to identify the main factors that influence design work and to build up a model of collaborative design work in practice. This model should describe the interaction of the different influencing factors with regard to different requirements in the design process, and should provide the basis for further development of systematic design with special emphasis on teamwork.

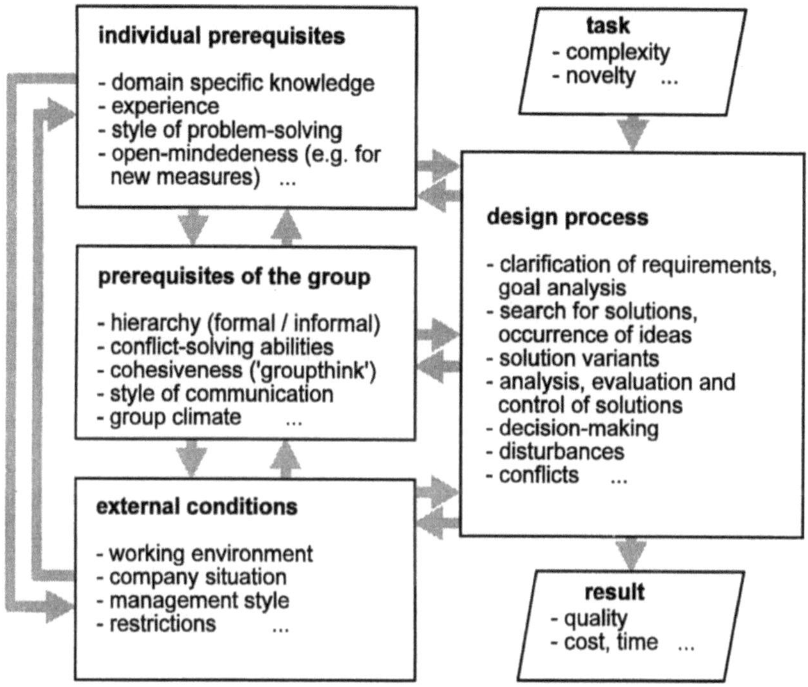

Figure 1: Examples of factors that influence the design process and the result.

3. Methods

A precise comprehension of design practice has to focus on a very detailed observation of 'single cases' over an extended period of time. Therefore a group of engineers and psychologists observed, documented and analysed four design processes in two companies for a total of 28 weeks without participating in the work. In order to guarantee valid results on the determining factors and their impacts on design work, a broad variety of investigation methods was chosen, which will be briefly described in what follows.

3.1 Compiling Data on the Design Process and the External Conditions

The *external conditions* summarise characteristics such as 'branch' and 'economic situation of the company', its 'culture' and 'flow of information', and the 'direct working surroundings' with their limitations in the particular working place of the design group. These data were assessed by interviews and questionnaires as well as by our own observations and protocols throughout the entire observation time.

In contrast to this, the dynamic course of the *design process* requires a very detailed observation. Therefore, both direct and indirect methods were used to compile data on the design process (see Figure 2).

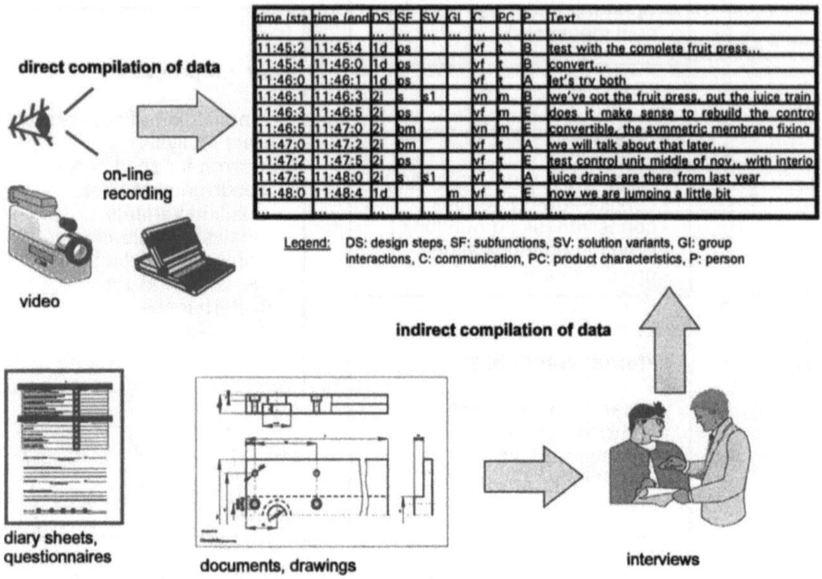

Figure 2: Compiling the design process using direct and indirect methods.

The primary *direct* method is the continuous non-participating observation of the design work. Sitting in the same room, a mechanical engineer observed the technical activities of the designers (e.g. working steps, sub-functions/components, ideas and solution variants), while a psychologist concentrated his observations on the social aspects of the design process (e.g. ways of decision-making and group interactions). Furthermore, video recordings of all phases of teamwork and relevant phases of individual design work were made in order to review the description of the design process in specific phases afterwards.

In order to fully account for the design process, it was also necessary to evaluate the non-observed work of the engineers involved. Therefore, *indirect* methods such as diary sheets were used. Diary sheets are papers with defined categories where the designers noted each evening the sub-problem(s) they worked on, how they solved problems or when/why they contacted colleagues or customers, etc. during the day. Additionally, we analysed the documents and asked the designers about their work and their personal opinions on the elaboration of solutions. Such 'indirect' data enriched and completed the online protocols of the design process.

3.2 Compiling Data on the Individual and on the Group Prerequisites

The aim of this investigation was to comprehend the rules and deciding factors of cooperative engineering design work. From studies concerning individual and group-action-regulation behaviour in complex situations [2–4], appropriate investigation methods are known which yield statements on relevant aspects of individual and group processes. Regarding these prerequisites, the methods chosen for compiling the different variables as listed up in Table 1.

Table 1: Variables and methods used to compile individual and group prerequisites.

Field of data	Variables	Methods
Biographical data	- age - professional education, career - qualification and experience	- semi-structured interview - questionnaire
Working surroundings	- motivation, job satisfaction - evaluation of the organisation - evaluation of the actual project - relationship to colleagues	- semi-structured interview - questionnaire
Ability in dealing with novel and complex problems	- analysis, information gathering - action planning - dealing with time pressure - dealing with stress	- microworlds: fire (individual) machine (individual) Manutex (group)
Special competencies	- heuristic competence - social competence - leader competence	- questionnaire - observation and analysing the interactions of the group
Abilities concerning the design process	- clarification of the task - search for conceptual solutions - selection and control	- diary sheets/marks on paper - on-line-protocol

The biographical data and designers' evaluations of the working conditions were compiled mainly by means of semi-structured interviews, whereas the ability of dealing with complex problems was assessed by asking the designers to solve three different computer-simulated problems. In contrats to design tasks, these computer-simulated problems can be solved without any specified knowledge. They are used to study the ability of subjects to tackle novel problems in a complex, dynamic and intransparent environment [5]. Dealing with these standardised problems, which are isolated from external conditions and pre-knowledge as far as possible, heurisms and strategies become obvious and allow statements to be made about the abilities of a person in dealing with complicated and new situations. In addition to this analysis of behaviour strategies, the particular competencies of the designers were compiled during the design work. In a similar way to compiling the specific individual prerequisites and the analysis of the subjects' behaviour, the different characteristics of the groups were aggregated. On the one hand we intended to elucidate the structuring aspects of the group, such as role-taking behaviour and leader functions; on the other, group interaction processes during the design processes were used to describe the group behaviour patterns with regard to group communication, group organisation and group climate on the basis of specific event protocols.

4. Evaluation

4.1 Distinguishing Between Critical Situations and Routine Work

The preparation and evaluation of the extensive data called for a new approach that would connect the data from the different fields (design process, external conditions, the individual and the group) and allow for generalisation of the findings. The basic idea of our method is the reduction of the documented design process to phases of routine work on the one hand, and critical situations on the other, where the design process takes a new direction on a conceptual or embodiment design level.

4.2 Types of Critical Situations

This method of critical situations seems similar to the 'critical incidents' described by Flanagan [6] or the 'critical moves' described by Goldschmidt [7], but the identification of critical situations in the design process follows defined rules that fit the requirements of general problem-solving processes [3, 8]. We distinguish five types of different critical situations: goal analysis, solution analysis, solution search and additionally disturbing and conflict management, as is shown in Figure 3.

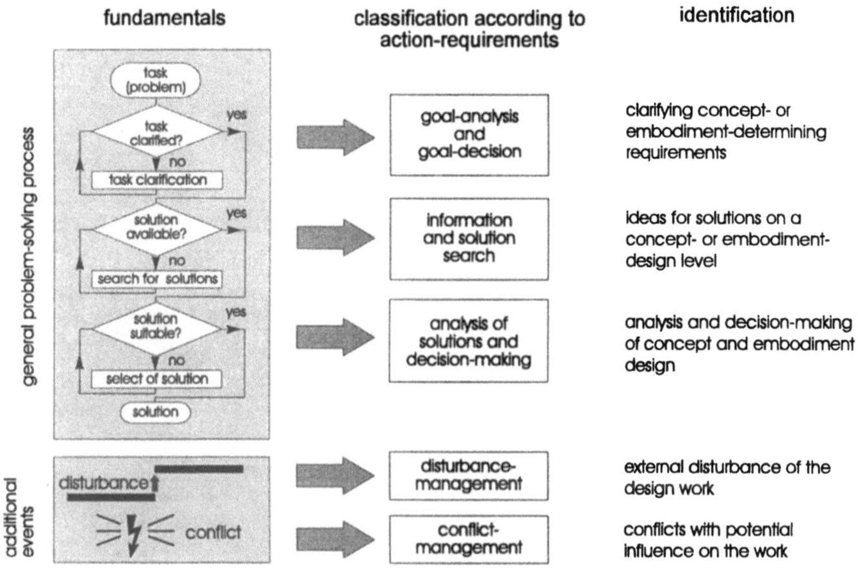

Figure 3: Division of 'critical situations' according to the general problem-solving process.

4.3 Explaining Critical Situations by Influencing Factors

An example of a critical situation is given in the initial episode, where designer B consults his colleague F, but did not succeed in getting the spatial requirements. This situation was classified as an unsuccessful *goal analysis*, because the analysis of important goal information failed. The questions to be answered then were: How did this situation occur, why did it happen in this way, and what was the effect of this situation in terms of quality, cost and time for the later process and product? We tried to answer these questions by explaining the critical situation in terms of obviously influencing factors, and their relations that showed up in the particular situation.

In this example, the communication between B and F is obviously caused on the one hand by the division of work (colleague F is the responsible project partner in the planning department) and on the other by the novelty of the problem. The lack of information in the situation is induced by the insufficient co-ordination of duties between the planning and the design department, and by the lacking self-assertion of designer B. The consequence of this unsuccessful goal analysis was a blocked solution development, with the effect that time was wasted for preliminary estimations of the dimension. Figure 4 shows the sub-model explaining this situation by relations between influencing factors and process characteristics.

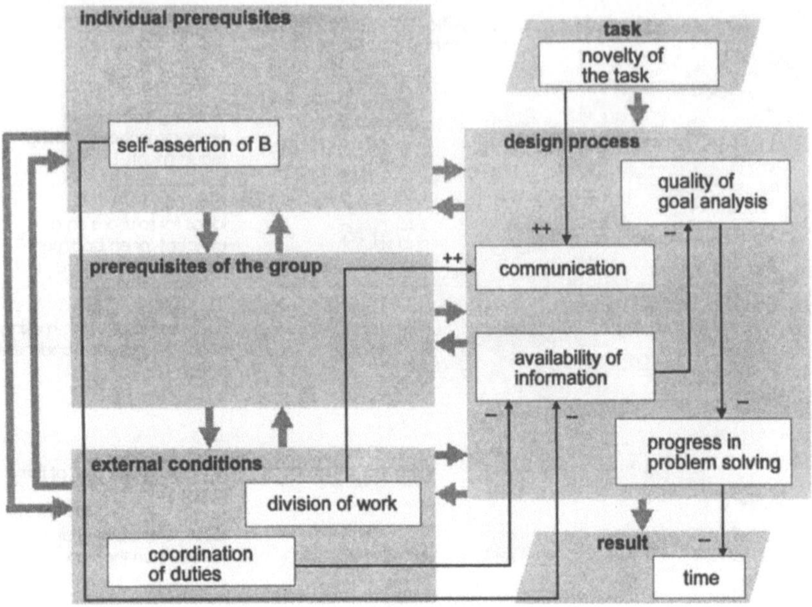

Figure 4: Sub-model explaining a critical situation by influencing factors (situation B50).

Each identified factor and its relation on the process and the result are explained separately. We identified in the four observed design processes a total of ***265*** critical situations of the different types, and explained the course of work by more than ***2200*** single interrelations between ***34*** factors, process characteristics and result characteristics.

5. Results

How can we now derive statements from this mass of data? The key to findings from the analysed critical situations is the assumption that the **frequency of occurrence** of the factors and relations represents their **importance** in the four design projects. Thus, by summing up the models of one type of critical situation (see 4.2), we can identify the main mechanisms responsible for positive or negative outcomes of the different types of situation. On the basis of this analysis we can answer questions such as, 'what are the main factors responsible for a deficient analysis of goals?' or 'what are the mechanisms leading to a successful conflict management?' Moreover, we can quantify the crucial factors and relations that determine quality, time and cost of the result with positive or negative consequence [9].

In this chapter, we want to introduce and discuss some central mechanisms that lead to success or failure in different *types of critical situations* in team design work.

5.1 Examples of Mechanism in Critical Situations

What is the basis of our analysis of mechanism in types of critical situation? Figure 5 shows how often the different types occurred with positive or negative consequences:

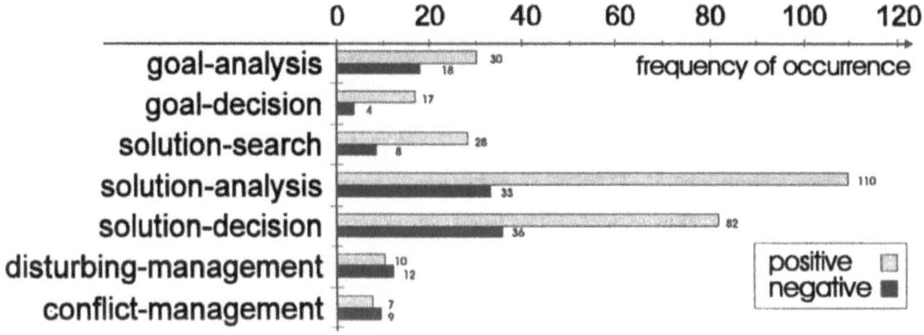

Figure 5: Frequency of occurrence of the different types of critical situations summarising the four projects (the sum is more than 265 because analysis and decision often occur together in one situation).

How important are the different types of critical situations for the result in terms of quality, time and costs? Figure 6 depicts for each type how often the situations influenced the result in a positive or negative way.

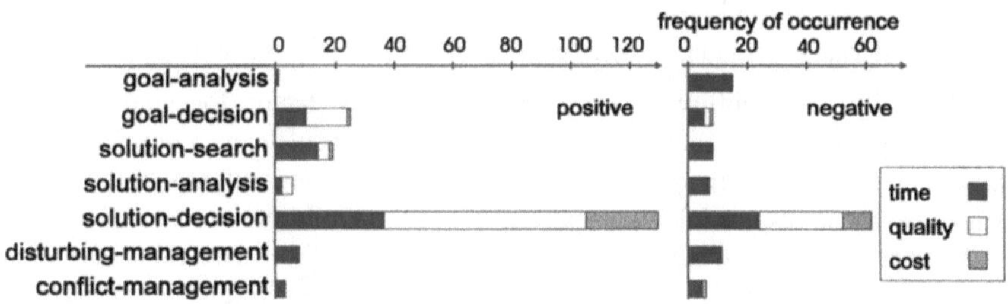

Figure 6: Importance of the critical situations for the result.

Obviously, *deficient goal analysis* is particularly time-consuming, *successful solution search* is time-saving, whereas *decisions of solutions* are of major importance for time, quality and cost of the result. These results are not very surprising, and the mechanisms that cause deficient or successful management of critical situations are more interesting. In the following we introduce the underlying mechanisms of deficient goal analysis, successful solution search, and adequate and insufficient solution decisions.

5.1.1 Mechanism of Deficient Goal Analysis

Figure 7 depicts the mechanism in 18 situations of insufficient goal analysis. The **thickness of the arrows** represents the frequency (in per cent) of the relations occurring in this type of critical situation. The **thickness of the frames** depicts the frequency (in per cent) of the factors identified in all critical situations of 'unsuccessful goal analysis'.

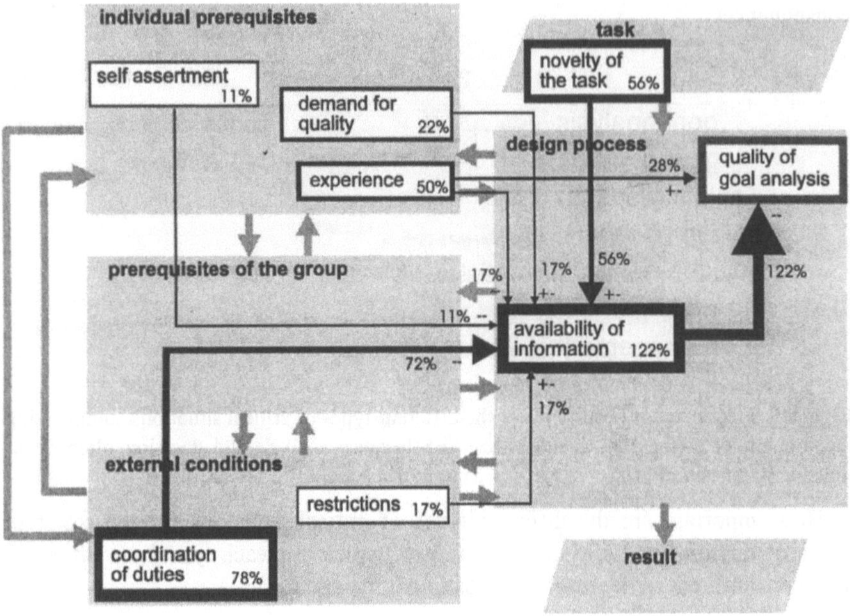

Figure 7: Mechanism leading to unsuccessful goal analysis (in 18 situations).

Obviously, 'availability of information' is a key factor for goal analysis. 'Availability of information' is mainly reduced by insufficient *coordination of duties* and by the *novelty of the task*. Moreover, *restrictions* and lack of individual *assertion* can be responsible for a sparcity of required information.

Another aspect is the individual *demand for quality*: if this demand is very high the designer will not hand out preliminary information to colleagues, because he is afraid of being wrong.

It is not surprising that *lack of experience* reduces the information availability but, in contrast, great experience leads directly to poor goal analysis in nearly 30% of the situations. The reason seems to be the 'same-procedure-as-usual effect': designers reduce their effort in analysing requirements, assuming they will know all the requirements from experience (see also Cross & Cross and Günther & Ehrlenspiel in this volume).

5.1.2 Mechanism of Successful Solution Search

Designer B is searching for a fleece-surface-control-system concept for the height-adjustable equalising-device. He thinks about a swivelling strip of plate and asks his colleague C beside him to generate an idea for the strip-support: '... I thought I'd support this near the centre of gravity, and a very small force will swivel the strip ...'. Together they develop two alternative solutions in an engaged discussion: a swivel support by an edge and by a pivot pin. Designer B: '... we need a very flexible axis, which means the smaller the diameter is, the better. I'll take a small DU-sleeve, 6 mm ...'. But designer C is warning: '... don't make it too small ...'.

In this successful solution search, designer B is chatting with his colleague at the next desk, with whom he has a good relationship. This check with the colleague is also driven by the demand for quality. Both designers exchange information and ideas, and consequently various aspects are taken into account. Figure 8 depicts the main relations leading to successful solution search in 28 situations of this type.

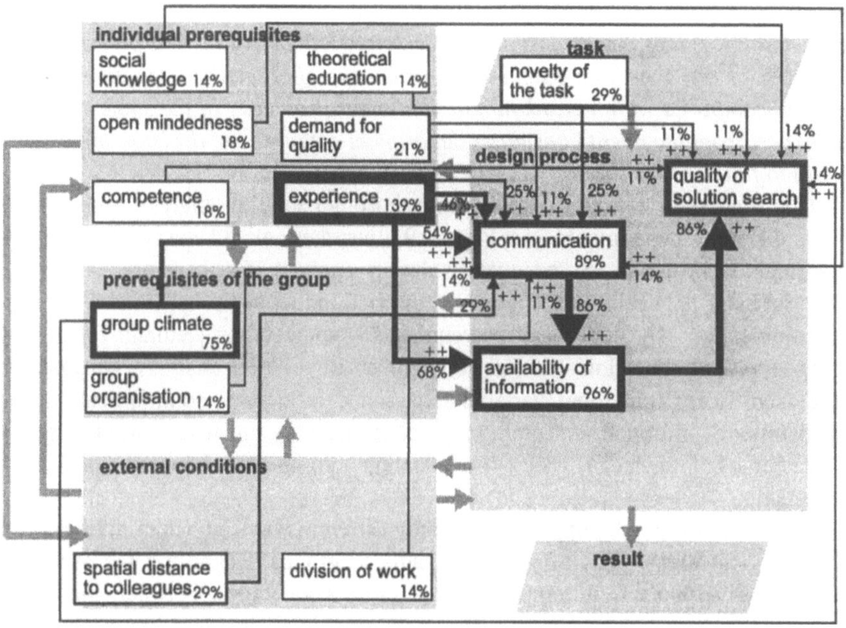

Figure 8: Mechanism leading to successful solution search (in 28 situations).

Obviously, the quality of solution search depends heavily on the *availability of information* concerning the requirements and possible solution principles. This information is often based on the experience of the designers.

Moreover, the *open-mindedness of the single designer* towards new solutions and a high *demand for quality* lead to an intensive and a successful solution search. A systematic search with the use of design methods such as morphological matrix or classification schemes occurred very seldom, but in the observed cases this *theoretical education* in systematic design was an important support for successful solution search. Furthermore, the *competence* of a designer shows up via a flexible adjustment of his procedure in *novel* problems.

Regarding the prerequisites of the group, a good *group climate* supports the open exchange of ideas between the designers. Actually, most of the successful solution search took place in cooperation with colleagues: *communication* was the main prerequisite for a satisfying *availability of information*. A 'creative' solution search is much more based on pleasant group climate than on good group organisation!

Keeping in mind that communication is an important factor for successful solution search and thus for time-saving, a good group climate and the opportunity of an informal talk with colleagues nearby become of economic value! How important are informal contacts for solution analysis and decision?

5.1.3 Mechanism of Successful Solution Analysis and Decision

Redesigning a pneumatic fruit press, inexperienced designer C asks electrician K for the required components of an electrical juice trough drive. The experienced electrician K points out the problems of an electrical drive, such as power failure, blocking of the trough or pulp on the guide, the effort needed for an electrical variant compared with a crank handle and other problems. In the same situation, an expert from the client-consulting department confirms the decision of designer C not to fit the electric drive.

This successful decision is based on a thorough solution analysis with experts from different departments. As Figure 9 illustrates, all 82 successful solution decisions were based on adequate solution analysis. Furthermore, successful analysis itself is heavily influenced by information that again is available through communication. The individual prerequisites – competence, demand for quality and theoretical education – are directly linked to solution analysis, whereas the most significant individual variable experience influences communication and availability of information. Communication is also very much supported by group characteristics, such as group organisation, group climate and quality of leadership.

Work groups may be organised in many different ways, considering that there are many different types of working teams or groups. Regarding the aspect of solution development, it turned out that close cooperation between the group members is very helpful because the main principles are then known by each group member. Group organisation thus means that group members may substitute for each other up to a certain level.

Whereas the organisation of the group seems to be a formal aspect that can be arranged and designed to a certain extent, group climate involves an emotional role, which is not easy – if it can be done at all – to organise. But in 20 % of all situations a good group climate is responsible for communication, and thus leads to successful solution analysis and successful solution decisions.

Although quality of leadership also plays a role, this factor seems to be less important than the other two group factors, group climate and group organisation. And there is yet another inter-group aspect, which refers to the external and organisational characteristics: division of work influences communication and later on successful solution decisions. Experts from different departments (e.g. the planning department, the buying department, etc.) have to communicate in depth to match and coordinate the different requirements.

In addition, the novelty of the task reinforces the necessity of coordination and communication to achieve successful solution decisions.

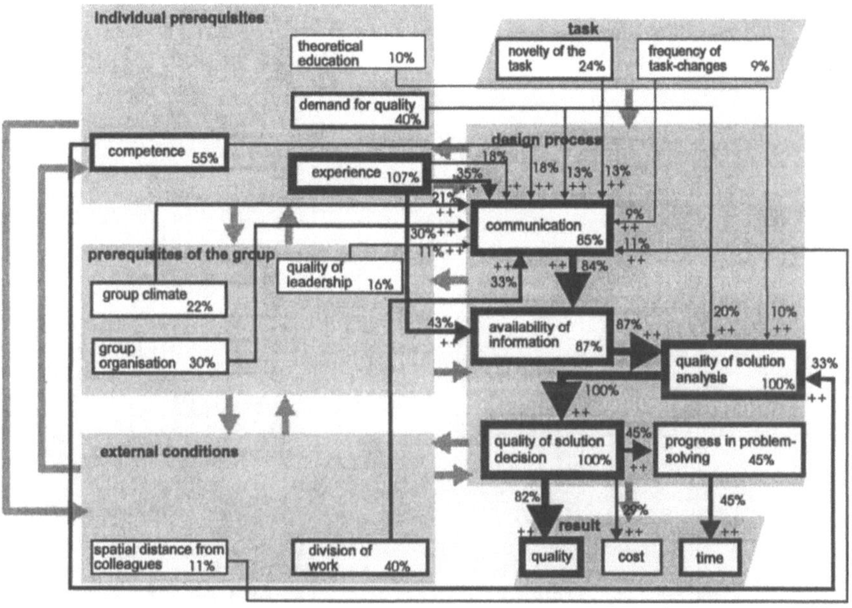

Figure 9: Mechanism leading to successful solution decisions (in 82 situations).

5.1.4 Mechanism of Deficient Solution Decisions

In the fruit press project, the motivated young designer C asks his experienced colleague B, during a team session, if he could not use a rack of plate instead a rack of tube, which has several disadvantages. The solution analysis reveals that the tube rack causes difficulties in manufacturing (colleague D who directs operations on the shopfloor, mentions '... fiddling around with blind rivets ...') and furthermore they estimate costs five times higher. Even knowing these problems and not being a supporter of the tube rack in the past, designer B becomes angry in face of the time pressure, and insists on a tube rack. The powerful designer B finally suppresses further discussion on this point, although all participants have realised the problems: under time pressure the decision for a tube rack is confirmed. The formal leader of the design department, who did not have specific technical knowledge about the problem, had left the room shortly before the important discussion of this point took place.

Surprisingly, in this documented situation, a faulty decision was made against the advice of a qualified solution analysis; rigid time constraints decreased the demand for quality of the designers, so much that the group decided to stick to the apparent wrong decision. Furthermore, the weakness of the formal leader of the design department, together with the great gaps of experience in the department, led to an informal hierarchy and thus to the powerful position of one designer.

This man was able to push this and other decisions through against the firm conviction of his colleagues. However, most bad decisions are based on wrong or inadequate analysis, as is depicted in Figure 10.

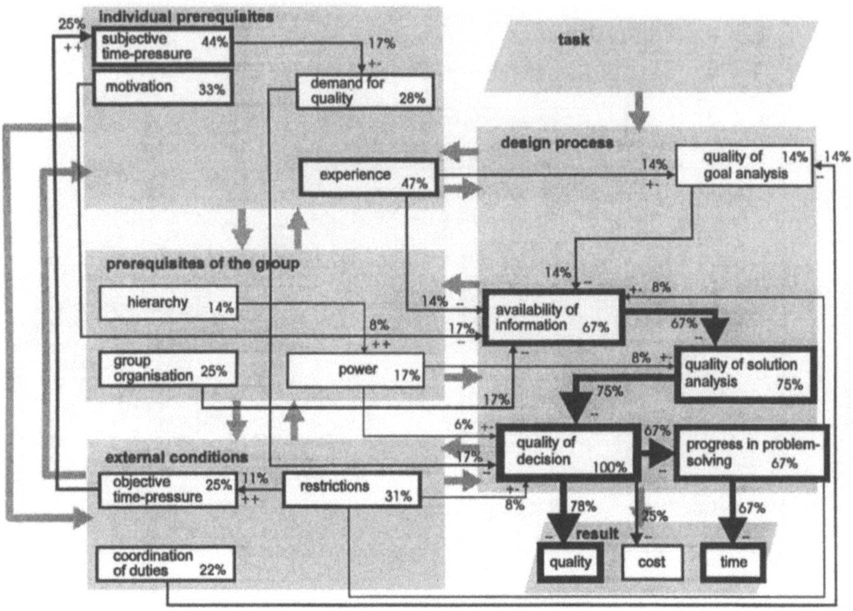

Figure 10: Mechanism leading to deficient solution decisions (in 36 situations).

At this point we do not want to discuss each relation in this network but we do want to stress five important mechanism that lead to deficient decisions about solutions:

- The consequences of **experience** as a prerequisite of the individual designer have already been pointed out in the context of deficient goal analysis. Thus, experience has to be seen as a two-edged weapon: (1) low/high experience can lead to less/more information availability; and (2) high experience can lead to an ignoring of the goal analysis.
- **Power** caused by an informal hierarchy is an important structure of the group in our investigations that can lead to three consequences in the design process: (1) less availability of information; (2) inadequate solutions analysis; and (3) wrong solution decisions.
- Low **group organisation** can lead to low availability of information. As pointed out earlier, the adequate organisation of the group is important for successful coordination and communication.
- **Time pressure** is one important external condition that can lead to decreasing demand for quality by the single designer, and thus the designer will accept solutions knowing they are sub-optimal or even wrong.
- In addition, **restrictions**, limitations on the daily design work caused by different types of problems, such as an unexpected increase of delivery time,

lead to less information becoming available, with the consequences of bad solution analysis and finally poor decision making.

At first glance the overall number of 36 insufficient solution decisions in four projects from a total of 265 critical situations seems not to be too bad. On the other hand it is important to consider the fact that faulty decisions on solution concepts might lead to far-reaching changes and modifications, with long-term consequences in the later phases of the project, perhaps in some cases even to legal confrontations (see Hales in this volume).

Apart from this, the number of visible faulty decisions is not the whole story; often decisions are delayed by the head of the design or another department. In an interview one designer told us: 'The worst thing is if *no decision is made*. Then you are in a situation where you don't know how to proceed. I as the designer have to do something because I have to keep to the deadline'.

6 Conclusions and Discussion

We have illustrated some important mechanisms for successful and deficient design activities in the different phases of the design process. We have tried to emphasise the importance of cooperation and communication for the whole process, from the goal analysis to the decisions. This thesis may be underlined by the observation that, in spite of the fact that designers worked individually for 85% of their working time, critical situations occurred in 88% of situations even with cooperation!

But what is the benefit of knowing the interactions of various influencing factors in different situations for both design education and practice?

Understanding the role that positive and negative mechanisms play in particular situations of design work helps to develop suitable precautions in the company and also allows relevant design education to be received at university. Many episodes of observed group interaction patterns confirm that work teams need support for daily work in the design process, and this support has to be really effective, not just cosmetic.

However, during the two investigations, the enormous effort that was necessary to compile data by the described methods became very apparent. Consequently, one aim of the further research is to develop investigation methods to compile the relevant data of the design process and the influencing factors with less effort to a suitable accuracy. One important idea is to educate the single designer himself to be aware of important mechanisms in the different types of critical situations.

Moreover, we want to point out that the empirical basis is broad because of the high number of 'critical situations', but the investigation included only two companies. Therefore, in order to generalise the model to a larger number of design situations, we are going to extend our basic data by additional investigations in the field of teamwork in engineering design practice.

Acknowledgement

The project 'Teamwork in Engineering Design Practice' is sponsored by the German Research Council (DFG).

References

1. Badke-Schaub P., Frankenberger E. 1997: Modelling Design Processes of Groups in Industry: An Empirical Investigation of Cooperative Design Work. In: D. Harris (ed.) *Engineering Psychology and Cognitive Ergonomics. Vol. 2, Job Design and Product Design.* Ashgate: Aldershot.
2. Badke-Schaub P. 1993: *Gruppen und komplexe Probleme*. Peter Lang: Frankfurt a.M.
3. Dörner D. 1996: *The Logic of Failure*. Metropolitan Books, New York.
4. Rasmussen J., Brehmer B., Leplat J. (eds.) 1991: *Distributed Decision Making*. John Wiley & Sons, New York.
5. Badke-Schaub P., Tisdale T. 1995: Die Erforschung menschlichen Handelns in komplexen Situationen. In: M. Kleinmann & B. Strauß (eds) *Computersimulierte Szenarien in der Personalarbeit.* Hogrefe, Göttingen, pp. 43–55.
6. Flanagan J.C. 1954: The Critical Incident Technique. *Psychological Bulletin* 51, 327–358.
7. Goldschmidt G. 1996: The Designer as a Team of One. In: N.Cross, H. Christiaans, K. Dorst (eds) *Analysing Design Activity.* John Wiley & Sons, New York, pp. 65–91.
8. Ehrlenspiel K. 1995: *Integrierte Produktentwicklung. Methoden für Prozeßorganisation, Produkterstellung und Konstruktion.* Hanser, München.
9. Frankenberger E., Badke-Schaub P. 1997: Influences on Design Productivity – Empirical Investigations of Group Design Processes in Industry. In: A. Duffy (ed.) *The Design Productivity Debate.* Springer, London.

Résumé of Section III

In the final analysis, a team is singular and thus the word 'team' can mean different things to different people. Teams are ubiquitous and heterogeneous on different levels: work teams, supplier teams, design teams, planning teams, functional and cross-functional teams, task forces, *ad hoc* teams, senior-level teams, etc. Moreover, team efficiency is a widely held belief that must be proved. Several investigations in group psychology show how careful we must be about unwarranted assumptions in this field.

In any case, investigating teams may be motivated by different philosophies, for example an unstructured ethnographic view with the emphasis on communication or a comprehensive view of a whole design project with a broad variety of predefined categories.

The advantages of the ethnographic view are aligned with the opportunity to observe processes without rigid commitment on categories. With the intention of investigating influencing factors and explaining processes, the necessity increases of establishing a pre-model and of defining the scope of the compilation of data. This procedure is of major importance especially when the solution development needs to be analysed and understood in content and with regard to the evaluation of results. Evidently the *post hoc* analysis is often reduced in explaining the chain of events, because a further examination of the influencing factors in the context of the situation is often not feasible. An extreme case may be forensic analysis, where people concerned with the accident have a strong motivation to hide minor or major mistakes of their own.

Another interesting field of investigation is the education of individuals and groups with regard to group working and to the goal of improving efficiencies in the group. With the increase of new information technologies, it seems important to use these new media in the context of education in order to support group communication and the decision-making processes. The overall aim of these technologies is to increase the transparency of communication in the group itself as well as to other groups.

Regarding forensic analysis of the engineering design process, Hales reveals communication also to be the most crucial aspect in the analysis of failures. Although the way of proceeding is different from *post hoc* analysis, Frankenberger & Badke-Schaub found *'information availability'* to be a key factor in different critical situations, for example in successful decision-making.

In general, all studies illustrate that the investigation of purely group characteristics is a too narrow street. It is necessary to broaden the view into external and organisational aspects which themselves influence group processes to a great degree. On the other hand, the observation and analysis of all likely potential influencing factors, starting with organisational culture and closing with the working environment of the designers, may fail to reveal the most important influencing factors.

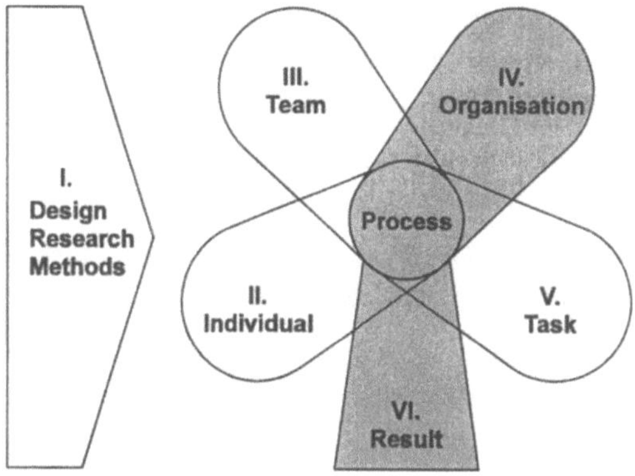

Section IV:

Organisational Influences

As the results of Hales and Frankenberger & Badke-Schaub in the previous section already indicate, the external and management conditions of the design process are of major importance in industrial design processes. Considering the expression 'external conditions' of design work, it is obvious that external conditions comprise almost everything from the direct working environment to the financial situation of the company or even the society. Thus, every empirical investigation of influences from this wide field has necessarily to focus on particular factors and on a limited period of time in order to generate adequate findings.

This section deals with the various *organisational* influences on design work. The main questions addressed are:

What type of group organisation is likely to lead to effective leadership?

How can management concepts be installed and adjusted in practice?

What are the requirements of specific processes in terms of organisation and management?

The course of the 'organisation section' starts with results from experiences from design practice and leads to a description of some results of a laboratory situation:

- Organisational and management influences on the steps of change processes are discussed by *Lindemann, Kleedörfer & Gerst.*
- *Andreasen & Hein* illustrate the differences between two management concepts in two cases.
- *Ottosson* presents the concept of a planetary organisation and reports on an empirical investigation in a laboratory setting.

The Development Department and Engineering Change Management

Udo Lindemann, Ralf Kleedörfer & Manuel Gerst

Abstract

Developing machines, plants and similar products has been the most important task for the department of design throughout the whole company in the past.

Increasing demands of customers and society, rising complexity of products and processes as a result of this, and ever tighter time constraints are now important parameters, which have necessitated a reorientation of product development.

As an important step to solve this acute problem, ***C***oncurrent ***E***ngineering has been successfully established. By team-working and changing from sequential to parallel processing, development processes have been speeded up considerably, even if current analysis demonstrates considerable possibilities of improving CE activities.

Using an example of change of products during development, with its resulting interfacial problems, it is shown, how methods and tools can be efficiently developed and utilised and the means by which suitable strategies can help to gain control over such processes. Apart from the technical measures used, it is still man who determines whether the processes are successful or not. For that reason the sharing of central control and decentralised moderation are other important aspects that remain to be considered.

1. Introduction

Changes in technical products are unavoidable and an essential part of every overall product development process. Since they have significant influence on the factors of time, cost and quality, the ability to control the processes of change in industry is of decisive significance. Because of cost and quality, changes should be carried out with minimum effort and with maximum efficiency, including safety and planning. Changes should also be realised as quickly as possible, because today's dynamic market not only induces changes to an increasing degree, but also requires faster development.

Because of the demand for an ever faster, more economical and safer implementation of changes, engineering change management is especially important for the control of change processes with regard to such complex products as automobiles. Management also has to consider decentralised organisation with delegation of decision and responsibility, as practised in today's 'Concurrent Engineering' (CE) and especially in decentralised development.

Until now, concepts for engineering change management were of purely administrative nature and adapted to a centralist and hierarchic company organisation (see Saynisch [1], Pflicht [2]). As a result, changes often cause time, cost and quality problems. Therefore, they are seen as interference, so that innovations in development are delayed or even prevented because of the high demand they incur. But changes not only offer a chance for innovation, they also contribute to a continual learning process within the company.

The concept of 'Integrated Engineering Change Management' (IECM), which was developed at the Technische Universität München by the Department of Design in Mechanical Engineering, takes into account decentralised forms of organisation and avoids the problems that frequently occur in changes by applying a methodical execution.

2. The Problem

2.1 Engineering Changes

From the point of view of design methodology, iterations are part of every product-development process. If it turns out that results and solutions as produced do not meet the existing demands on the product, these results or the even defined aims of the product-development process have to be modified. A change induced in this way always signifies a change of the product model as well as of the process. In order to control this process, functions such as planning, coordinating or supervision – in other words 'Engineering Change Management' in the fields of technology (design solution), structural organisation (competences and interfaces) and process (final dates and dependences) – are of decisive importance.

Classic change management, which arose as a form of configurational management on the basis of project management in the 1950s and 1960s, assumes a centralist management of change processes. Planning, coordinating and supervisory tasks are clearly separated from the executive tasks, which are not considered in classic change management. One result of centralist management is large circles within the process, where many organisational interfaces have to be overcome. Interfaces pose problems in the flow of information, especially of informal information, and are therefore a cause of problems with time constraints, costs and quality during product development (also see Ottoson, in this volume).

Normally, impulses for changes occur unrelated to time and content. Accompanied by inadequate control of interfaces, these circumstances lead to an uncoordinated treatment of the cause of the problem, or only of single symptoms. Especially in complex product-development processes in combination with defective coordination, the overview on changes, their momentary status and the employee or organisational units involved is impeded, so that effects on cost and time constraints are hardly predictable. In the extreme, problems in the product or in its models – subsequently referred to as reason for change – are solved differently from different viewpoints. The results are money- and time-consuming alterations to a change, or a decrease in quality of the product.

2.2 Process- and Interface Management of a Change

The first objective of effective and efficient engineering change management must be the prevention of ill-considered and unpredictable changes. At the same time, the change process has to arrive at its target by mean of a methodical procedure as quickly and safely as possible with the least possible effort.

For the control of the process it is especially important to weigh the possible gain from a change against the often high expense. The choice of useful and desirable changes results from consideration of technical and economical risks, time constraints and opportunities. The expense of a change for a complex product-development depends on several factors. In particular the extent of the change, as well as the point of time in the overall process where a reason for change occurs, play a role.

The extent of change represents the parts or components of a product that are affected by the change. Frequently it is not just one component that is affected by a change ('local' change), but several components (interface-overlapping change), especially in complex products with a high connectivity of parts. As Kleedörfer & Negele [3] claim, it is important to differentiate between a change and its cause with relation to the interface reaction (component interfaces). A reason for change can involve only one component or the relation between different components, e.g. a collision of components. The problem of collision (interface-overlapping reason for change) can be solved by a local or interface-overlapping change.

Since the working results of the single stages of a development project are interdependent, the whole network of results is endangered by the change of one working result. It is therefore necessary to calculate the entire extent of change as soon and as precisely as possible, because the expense of a change increases progressively according to the rule of ten (see Clark & Fujimoto [4]) during the course of the overall development.

Of decisive importance for process management of changes is, as described above, the control of structural organisational interfaces and related problems in the flow of information. In recent years different integrational procedures to overcome interfaces have established themselves in companies with complex processes. Concurrent Engineering (CE), to be mentioned foremost here, uses primarily the idea of interdisciplinary teamwork in CE teams apart from project management. CE teams are oriented to aspects of the product and its components. As far as possible, they work according to the principle of subsidiarity, according to which functions like planning, coordinating or supervising are to be carried out by a CE team as well.

Through the introduction of CE, structural interfaces throughout the overall development process could be minimised or even eliminated. This of course leads further to an efficient engineering change management, especially if the change involves only one single component (local change).

Within complex processes, despite CE, there are still several interfaces between the different CE teams of a development project. New walls in the flow of information, as we know from the example of departments in classic functional organisations, can arise exactly there (see Figure 1). Because of the specific orientation of CE teams to components, problems are especially apt to arise in interface-overlapping changes.

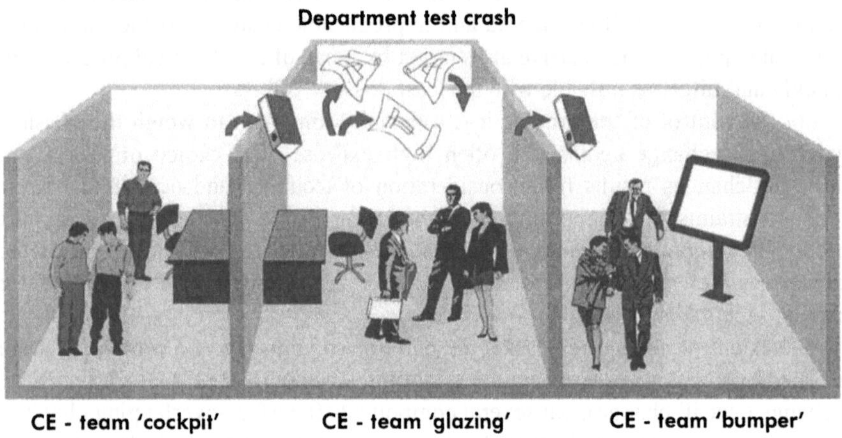

Figure 1: Walls in the flow of information between CE teams.

The concept of classic change management does not include the premises of integrated product development, such as the principle of subsidiarity and the minimising and controlling of interfaces in order to optimise the flow of information. In the following a concept of 'Integrated Engineering Change Management' (IECM) based on integrated procedures like CE and the related principle of subsidiarity will be presented.

3. The Concept of Integrated Engineering Change Management

IECM is an organisational concept for handling changes in technical products. It can be applied to interface-overlapping as well as to local changes. The concept is divided into the following areas:

- the process
- structural organisation, and
- computer support.

3.1 The Integrated Change Process

Since the technical orientation of the solution determines the change process, the technical solution process and the organisational and administrative procedures have to be considered and linked up, in order to control the change process. In the concept presented, integration of management functions into the solution process of design methodology with its typical procedures is of primary importance when implementing a change. The solution process represents an 'early stage' or 'planning stage' of a change and therefore has to be given high priority.

The structuring of the change processes is based on the procedural cycle according to Ehrlenspiel [5] and on the problem-solving process of 'Systems Engineering' according to Daenzer & Huber [6]. Both papers break down the problem solution into the following stages:

- clarification of the task,
- the search for solutions, and
- selection of a solution.

These stages of the procedural cycle can be used for engineering change management. However, for the implementation of a change, especially with regard to practice within companies, the abstract procedural cycle has to be brought down to a concrete level. Typical tasks of engineering change management have to be examined and assigned to certain steps. Concepts for the solution of interface problems will be especially considered in the change process in order to achieve a holistic treatment of the reason for change. Finally, the complete process has to be documented, if possible with the support of a computer-tool (see below – Computer-supported Documentation). The reason for documentation is the need for information for outsiders and employees who have been integrated into the process at a later point of time, the comprehensibility of decisions, and to allow for data-evaluation for project and change procedures.

3.1.1 Clarification of the Task

The first step in clarifying the task is identification of the reason for change, meaning the recognition of the problem and definition of the required stage (i.e. target). The integration of this step into the change process is important to avoid an interface between the recognition of the problem and the implementation of a change *per se*. For the systematic development of a product, the task of recognition of the problem has to be provided for appropriate moments in the development process in order to integrate changes specifically into the development process. The use of methods for the early recognition of product properties as described by Bernard [7] is therefore especially important. By early recognition of problems within a product or its models, time and costs can be saved (according to Ehrlenspiel [5]). During the development stage, the findings from the design and assemblage of prototypes (testing tools, model and prototype design) are important for the recognition of the reasons for change. But also other findings – from the production, the application of previously developed products on the market, and from the activities of competitors – contribute to changes, as well as new regulatory legislation. Lastly, late 'demands' of the management or customers are often a reason for change.

From the long enumeration of possible influences it can be deduced that it can be problematic to catalogue the necessary changes, especially in the case of large organisations and complex products. Braunsperger [8] addresses this problem and names a so-called 'control list' as a methodical measure for safeguarding the flow of information between all departments of the overall product development process (product planning, design and development, production, assembly, etc.).

For safeguarding the flow of information it is important to bear in mind that the responsibility for the problem, which rests on the one who recognises it at the time of identification, is consigned on the basis of clear agreements. The person subsequently responsible for the change (the designer or moderator of a change, depending on whether it is a local or an interface-overlapping change, see below – Moderation of a Change) takes over the reason for change and brings it to a solution.

First, the analysis of the problem has to be made by the person responsible for the change. The aim of this step is the clear ascertainment of the cause for the reason for change, in order to avoid alterations to a change. The objective is to attack the cause as efficiently as possible rather than the symptoms. This step represents the organisational and technical version of the step 'clarification of the task' of the procedural cycle.

On the basis of the analysis, components affected by the change can be ascertained. From the number of affected components and the complexity of the reason for the change it is now possible to make a rough draft of the change and the appropriate process. The desired deadline for conclusion of the change has to be ascertained and judged in a first assessment. Access to a current process model of the concrete development project is necessary for this. Especially in larger organisations and decentralised forms of organisation it has to be kept in mind that the problems resulting from a single cause can be bundled and coordinated as one change. The triggering of several changes by different organisational units due to one single cause, possibly with various symptoms, is to be avoided. In those cases, coordination is extremely urgent, so that one and the same problem is not solved differently by different sides. A safeguard has to be provided here (see below – Matrix Teams of IECM).

For coordination of the change, all organisational units actively participating in the technical solution process have to be informed. The identification of these units to be integrated is done by means of the affected components or properties of the entire product (vibration characteristics, aerodynamics etc.) (see Figure 2).

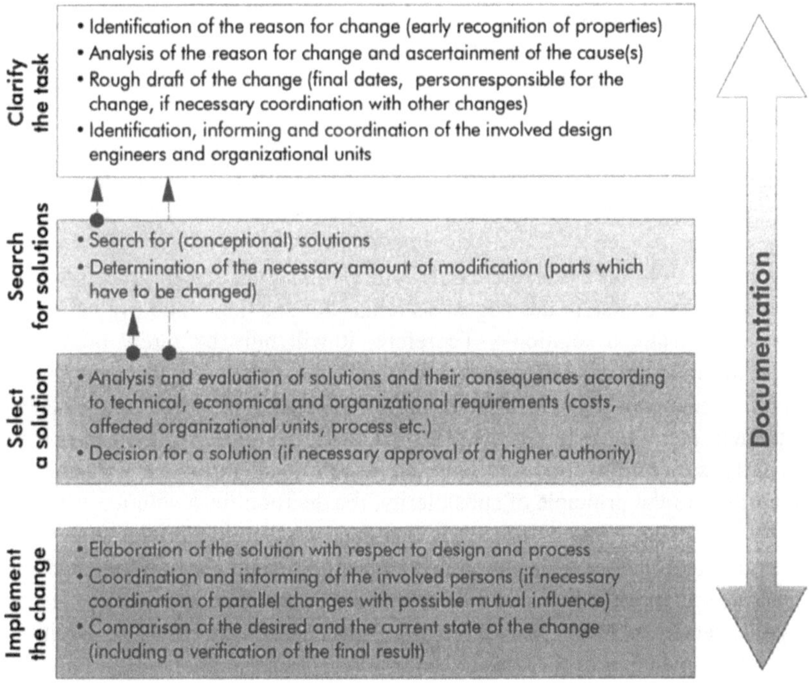

Figure 2: Steps in Integrated Engineering Change Management.

3.1.2 The Search for Solutions

This step in the processing of changes corresponds to the identical step in the procedural cycle according to Ehrlenspiel [5]. This is where conceptional alternatives for solutions are sought. In the case of a local change, the search for the solution takes place within one CE team. The expertise can be used here for finding and evaluating the solution on the basis of teamwork. Because of the special problem of interface-overlapping changes, interfaces have to be considered while searching for a solution. Organisational concepts for this are presentedbelow – see Matrix Teams of IECM.

In the search for solutions, it is possible to revert to certain strategies and methods. Ehrlenspiel [5] suggests as strategies to 'start from the existing and familiar' and 'from the preliminary to final', and, as methods, techniques of creativity, design catalogues, checklists, etc. (see also Pahl & Beitz [9]). The extent of change, which includes the number of parts that have to be changed and the difference between the old and new form, is directly linked to the alteration.

3.1.3 Selection of a Solution

The different solutions for a reason for change, worked out in the previous stage, are now in the third stage of the change process, first subjected to analysis and evaluation. Important for the complete evaluation is consideration of the solutions and their consequences from different perspectives. An assessment as to technical quality and feasibility, as well as to the costs and the process effects of the solution, has to be made. Further, the affected organisational units and especially the consequences for the development process have to be regarded on the basis of the amount of change. For this purpose, access to a current process model is necessary.

It is known from the observations of Dylla [10] that the design process follows a highly iterative course in the stages 'searching for solutions' and evaluation and decision ('choosing a solution'). Therefore, it will only be rarely possible in practice to make a clear distinction between those steps. For a methodical, systematic treatment of a problem, the single steps of engineering change management (see Figure 2) for the assigned tasks should be fulfilled in order to avoid hasty, supposedly good solutions and associated alterations of a change.

According to the principle of subsidiarity, the decision for a solution is usually made by the designers or the CE team. Company practice has proven that, within a certain range of cost, a decision on this level is sufficient. Only when exceeding the cost limit is the approval of a higher authority (superiors, sub-project leaders, project leaders) necessary. In this way, in more important cases, an additional control of the planning of the change can be made before implementing it.

3.1.4 Implemention of the Change

The final stage of the change process is the implementation of the change in that the conceptional solution is carried out and verified. By detailing the resulting solution, this step is an extension of the procedural cycle. In it, the design solution process is carried over as an implementation of a highly administrative nature, in which the mechanisms of classic change management, or alternatively, configurational management, take effect. Through an integrated consideration of both areas, the technical solution and the (administrative) implementation, a basis for a thorough processing of the change is given. Interfaces within the system are avoided and an easy flow of information is guaranteed.

Before implementation of the change begins, the planning of the change, roughly drafted in the previous steps, should be worked out in detail. To this end, the form of the part or components that have to be changed is modified, and the current planning process of the affected project is extended on the basis of the projected task. Based on the analysis of the effects of the change and the planning process, the tasks of the affected employees have to be coordinated. For that, all involved employees have to be informed about the state of the change. Lastly, a comparison of the desired and the actual state of the change, including a check of the final results, needs to be made.

Analogous to the procedural cycle, the change process is marked by an iterative procedure in the treatment of subtasks. If informational deficits are recognised at one stage, it is necessary to go back to the previous stage where the information was produced.

3.1.5 Computer-supported Documentation

For the documentation of data during the change process, computer-based support is needed. Therefore, a data-based tool was developed according to the process represented in this paper. The main form is pictured in Figure 3.

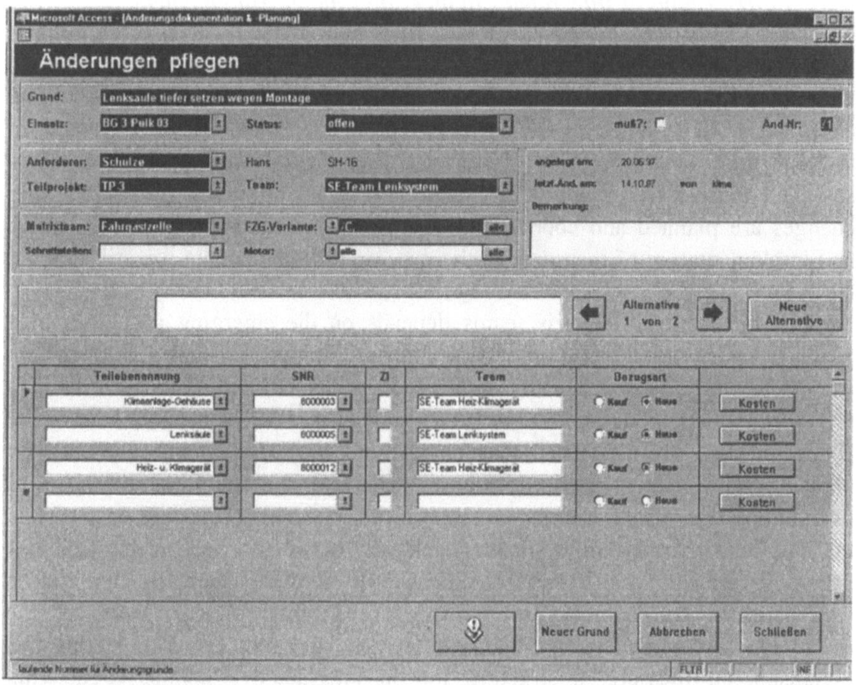

Figure 3: The main form of the Alteration Documentation & Planning Tool.

The top of the form is mainly for the first step of the process. The alternatives for a reason for change can be stored. They are described by the components, their ID and their version number. The technical solution could be described in a text field or documented in the CAD System.

For the evaluation of alternatives, it is possible to store the costs of the change of each part of the alternative. Furthermore, it is possible to link a project planning tool to the system. This link was accomplished with the system described by Stuffer & Kleedörfer [11]. It allows an evaluation of the effects of the change in the recent process model. This leads to a holistic evaluation of the alternatives of one reason for change.

3.2. Structural-Organisational Concepts for Managing Changes

The problems of interface-overlapping changes have already been addressed in 2.2. In order to ensure fast, economic and safe treatment, it is further necessary to define a new concept of structural organisation that does not contain those problems. Especially in interface-overlapping changes, the question arises of how the shared responsibilities and development processes of the CE teams can be coordinated. To this end, another form of organisation in addition to the CE teams is necessary.

3.2.1 Matrix Teams of Integrated Engineering Change Management

A new unit is suggested for the coordination of interface-overlapping changes with shared and component-specific responsibility in the CE teams, permitting coordination of designers or CE teams in more complex product development. In this unit, representatives from different CE teams and departments are united in so-called **matrix teams**. Only CE team-overlapping problems are solved here; changes are planned and coordinated. But the component-specific coordination and implementation of measures in the process of a component are still done in CE teams.

The organisation of matrix teams depends on the emerging properties of a product. Every product has properties that become apparent only in interaction with different components. Parts of the product are interlinked to each other in this way. When changing one part, attention has to be paid to whether other connected parts are influenced by such relations (i.e. component-interfaces). By focusing the respective matrix team on one characteristic, the possibility arises of overcoming the interfaces between the components. This procedure ensures that the interfaces between the designers, especially between the CE teams (see also Figure 1) and the specific departments (test or simulation departments) can be overcome. In automobile development, for example, matrix teams for the assembly, aerodynamics, etc. have to be set up (see Table 1). With this organisational measure, a structure for an interface-overlapping engineering change management and a controlling authority for properties of the entire product can be built up.

Through institutionalisation these new units also function as a basis of information for the participating developers. To ensure this function, two prerequisites have to be fulfilled. For good teamwork within the matrix team, the leader has to be qualified and experienced in the technical, methodical and social areas. Secondly, the team must hold a clear-cut position within the project organisation.

The specific matrix teams are important for interface-overlapping changes. This is where the question of the coordination of designers or CE teams arises. The responsibility for the implementation of component-specific adaptation is divided up between different designers or CE teams. Therefore, coordination of the responsible units is necessary. But often the coordination is casual and depends on the relationship between the employees. For efficient coordination it is necessary to provide an exchange of information between the partners that is as direct and reliable as possible. The responsibility has to be borne by one unit, into which the partners are integrated. The problem could be solved by combining the structure of the matrix teams into the methods of teamwork.

Table 1: Matrix teams for overcoming interfaces in larger project organisations (according to Lindemann & Kleedörfer [12]).

Product Development Team (oriented to product components and process)

Matrix-teams (management of properties of the entire product)	CE-team front gate	CE-team glazing	CE-team bracket	CE-team bumper	CE-team cockpit	CE-team seat	Test aerodynamics	Test crash	Pattern making
CE-team/unit sends employees to Matrix-team									
Team aerodynamics	X	X		X			X		X
Team passive security			X	X	X	X		X	X
Team acoustics	X		X		X		X		
"Assembly"-team engine compartment		X	X	X	X				X
"Assembly"-team passenger compartment		X			X	X		X	X

Further, the principle of subsidiarity is applied to the matrix teams. For the elaboration and selection of solutions, up-to-date expertise and knowledge of details are required, which usually only specialists possess. According to integrated product development, the executive levels are responsible for the problem solution in both content and process. The middle and top management will therefore only be called in if the task cannot be solved by the employees alone. In general, direct settlement at the specialist level is faster than the detour via hierarchies, which would present just another interface in the process. This connection speeds up the change further.

Another effect of using matrix teams is that information is no longer random and orientated to the subjective need of the individual, but can be interchanged with a high level of reliability. Informational deficits are recognised by good team management and can be cleared relatively quickly. Further, peripherally affected designers will also be informed about the state of development of their colleagues. Finally, the documentation of the results in a team by means of written records is usually more consistent than documentation in random bilateral agreements. For this task a computer-based system is developed and adapted to the organizational structure (see above – Computer-supported Documentation).

3.2.2 Moderation of a Change

Several problems can turn up in the implementation of a change process in view of the premises of integrated product development. These problems can be neutralised with the help of the concept of moderation.

Seifert [13] states that in a moderation the 'team leader' tries to lead the team to the solution of a problem, and does not try to solve the problem by himself. At the same time, he has to keep a neutral position. His task in supporting the process is to ask the right questions at the right time. This permits a bi-directional, interactive flow of information within the frame of teamwork. Thus, he allows the team to work out a solution on its own and to make common evaluations and decisions.

In contrast to current standard procedures of engineering change management, there will be a networked flow of information and not a control of the change (see Figure 4). In hierarchical and centralist change management, 'opinions' of the affected units are taken. As shown in 2.1, large circles evolve, which should be avoided in integrated product development. Further, the 'controlling units' are bottlenecks in processing information because of the centralist structure. It is known from observations in practice that this form of organisation regularly leads to delays in implementation of changes. These problems can be avoided by handing over the technical and process planning of the change to matrix teams. The team leader is not the bottleneck in the exchange of information, but rather the trigger for a direct and interactive exchange of information. This speeds up the implementation of a change and improves the quality of the solution.

In methodical handling, the solutions for a reason for change have to be developed without fixation on the sphere of the responsible employee. A systematic tree of solutions has to be designed, from which then the best one can be chosen. Especially with interface-overlapping changes, the solutions of the responsible designers have to be compatible. Therefore, if a moderation of a change is made, the team will try to develop several alternatives in a large field of solutions by integrating all affected members. If regarded comprehensively, better solutions can be found than by fixation on one single designer or component. Further, the 'Not Involved Here' (NIH)-syndrome can be prevented, because it is more likely that alternatives developed within the team will be discussed more openly than if dictated by an external unit.

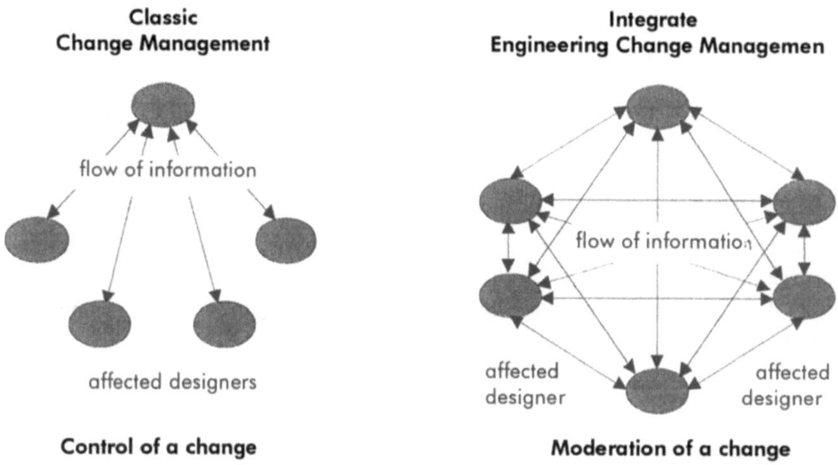

Figure 4: Moderation instead of control of a change.

For efficient moderation there are special tools and methods available. An important method, according to Seifert [13], is the visualising of problems and tasks in the solution process. Designing teams, for example, need product models in order to work efficiently, according to Stetter & Ullman [14] and Hacker & Sachse [15]. They are helpful in analyzing and searching for the solution and serve an enhanced exchange of information, therefore also providing better understanding. The team leader must be aware of these coherences and should use the methods of moderation to lead the matrix team to an effective and efficient procedure.

4. Conclusion

With IECM a methodical and holistic approach is introduced to handle engineering changes in the overall product development process, especially for highly complex products. Implementation in a company in the automobile industry is markedly speeded up. Furthermore, alterations of a change are reduced because one problem and its effects on the product and the process are handled as one system.

For introduction into companies, three areas must be considered and adapted, as described above. These main areas are:

- process,
- structural organisation, and
- computer support.

These three areas should be introduced into the company as one system. The process and the structural organisation have to be adapted to the product and structure of the enterprise. Based on this, the computer tool must be adapted and introduced into the development projects. Last but not least, an open attitude regarding errors and changes has to be established within the company. Only on this basis can IECM be installed as a system to realise optimisation, innovation and, therefore, better products.

Acknowledgements

The contents described are results of a project supported by a grant from the BayFORSYS and carried out in cooperation with a Bavarian automobile manufacturer.

References

1. Saynisch M. 1984: *Konfigurationsmanagement: Fachlich-inhaltliche Entwurfssteuerung, Dokumentation und Änderungswesen im ganzheitlichen Projektmanagement.* Verlag TÜV Rheinland, Köln (Schriftenreihe der Gesellschaft für Projektmanagement).
2. Pflicht W. 1989: *Technisches Änderungswesen im Produktionsunternehmen.* vde-Verlag, Berlin.
3. Kleedörfer R., Negele H. 1997: Planungssicherheit von Änderungsprozessen. *Endbericht FORSYS*, München.
4. Clark K., Fujimoto T. 1992: *Automobilentwicklung mit System.* Campus, Frankfurt a. M.
5. Ehrlenspiel K. 1995: *Integrierte Produktentwicklung.* Hanser, München.
6. Daenzer W.F., Huber F. 1994: *Systems Engineering.* Industrielle Organisation, Zürich.
7. Bernard R. 1997: *Rapid Prototyping in der Konstuktion.* Herbert Utz Verlag Wissenschaft, München (Seminarberichte 27).
8. Braunsperger M. 1993: *Qualitätssicherung im Entwicklungsablauf.* Hanser, München (Konstruktionstechnik München, Band 9).
9. Pahl G., Beitz W. 1986: *Konstruktionslehre.* Springer, Berlin.
10. Dylla N. 1991: *Denk- und Handlungsabläufe beim Konstruieren.* Hanser, München (Konstruktionstechnik München, Band 5).
11. Stuffer R., Kleedörfer R. 1997: Prozeßmanagement im Wandel. *EDM-Report* 2, 25–31.
12. Lindemann U., Kleedörfer R. 1997: Erfolgreiche Produkte durch Integrierte Produktentwicklung. In: G. Reinhart & J. Milberg (eds) *Mit Schwung zum Aufschwung: Information – Inspiration – Innovation/Münchner Kolloquium '97.* mi, Verl. Moderne Industrie, Landsberg/Lech, pp. 115–136.
13. Seifert J. W. 1996: *Visualisieren, Präsentieren, Moderieren.* Gabal, Offenbach.
14. Stetter R., Ullman G. 1996: Team-Roles in Mechanical Design. In: *The ASME Design Engineering Technical Conference and Computers in Engineering Conference.* ASME.
15. Hacker W., Sachse P. 1996: Unterstützung des Problemlösens im konstruktiven Entwurfsprozeß durch Prototyping. In: H.-J. Franke (ed.) *Bild und Begriff 4.* TU Braunschweig.

Innovating the Product Development Organisation

Mogens Myrup Andreasen & Lars Hein

Abstract

The organisational innovation of the product development function is a doubtful affair since we can hardly describe why a specific organisation works. In this paper, two comprehensive innovation campaigns in Danish industry are described with reference to the nature, content and results of innovative processes, and as the crux an attempt is made to explain the principal elements in the changes of attitudes and behaviour in companies.

1. Introduction

Organisational innovation, which within the area of product development means new values, mindsets, methods, procedures and perhaps a change of organisation, is an area in which companies are in trouble. Management only has a few instances from which to draw experience. Insight into models and ideal solutions may not be present and the process is so manifold that it is difficult to direct its course and results.

The authors are consultants within product development and have in this capacity carried out a number of organisational innovations in Danish companies. An account of two comprehensive innovation campaigns is given in what follows. Focus is on what companies do and how the parties involved react, and an attempt is made to explain the supportive elements in the changes of attitudes and behaviour of the parties involved.

The authors' aim is to illustrate the nature of the innovative processes, and particularly to explain how it, apparently, differs from the engineering perception of manageability.

Experience reported here has neither been gathered nor worked up properly researchwise. Hence this paper does not differ from many generalisations in European design methodology research, but we hope that the observations can be applied to certain types of participatory research. Following the results referred to in this paper, the authors' research into innovative processes and implementation of methods has continued within specific areas like DFX [1], 'setting the stage' in product development or design preparation [2], integrated manufacturing systems [3], choice of method [4] and conceptual design of manufacturing systems [5].

2. Implementation of Integrated Product Development

Integrated Product Development (IPD) is a general procedure for product development with focus on integration between market, product and production aspects when creating new business [1]. The concept seems to have arisen in a kindred form in several places at once. The Danish version had been inspired by Swedish ideas and consists of a general activity model with optimum simultaneous activity patterns in three parallel courses with common milestones.

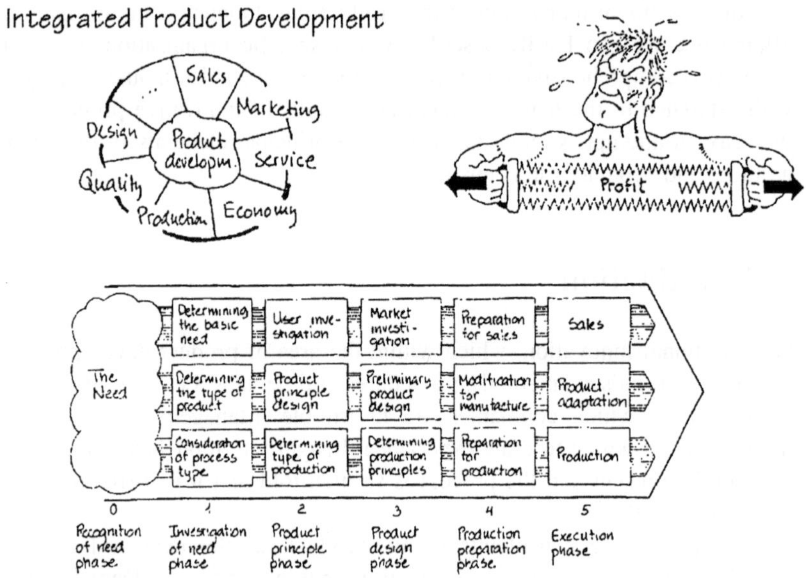

Figure 1: Integrated Product Development is a comprehensive procedure and methodology for product development [6].

The IPD concept was introduced from 1984 to 1988 into Danish industry (mechatronic products) in a country-wide campaign and was supported by a publication [6] (see Figure 1). At open meetings, the fundamental idea was explained together with its application to typical problems and shortcomings in existing working patterns in industry. For a specific company, implementation of IPD meant that the procedure mentioned and its milestone activities were established in a form that had been adjusted to the company, accompanied by selected tools and often a changed work approach, i.e. elements from this list:

- A procedure/milestone plan
- A set of tools
  - Product specification
  - Business specification
  - Project start seminars
  - Planning based on experiences/tactical means
  - Analysis of competitors
  - Design for *X* tools, where *X* could be assembly, manufacture or cost
  - Measements at the milestones
  - Business thinking
    - Cash flow
    - Dispositions
    - Optimising the business
  - Project organisation
    - Team
    - Project leader
    - Product steering group
  - Systematic search for new products
- A working attitude
- A leader style or approach

Typically, implementation of IPD was carried out using consultant assistance, but some of the companies established study groups based on the publication and themselves found a suitable form of IPD. The consultant assignment was solved in a characteristic sequence of activities:

- *Management meeting* between the consultants and leading employees of the company. Here the current need for change within the company and the role that the product development function must play during the realisation of the company strategy were identified and formulated.
- *Planning and structuring* of a course activity, with a typical duration of three days, aimed at giving information about and inspiring new working methods. Typical elements of the course were:
  - Management contributions concerning the interpretation by management of the tasks and performance of the development function, interpretation of strategy and results, and formulation of the aim of the course.
  - Contributions from the department managers involved: production, sales, quality, marketing, R&D, documentation, etc., with interpretations of the role of the development activity for their area and their expectations in participating.
  - Contributions from the consultants with presentation of ideal courses, ideal cooperation, explanatory models and presentation of new methods. The material was chosen and adapted to current company conditions and problems.
  - Discussions, group work, role playing and gradual gathering together of pro-posals, goals, remarks and so on.

- *Course completion*, typically with a plenum part with manager participation, and a more operational part for the line organisation during which technicalities were discussed.
- *Elaboration of an action programme* based on management decisions and carried out in groups of employees that, together with a consultant, are to carry out activities such as:
  - formulation of a company-specific procedure;
  - preparation of milestone activities;
  - implementation of chosen tools;
  - preparation of and participation in a pilot project which is carried out under the new conditions.
- *Implementation*, often with moderate consultant assistance, run by employees and managers themselves.

This pattern is typically based on a planning approach: problems can be identified and solved by means of tools, and tools can be introduced in a planned activity. Progression and results in the implementation process were attributed by us as consultants to the outstanding nature and topicality of the inherent message.

2.1 Some Important Features of IPD Implementations

Before discussing the psychological and social conditions within the changes, we shall accentuate some important aspects of the implementation:

- The process activates participants, but is organised by a consultant.
- In the course of the process, a consensus on new values and new working patterns arises, especially the necessity for improved integration.
- Normally, the role of the development manager is modest; he focuses on results, not methods or organisation.
- In a significant number of cases, the development manager left the company during or after the process.
- There is often an enhanced ability within the company to see and admit existing bad practices.
- Often, the mission and the goals of the reform process are carried out by a few key persons.

If we are to try to summarise the results of IPD implementations, the following statements, based on observations, not measurements, are typical:

- The company has difficulties in maintaining formal elements (procedures, roles, tools, etc.), whereby the integration effect slowly disappears.
- Employees say what they think in a new way: 'it's in our blood', 'we all believe in IPD'...
- There are clear improvements connected with the changes, whereas formalism weakens the effects.
- Sales/marketing and production are most enthusiastic about IPD which provides them with professional cooperation.

- Companies fail to accompany new working patterns by ambitious goals. Therefore, in certain companies the effect has been less than expected.
- IPD implementation requires that basic working patterns and norms (on the 'nuts-and-bolts' level) are in place.

2.2 An Interpretation of IPD Implementation

An analysis of the changes in values and attitudes that took place during a characteristic course shows that the following features can be identified:

- Confrontations between departments concerning cooperation patterns and interdependency lead to the acknowledgement of a need for and possibilities of integration. Here company internal contributions are very helpful and the consultants can accentuate and appeal for improvements.
- When there is plenty of time for discussion, the consultants are 'de-clutched' and the participants discuss reciprocally. Thus a new consensus arises. This means that the course has initiated an organisational innovative process which is not formulated and governed by management (see Figure 2).
- When the new working pattern comes into existence on the basis of consensus, each employee sees his/her role, competence and task outlined. For some people it is a challenge and for others a shock and a defeat (see Figure 2).
- The innovative process may be so strong that the course may pass 'the point of no return', which is to say that management will face considerable obligations to follow up and deliver.

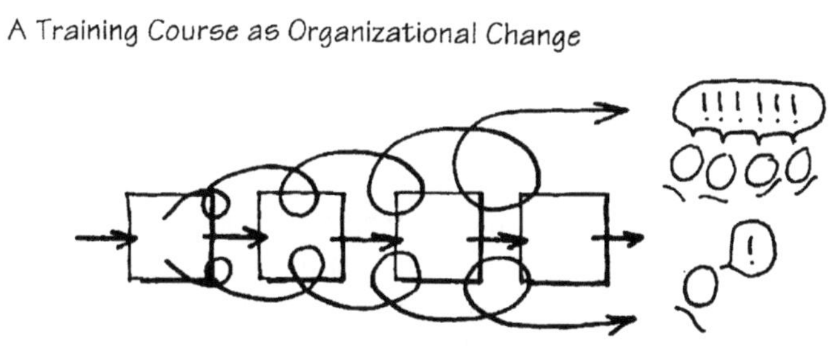

Figure 2: Teaching and discussing integrated product development during a company internal course becomes a process of group consensus and individual role identification.

We conclude that what we as engineers naively saw as a logical top-down process with considerable scope, has through the participants become an innovative process with results that were not wholly foreseen but which are mainly positive.

3. The Unic Processes for Innovation

UNIC stands for 'Innovating the Product Development Ability' and is a research and consultant project, the aim of which is to provide Danish industry with tools to innovate the development of an organisation. The initiative originated in acknowledgement [7] of the fact that many development departments are unequal to the task of doing something new, and thus do not constitute the important power source that they ought to be. The project came into existence in a cooperative effort between companies, consultancy firms and our university, and in a campaign-like activity the results were offered to and used by Danish companies. This research was done in the years 1986 to 1988 (GBP 800,000), and since then the developed tool has been used in a considerable number of companies; the authors have been involved in 45 organisation innovation projects.

3.1 The UNIC Concept

Whereas Integrated Product Development is to be seen as a concrete set of tools, UNIC is an organisational change in product development that leads to a set of tools not determined in advance. The goal of the research was to make possible a 'design of the product development machine'. This assignment demands humility, because through our work with product development organisations we have acknowledged that it is difficult to understand and explain how development tasks are being solved. This research has provided us with tools and knowledge that may form the basis for the following important preconditions for and constituents of innovation:

- Diagnosis, i.e. draw up appropriate visions and targets that will ensure the future of the company, and then measure the existing development system against these, and make a diagnosis as to whether the existing development system is based on sound and productive mechanisms.
- Specification of a new or modified development system, based on the existing one but with enhanced productivity and performance, by setting new tasks.
- Design, i.e. choose means – in the shape of an organisation, procedures, manning schemes and so on – that correspond to the specification, and which with a reasonable degree of probability can lead to the desired results within the time allocated.
- Implementation, i.e. introduce the new 'development machine', based partly on the old development system, but with a reasonable guarantee that the new one will work as intended.
- Measurement, i.e. set up qualified metrics for productivity and performance, and measure whether these targets are reached after the new development system has been set up. A particular area of interest is measurement of whether things have fallen into place, by means of which the actual establishment of the new development system can be checked.

Innovation is not only about implementing new tools, but about increasing coherence between the conditions that prevail in a company and the influence of the development department. This may lead to new visions and development strategies, new targets, formulated both as product and business targets, and new performance and innovation targets. The revised targets might crystallise in a new formulation of the design commission, its qualitative and quantitative technological content, and time and aspirational targets for each product area [3]. The development assignment forms the basis for design of the development system, i.e. the organisational, methodical and managemental solution. The product development system is more than the formal organisation matrix. Current projects have shown that it is useful to work with the following seven elements or subsystems (see Figure 3):

- The organisational structure, i.e. the functional units and areas of responsibility that together make up the development system.
- The decision structure, i.e. the relationship between targets and strategies, plans and project targets, and the decision makers who play a role in this interdisciplinary pattern.
- Groups and individuals, i.e. the social system in the form of formal and informal groups and grouping mechanisms, who – by their knowledge, abilities, attitudes and sets of values – determine the performance, dynamics and ability to introduce change into the development system [3].
- Methods and tools, i.e. the sets of procedures, planning, development and evaluation methods and design tools that are to be used in the product development.
- The knowledge structure, i.e. the way in which the company chooses to obtain, develop, store and channel knowledge in relation to key persons within the company.
- The measurement system, i.e. the set of metrics for productivity and performance, which are to be formulated and followed up, together with the tools to be used in this measurement and its follow-up [8].
- The physical framework, i.e. the actual workplaces, their design, and the physical, visible impression that they give for the different sub-activities of product development and that contribute to giving the impression of innovation and significance.

A critical element in renewal of a development function is of course access to dynamic solutions, both with regard to method, organisation, and in terms of the new 'world pictures' and mindsets of the participants. In the current projects, the consultants' knowledge and recognition of good solutions have been important success factors.

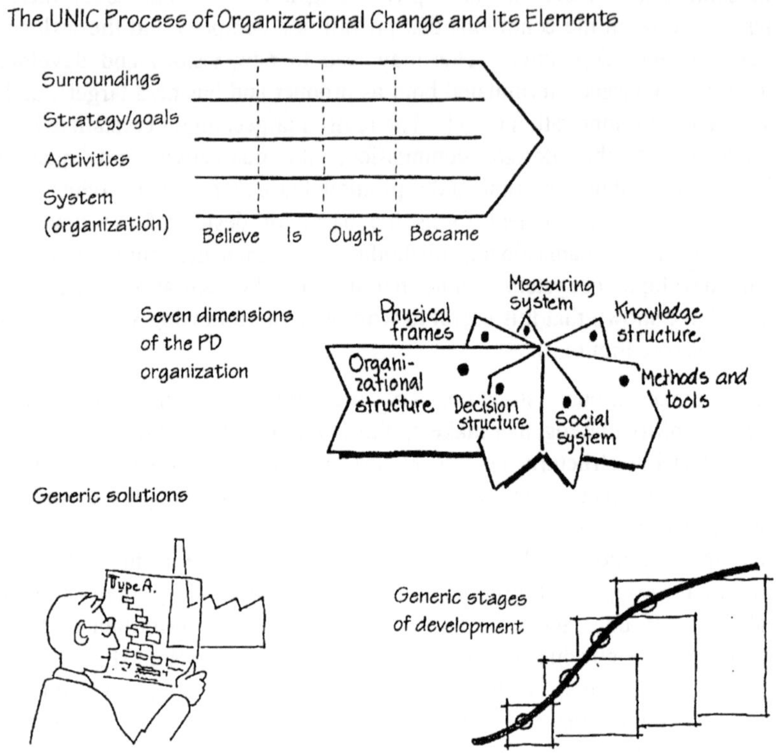

Figure 3: The UNIC process for organisational innovations has elements such as a basic stepwise procedure, a seven-dimensional view of the organisation, generic organisational solutions, and identified generic stages of organisational development [9].

3.2 The Innovation Process

The making of a diagnosis is, as mentioned above, the basis of innovation. It has turned out that the deficiencies of development ability and too low productivity can often be traced back to a few causal patterns. Thus, it has been possible to formulate a set of hypotheses that the innovation consultant can test out in the company concerned. Examples of hypotheses are: 'Previous product development has been so crude that the company is now faced with considerable maintenance and fault correction activities that require excessively high degrees of manning', or 'The R&D director's chair is effectively empty: he does not look after problems of technological strategy, product relationships or standards for development, but just allows things to grow wild'.

Such hypotheses can be used to determine whether or not the company suffers from a particular problem.

In the innovation process, which was created and which led to the production of a manual [10] in support of the company leaders of the renewal project and of the consultants, one should work with different classes of description for the four areas discussed above (environment/aims/tasks/development system), as follows (see Figure 3):

- 'believed to be', i.e. the view of the company on how the development system is built up and its performance, as these are described in organisational models, a plan and so on.
- 'is', i.e. the supposedly objective and 'true' picture that arises through diagnosis of the development system.
- 'ought', i.e. the chosen, desired development system and its performance to be used as a target for the activity of renewal.
- 'became', i.e. the result achieved after introduction of the new development system.

These different states that the process of change passes through reflect the difficulty of 'understanding' product development, as has been mentioned previously.

As has also been mentioned previously, the actual design activities are carried out using of the type 'modern' forms of organisation, such as Integrated Product Development and new tools like Quality Function Deployment (QFD), Design for Manufacture (DFM), etc. The seven dimensions serve to draw attention to some of the other things that need to be in place, e.g. goal formulation and result balance, and the need to measure whether the new organisation is indeed in place and working satisfactorily.

3.3 Typical Courses of Innovation

It is obvious that the UNIC process is very complex compared with the usual method-oriented implementation of Integrated Product Development.

A typical course that the UNIC innovation process follows consists of the following:

- Management meeting.
- Preliminary diagnosis based on the 'believed to be' of the management, and material in writing such as minutes of meetings and procedures.
- Planning and making of a diagnosis where problems are identified and where the necessary changes and performance improvements are identified.
- Identification of goals and tasks for the potential new product development function.
- Selection of solution elements and structuring of a new product development system.
- Planning of implementation activities.
- Implementation and measurement of current achievement.

In practice, the implementation could consist of introducing Integrated Product Development, creating a procedure, enhancing activities concerning goals and planning, organising product planning, creating cross-organisational teams, etc.

3.4 Important Features of the UNIC Process

Renewal demands a pivotal point in the organisation. Our experience is that one individual in management can be put in a situation in which he alone is responsible for and obliged to create a new power balance as the breeding ground for renewal.

The UNIC process has not been tried out without consultant assistance. The making of a diagnosis is almost an impossible task for a company employee, because it carries elements of disclosure, criticism and exposure. Design activities may also necessitate a generalist view that the company does not have. Finally, perhaps measuring may best be carried out when it is supervised by an outside organisation.

A decisive point is the feasibility of the process. We take the liberty of making the following conclusions concerning this point:

- It is possible to make a trustworthy diagnosis of the development organisation of a company and to obtain sufficient insight into it for the consultants then to create a new development organisation in a credible manner.
- It is possible, together with the company, to design, select and staff the new development organisation in a manner that is both inspiring and credible for the company.
- It is possible to initiate the introduction of the new development system in such a way that the company believes in it.

3.5 An Interpretation of the UNIC Process

Interesting questions can be asked concerning the UNIC approach: How and why does innovation happen? What causes changes in values, attitudes and behaviour?

Employees who are involved in the diagnosis, see it as the initiation of a process of change. The very fact that existing conditions and performances are queried is perceived as a token 'promise' of change by management and that such change will be initiated by management, visible and quickly applied. The diagnosis can result in action paralysis, as everybody anticipates the new organisation and the new working methods. This is partly the result of a resigned attitude to the existing state of things, and partly the result of the job rotation that will perhaps become part of the innovation. Pointing out weak operations and unexploited possibilities forms part of the diagnosis. The employees of the company must 'buy' the diagnosis. Here, the integrity of the consultants and the necessity of all participants recognising the situation of their company play major parts.

The next step in the UNIC organisation innovation process is to identify and formulate the goals and visions that the revised product development function shall meet. Here the individuals, with a focus on top management and key persons, will have to agree what needs to be accomplished in order to secure an attractive and reasonable future. In a specific company this particular ground may have been covered, and there may be a set of strategies already formulated, from which the goals for product development may be derived. However, in most companies we found that this exercise was a new, if attractive and stimulating, one for the company. The balance was often one between near-platitudes, like 'always first with the new products' on the one hand, and very specific, but unimaginative statements like 'to get the HM17 ready for the Hannover fair' on the other. A certain shift could be seen from individuals involved in the process: from an unharnessed declaration of ambitious (and unrealistic?) goals, through to the realisation that these goals would mean an impossible workload for many years to come, to the final precise formulation of reasonable but professionally and commercially attractive goals.

Designing the new product development organisation differs in nature from a standard design process for a number of reasons. The first is that you are playing with cards that have already been dealt, that you already have a number of employees on board, and that you will not suffer the consequences of having to take on completely new ones. The second is that we cannot (at the moment) be very normative in the design of product development organisations, since we do not know enough about why things work. Therefore, we tend to build (for good reasons) on the existing good elements and mechanisms in the existing organisation. In designing this new organisation it will become apparent (if it was not before) for the people involved that their future may be with new roles, new responsibilities and new patterns of cooperation. Here, some will imagine themselves losers and some winners, and sudden reactions to the new organisational design may be hard to explain until this is taken into account.

Implementation of the new organisation (or product development function) suddenly takes the process out of the hands of a small group of people in the company and into those of nearly all of the company, even in some instances into those of subcontractors and affiliated companies. The success of the implementation will depend on a number of considerations:

- that the factors that initiated the process need to be communicated (need for survival).
- the goals of the reform need to be communicated (what we want the result of the process to be).
- the process of the implementation must be understood (what will happen, and when).
- the concept must be perceived as sound, and shown to be meeting the problems that are perhaps already recognised at the operational level.
- it must be made clear that the management group is united in its intention to carry the process through.
- it must be understood that help will be provided to enable everyone to participate, both as individuals and in groups.

The final stage, measurement, has two important dimensions. One is to measure if and when the new organisation is in place, following the reform. A fair amount of this is concrete, but a very important part is 'soft', because it deals with measuring wether people have acquired new norms, new working patterns, new mindsets, etc. The other is to measure the performance of the new organisation, both after implementation, and then over time. As measurement of performance is virtually new in product development, the reaction is often one of incredulity. But since it can be illustrated and proved that meaningful measurements and indicators can be formulated, the situation is as a rule accepted without resentment.

Generally speaking, the UNIC approach has been welcomed by the professionals in product development. It has provided a new and more comprehensive framework for product development, it has resulted in modified and more meaningful roles and tasks in product development, it has given a name and a form to professionalism in product development, and finally it has given to the area of product development what has been the prerogative of manufacturing and production for many years – a manageable development process.

4. Discussion and Conclusion

The aim of this paper was to report case stories and participatory research observations from organisation innovation projects in product development.

The two types of innovation projects observed are quite different. The first is an implementation of a new product development methodology, the solution being given: Integrated Product Development. The second is a more complex innovation procedure – to establish the situations and needs of the companies, to search for a new organisation and to implement it. Our aim is not to comment on or compare the two approaches, but to look at the individual processes of change.

The projects in question have been led by engineers, but the most important preconditions for changes and success are of a psychological and social nature. A primary question is whether the necessity for change is accepted.

Organisational change cannot directly be compared with networking and learning, see Leifer & Mabogunje (in this volume). There are levels in organisational change related to 'doing the right things', later 'doing things right', and then comes the performance itself. It is our experience that monitoring the organisational change needs both a means of measuring whether the change has occurred (Have we introduced the change?) and a means of assessing specific new methodology-oriented performances (Do we now carry out Design for Manufacture with substantial cost reductions?).

Our approach to innovation may be seen as looking upon design performance under new influences, see Wiegers & van Breemen (in this volume). We manipulate influences and such things as procedure, integration means (teams, milestones), methodology and new goals (business focus, time). But our research has such a goal-oriented setup that we have 'to advance from a bad to a good organisation'.

Also the work of Duffy & O'Donnell (in this volume) on product development performance may be related to our work on organisational change, because the evaluation of performance should be seen as a very important metric, although not the only one that should be used for controlling organisational changes, as pointed out above. Normally, new tasks, new markets, new goals, new organisation and the problem of making them function are preliminary, performed in the hope of obtaining improved performance and subsequently enhancing it.

Product development shows such complexity that a description of the modes of operation of the organisations is almost impossible. Because of the authors' engineering backgrounds and their limited insight into the psychological and social nature of the change process, this paper poses the question: Who is qualified to draw up innovative projects? Engineers lack insight into what pushes innovation forward, whereas psychologists and sociologists lack insight into what is at stake with regard to technology. Danish industrialists, mostly engineers, expect the best solutions to come from non-engineers.

It is the authors' hope that the case studies described in this paper may contribute to the perceptions that changes of method in product development do not happen spontaneously, are not a top-down process, and cannot be implemented without focus on a number of soft aspects.

References

1. Olesen J. (ed.) 1993: *Proceedings of WDK Workshop on DFX.* Inst. of Engineering Design, TU of Denmark.
2. Hein L. 1993: Setting the Stage. *ASME Winter Annual Meeting (WAM),* New Orleans, December.
3. Andreasen M..M. 1996: Product Development in a Life Cycle Perspective (in Danish). In: J.O. Riis (ed.) *The Enhanced Production Concept.* Industriens Forlag, Inn.
4. Araujo C.S. 1998: *Assessment and Selection of Soft Product Development Tools: A Theoretical Contribution.* PhD Thesis, TU of Denmark, Lyngby.
5. Research project: Manufacturing after year 2000 (in progress).
6. Andreasen M.M., Hein, L. 1987: *Integrated Product Development.* IFS, Bedford.
7. Andreasen M.M. 1991: Design of a design organisation. In: V. Hubka (ed.) *Proceedings of ICED 91.* Heurista, Zürich.
8. Kirkegaard L. 1989: Developing Efficiency and Effectiveness in Product Development. In: V. Hubka (ed.) *Proceedings of ICED 89 Harrogate*, IMechE.
9. Andreasen M.M., Hein L., Kirkegaard L., Sant K. 1989: *Product Development - Fundament of Innovation* (in Danish), Dansk Industri. Copenhagen.
10. Andreasen M.M. 1987: Design Strategy. In: V. Hubka (ed.) *Proceedings of ICED 87 Boston*, ASME.

Planetary Organisation Offers Advantages in Project Work

Stig Ottosson

Abstract

A new organisational concept for project work is presented in this paper. The aim of this new concept is to increase flexibility and involvement of every team member in the project group as well as to shorten the time needed to accomplish tasks properly. The organisational form is called 'Planetary Organisation' with the project leader in the centre, sub-project leaders around him/her, and these being in direct contact with each other. Around each sub-project leader there are project members also in direct contact with each other. The management style for the project leader and the sub-project leaders should be Management by Walking Around (MBWA) for maximum efficiency, which is possible to perform as time is allocated for this under the new organisational form. In the system, experts and senior managers also act as comets, influencing the project members in a positive way.

1. Introduction

Flexibility, dynamic organisations and decentralisation are three popular terms used in management today. However the common line-organisations and matrix organisations were originally designed to attain opposite goals. They are difficult to reconstruct, in order to satisfy the new demands, without losing their meaning. A new organisational form called 'Planetary Organisation' is therefore suggested in this paper and explained for the first time at an international conference. The way the planets are kept together and influence each other is the metaphor for the planetary organisation. As in the solar system, comets – in the form of experts and senior managers – can move freely between the planets.

The model is to some degree implemented at Algon AB, which is a company involved with IT products. At Prosolvia the model is used fully. Prosolvia comprises experts in Virtual Reality and advanced simulation techniques. Both companies are listed on the Swedish Stock Exchange. R&D spendings for each company are in the 15–20% range and their product life cycles are 0.5–2 years.

The fact that this model has been adopted by at least these two companies suggests that it is especially useful in companies that have to come to market rapidly. Figure 1 illustrates the model.

Figure 1: The planetary organisation suitable for project work [1].

2. Simple Experiments Performed to Test the Method

To test the basic ideas of the planetary organisation and to compare it with the basic line-organisation ideas, different groups of students and engineers were given the same problems to solve in team against team competition. The only difference in these pairs of teams during competition was the organisational form (planetary vs. linear organisations). For both organisations, three management levels were chosen. The project leader had 2–3 sub-project leaders and 2–5 group members were coupled to each sub-project leader. Figures 2a and 2b show how the different groups were placed during one of the experiments.

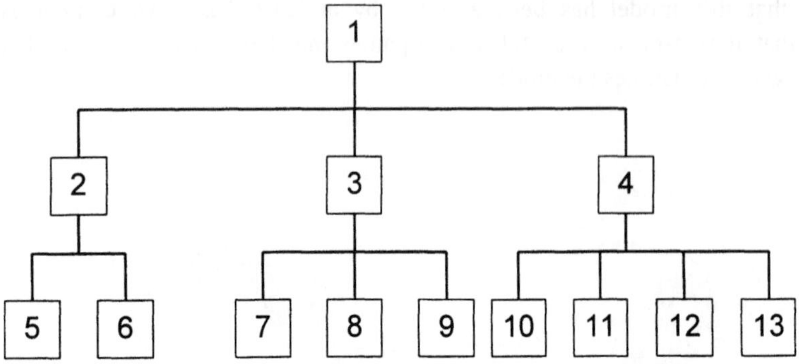

Figure 2a: The placement of project members during one of the experiments when 13 team members were used in a 3-level line-organisation.

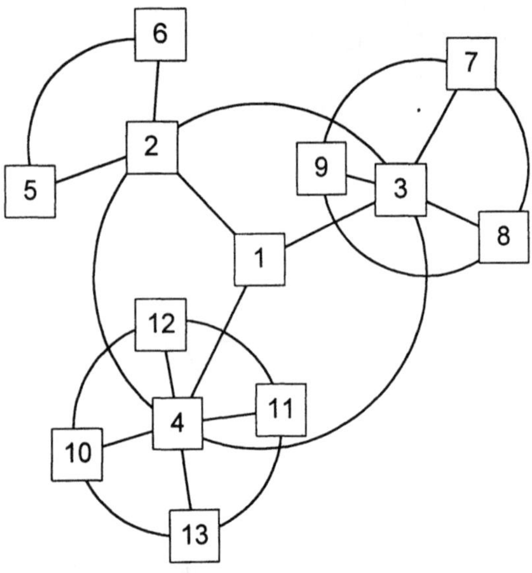

Figure 2b: The placement of project members during one of the experiments when 13 team members were used in a 3-level planet organisation.

Before the experiments started, each team member was given a bundle of small pieces of paper (5 × 10 cm), pencils and each received an equal number of playing cards. The playing cards were distributed at random and placed face down prior to the start of the experiment. The project leaders and sub-project leaders were given more paper as more information was anticipated to be sent from them than from those further out in the organisation. In the experiments, the small pieces of paper were to be used to write down questions, orders and suggestions.

Before the start of the competition/experiments, everybody was instructed that no communication, oral or otherwise, was allowed during the experiment other than such that was written on the pieces of paper, and that the paper as well as the paper and cards could only be exchanged along the lines (shown in Figures 2a and

2b) between respective team member. The papers and cards were to be passed with or without the help of one to five experiment assistants during every experiment. The experiment assistants had to be especially active when the table placements were such that team member could not directly reach each other. The assistants also had the duty of enforcing the simple rules. They were furthermore asked to commit to memory certain items and events as a basis for comments after each experiment had been completed.

The problem which was to be solved – the mission – was to rearrange the randomly distributed playing cards so that at the end each team member had a predetermined set of cards. This defined goal was given to each project leader on a spreadsheet when the experiment started. For example, team member 4 should have all four kings, or another combination of cards, when the experiment was completed. The special problem for each project leader was to manage the team and the team work by sending written orders and questions to get information for his/her subsequent decisions.

The participants were also instructed that two team members only could exchange 1, 2 or 3 cards each time with each other, only after they first had communicated, and agreed upon, the exchange. In addition, the communication was just to be one way and only done with the help of written information on the paper slips. Exchange had to be performed simultaneously, so that one team member could hold all the cards. Thus for example, the project leader could not order everybody to send in all their cards to him so he could rearrange the cards and send them back with information on who should retain which cards.

Seven experiments have been performed so far during 1997. Five of them have been done in Sweden, one in Estonia (Tallinn) and one in the Czech Republic (Ostrava). The planetary team organisations always fulfilled their missions but in two cases the line-organisation teams were stopped when no final solution was achieved after 60 minutes. Typical completion time spans for the planetary teams was 15–30 minutes. Completion times for the line-organisation teams were 0.95 (only one case!) to 3.0 times the planetary organisation's competition times. For 20 team members in the projects, the differences in times were much larger than when eight members were used. Thus the larger the teams were, the larger were the advantages for the planetary teams.

3. Discussion

Notably, the team members at the 'bottom' of the line-organisation were very inactive most of the time, while the project leader was always overloaded with work all the time. In the planetary organisation everybody worked all the time and the work load on the project leader was slight compared with the project leader in the line-organisation, giving him/her the possibility of controlling and correcting rather than commanding. Thus a harmonious and efficient operation was performed by the teams working in accordance with the planetary model.

In all the Swedish experiments it was noticed that destructive initiatives started to take place by those team members who sat idle for some time at the 'back' of the line-organisation. For example, they started to send questions to their sub-project leaders with the messages: ‘Are you sleeping?’, ‘How long do I have to wait to get answers?’, and even messages totally unrelated to the task at hand such as: ‘I am hungry!’, ‘What are you going to do this summer?’, etc. These questions increased the stress levels for the sub-project leaders and caused them even to spend time on, in written form, reprimanding those team members. When the sub-project leaders reverted to the same behaviour against the project leaders, the entire project performed at a very low efficiency level. The longer time that team members were sitting idle, the more their destructive behaviour was shown to be. Insinuating questions thus were a strong sign of too little to do and/or unwillingness to contribute to the work.

It was interesting to note that, in the former Communist countries, the destructive pattern did not show up in the line-organisation teams as it did in the Swedish teams. In the Estonian and Czech experiments the team members, with the exception of the project leaders, were simply passively waiting for orders, taking no initiatives whatsoever on their own. In the discussions afterwards they explained that they had for so long been taught not to question orders and not to take the initiative that they, when put in a line-organisation, reverted to such a trained behaviour. In the planetary organisation they however proved to be active in much the same way as in the Swedish experiments.

When very efficient and structured project leaders were chosen for the line-organisations, they could attain almost the same efficiency as the planetary organisations did with an ordinary leader. In one case, the structured project leader also managed to beat the planetary project leader in reaching the goal for the experiment. If the project leader in the line-organisation was not well structured, the result in two cases was so poor that the mission was aborted when the planetary organisation had fulfilled its task. The influence of project leaders who were not very well structured on the performance of the planetary organisations was slight, as the organisation appears to be ‘self-repairing’.

The discussion of feelings and experiences, to reduce the stress levels for the project and sub-project leaders, was felt to be important after the completion of each experiment. In both the non-Swedish cases and one of the Swedish cases, the project leaders of the line-organisations blamed themselves so heavily for the result that they were in a very bad mood for a long time. Not until they started to realise that perhaps the organisational form could also have had some impact on the result and that it is very important that everybody in an organisation supports each other, was this frustration alleviated.

4. Conclusion

From these simple experiments the efficiency of planetary organisation has proved to be superior compared with that of line-organisation when solving a complex task during which everybody had to be involved. The planetary organisation automatically seems to foster active work, individual initiatives and real team feelings. The quality of the project leader did not prove to be critical for the outcome and a 'self-repairing' pattern was built in the system. The work load for the project leader was decreased which meant increased possibilities to investigate and influence the team members rather than giving orders. Thus a management style symbolised by 'Management by Walking Around – MBWA' was established. Many of the disadvantages with line-organisations blossomed fully in less than a quarter of an hour, which is a powerful experience for those participating in similar experiments.

In real life projects, people are allowed to speak to each other. They also build their own internal networks in which they act independently of organisational form and set rules. Thus they to some degree act as the planetary organisation prescribes.

Project leaders who have experienced such experiments as the one described will better understand why their team members act in peculiar ways in real projects. The extreme situations that became evident in a few minutes during the experiments focus on problems that can take years or even hundreds of years to understand. The brigade headquarters of a military manoeuvre is, for example, often overwhelmed with work and the staff members do not have time to sleep or eat before they are more or less totally stressed out. At the same time the soldiers lie on their backs for days without getting information and having anything to do. So it has been going on for hundreds of years and very few signs are seen that the military organisations will try new ways of organising their operations for the typically project-based missions that war and peace are. For companies with short product life cycles, the planetary organisation model seems to be especially favourable.

Reference

1. Ottosson S. 1997: *Innovation och Företagsförnyelse (Innovation and Business Renewal)*. Tervix, Floda, Sweden.

Résumé of Section IV

External conditions in terms of 'isolated variables' are not the main issue in the contributions of this section, but they are treated in a more general way as 'management' influences. This *modus operandi* is not surprising. Whereas a laboratory situation allows the effects of separate variables to be investigated, this procedure is very much reduced in design practice – it seems almost impossible to define independent variables and to investigate their effects under isolated conditions. In addition, the 'effects' of variables may concern a wide range of factors, e.g. from the individual contentment of employees to the question of company culture.

It is superfluous to emphasise that 'management' is a necessary ingredient of each product development activity, for example in order to reduce friction and conflict between departments and teams, which occur inevitably in the collaboration of individuals and teams. Furthermore, the goals of management activities and organisational changes are various. On the one hand, it is the aim to destroy 'walls' between departments with the objective of increasing flexibility in dynamic development processes. For example, the introduction of self-organisation of single units (planetary organisation) should reduce the workload of design managers and increase the responsibility of the group and the individual. On the other hand, management activities are directed towards creating an optimal climate for innovation. This climate for innovation can, for instance, concern the management of changes during the development process. In terms of a long-term product planning, management of changes does not just involve the actual changes, it has in addition to ensure flexibility of the possible changes to fit the production of the product or product parts in the future.

According to these different goals, 'organisational change' consists of many varied and balanced steps. These steps may address specific external conditions, such as the location of team members with respect to each other. The changes themselves are related to further questions such as: at which time and how often are such changes introduced, who is affected by the changes, how are the changes prepared and communicated, and who are the winners and losers of changes? These questions mirror the sometimes contradicting and risky goals of management activities: the conclusion may be a balanced procedure on the sharp edge between the demand for security on one hand and the necessity of change to support flexibility and an innovative atmosphere on the other.

These organisational problems will increase further in the future. Modern trends like 24-hour development will demand the flexible adjustment of organisations to meet actual needs, including those of internal and external partners over long distances.

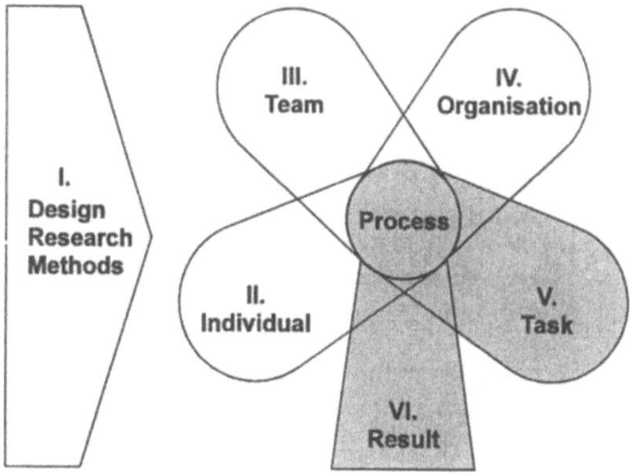

Section V:

Task and Process Variables

The design task as the starting point and objective of the design process determines subsequent design activities and is therefore of major influence for the method of procedure. As already mentioned, tasks in engineering design are varied. Therefore, this section opens with an introduction to the approaches for *classifying design tasks*. The characteristics of a design task affect the design process, which means that important variables in the design process are based on characteristics of the task. For example, the availability of information depends on the novelty of the task. This leads to a key issue of design work: *information* and its mental treatment, external management and availability. Designing can be seen as the development of product and process information. In this regard it is not surprising that the availability of information is a main factor for the success of nearly every design activity. On a general level, the suitable treatment of information is a basic activity of every individual problem-solving process and consists of basic cognitive processes. The support of these *internal* information treatment processes is an issue of the first contribution to this section. In addition, the following questions are addressed in this section:

How can we differentiate *design tasks* in terms of *requirements* on the designer?

How can we *support* mental processes, for example by *models* of the product and *tools*?

What factors influence the *need* for information in the design process?

What approaches should we use to support the designer via *information technology*?

The course of this section proceeds from the basics of design thinking to the use of models and information systems to support information processing and storage during the design process.

- *Hacker, Sachse & Schroda* start with a characterisation of design *tasks* and its effects on design thinking as information processing. Thus, the emphasis of the presentation is on *individual* aspects, but there are also other aspects discussed, such as solution-supporting mental representations, design procedures and the use of knowledge in the design process.
- *Radcliffe* identifies the use of different types of models and tacit knowledge in various design projects. He distinguishes these projects (and tasks) by different categories such as social dimension and duration.
- *Christiaans & van Andel* investigate the use of information in a team during an experimental design process. The difficulties in supporting this information gathering by a computer tool leads to the issue of information support by tools in the next paper.
- *Meerkamm* gives a survey of the problems and approaches of information technology that solves these problems. He also discusses the influences on the demand for different types of information in the design process.

Design Thinking – Possible Ways to Successful Solutions in Product Development

Winfried Hacker, Pierre Sachse & Frauke Schroda

Abstract

Engineering design is analysed in terms of design problem-solving. Engineering designers of differing productivity hardly differ in mental abilities, as measured by intelligence testing, but in their strategies concerning the analysis of requirements and of the problem, in the search for general principles that solve the relevant problem, and in the procedures that develop specific solutions. Experimental and interview results stress that the interrelationship between thinking and sketching, as well as other kinds of early low-cost prototyping, are of crucial importance for the efficiency of the problem-solving procedure and the result. Possibilities of assisting design problem-solving and of improving the training of designers are discussed.

1. Introduction: Engineering Design as Thinking and Problem-solving

Engineering design – like other designing activities, for example the design of technological processes or software design – is a working activity with a decisive impact on two key problems of economic development. The first issue is the design of innovative products to gain access to new markets; the second concerns the competitive prices of these products. Thus, engineering design will have a high impact on competition within regions and between firms.

What are the characteristics of 'engineering design'?

According to Steuer [1], engineering design means 'thinking ahead creatively and completely in order to create a technical object fitting with the requirements of the historically determined level of technological trends and the development of all documents necessary for its physical implementation. It consists of designing and shaping'. This definition concentrates on two types of design tasks only: on the design of completely new objects and on the adaptation of given solutions to new requirements. It more or less neglects tasks that require the recombination of already elaborated modules. This paper will confine itself to tasks covered by the cited definition.

According to the definition, engineering design is a thinking activity, more precisely the most complicated type of the manifold types of thinking and problem-solving. It just does not imply reasoning on an already given issue, but thinking ahead in the sense of developing the mental model of a future object that has never existed so far. Furthermore, this thinking ahead should be creative because the future object should have useful new parameters.

This 'design problem-solving' process [2] contains some important phases. The main ones are:

1. identification and clarification of the problem,
2. development of a frame conception of solutions,
3. design of the favoured solution, and
4. working out the details [3].

The main point is that the least observable and least formalisable 'early phases' of task identification and the development of a frame conception of solutions will have the highest impact on innovativeness and on manufacturing costs. Ehrlenspiel [4] illustrated this decisive function for the possibility of influencing costs. This influence will be much higher within the early phases of design problem-solving than in later ones, but is only vaguely calculable. To stress the point: a reliable calculation of the costs of a design result only becomes feasible when it is too late to do anything about them.

Consequently, in its early phases, engineering means problem-solving with high uncertainty concerning its results. On the one hand this uncertainty offers a challenge to proceed to innovative solutions, on the other one, however, it also implies the risk of detours and failures. Thus, uncertainty may challenge or worry the designer and in the final analysis simply inhibit a creative procedure. Therefore engineering design is design problem-solving with an internal contradiction between challenging options and stressing risks. At this point the question arises whether possibilities exist to support design problem-solving. Working activities need strategies that offer information – what options should be used and what detours should be avoided. Such strategies or technologies may support manufacturing as well as information processing activities. There are no convincing arguments why it should be not feasible to develop a technology of efficient design problem-solving. First modules are offered and are successfully applied in design education and daily work. Examples are guideline 2221 of the Verein Deutscher Ingenieure [5] offering a strategy of engineering design, or the textbooks on design strategies - in Germany for example Hansen [6] or Ehrlenspiel [4], both integrated into the frame approach of Pahl & Beitz [3].

What might be contributed by Psychology to the further development of supporting strategies or even of a methodology of engineering design?

First of all, highly and less efficient designers may not be distinguished significantly in terms of their intelligence as measured by traditional tests [7]. Spatial imagination, too, will contribute only marginally to an explanation of excellent versus less convincing solutions. Even job experience does not explain the quality of solutions sufficiently.

A more promising approach offers the analysis of working strategies of designers. Before, however, discussing these mental working strategies it is important to bring into mind a few essential characteristics of thinking within engineering design. Thinking strategies should not be misinterpreted. Thinking may follow logic or other systematic rule systems in some cases, but in others it might not. It may apply rational and abstract concepts, but can also be based on pictorial imagination. The thinking process may be consciously controlled and verbalisable, however more often an unconscious, implicit or tacit processing takes place which in some situations may converge into sudden conscious insight. Finally, thinking as design problem-solving mostly combines the development of actual new solutions with an activation of conscious or tacit knowledge of already known ones.

In order to support and optimise engineering design processes, design problems of different requirements must be studied and compared. Unfortunately, however, an agreed and valid procedure to analyse and rank the requirements and, thus, the difficulties of engineering design problems is still lacking. Therefore we have tried to develop a first draft of a requirement analysis.

2. Requirements of Engineering Design Problems

Problems are characterised by subjective and objective features. Subjective features refer to the individual mental representation of a problem including individual ability, experience and the knowledge of a problem solver about the field and about possible operations. This is often described as 'problem space' [29]. The objective features refer to characteristics of the task structure that are independent of the concrete problem solver. A general and formal classification for engineering design problems should focus on this task structure. There are several existing taxonomies for engineering design problems [8–11]. Based on these ideas we propose the following criteria and operationalisations [29, 30].

1. **Multiplicity of goals** (number of goals, number of conflicting goals and strength of conflicts between the goals)
2. **Complexity** (number of subfunctions, number of interactions between subfunctions, strength of the interactions)
3. **Transparency** (information about the situation, information about the method of solution, information about the solution aimed at)
4. **Degrees of freedom** (number of possible variants of solution, number of possible ways of obtaining solution)
5. **Dynamics** (changeability of the situation, predictability of the consequences of the problem solver's action and desicions, influences from outside)
6. **Necessary knowledge** (specific knowledge about facts, specific knowledge about methods, general problem-solving strategies)

These criteria are differently weighted by different design experts [22, 30–31]. They attribute high importance to the multiplicity of goals and problem complexity, medium importance to transparency, degrees of freedom and necessary knowledge, and low importance to the dynamics. There is a satisfying reliability and concordance of the averaged expert rating judgements for nearly all criteria (Table 1).

Table 1: Reliability and concordance of the criteria ($N = 58$ experienced designers).

Criterion	Reliability (Cronbach's Alpha)	Concordance (Kendall's τ)
multiplicity of goals	.85	–
complexity	.81	$.59^{(*)}$
transparency	.73	.48
degrees of freedom	.78	$.71^{*}$
dynamics	.77	$.77^{*}$
necessary knowledge	.54	$.66^{*}$

* ... $p < 0.05$; (*) ... $p < 0.1$

Thus, based on this requirement assessment an overall problem difficulty may be calculated by a weighted sum. Moreover, the individual requirement dimensions may be represented as a profile of task difficulty. Figure 1 shows the different design tasks for two examples.

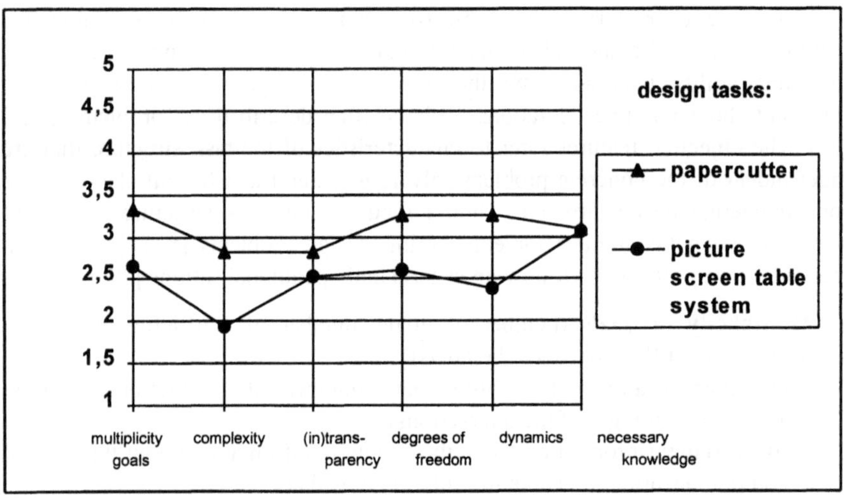

Figure 1: Profile of task difficulty.

These first results encourage further research in this topic. A valid checklist for requirement analysis might be highly useful in several directions. From an theoretical point of view it would be an essential instrument for problem-solving design research. From a more practical point of view, an instrument for requirement analysis might help with both, engineering design education and the organisation of engineering design tasks in industry. In the first case the instrument might ease the difficulty of sequencing learning tasks, in the second it might offer a rough estimate of the necessary working time as a prerequisite of planning.

3. Characteristics of Successful Strategies in Engineering Design

The following review draws on unpublished results of our Dresden research group and on the literature [12–19, 22, 30, 31]. Successful procedures are based on characteristics of the development and implementation of intentions on the one side (the intention aspect), and on the procedures and tools applied in implementation on the other side (implementation aspect).

Seen from the *intention* side, successful strategies are goal-directed and planning procedures. The task, internalised as the intention or goal of the design process, is clarified extraordinarily thoroughly, the individual goals are integrated into a plan but nevertheless they are dealt with in a flexible manner.

Seen from the *implementation* point of view, in successful strategies first of all a couple of different solution principles are developed and compared, before the 'search space' is efficiently focused on the one most promising version. Moreover, problem-solving and knowledge application are efficiently combined.

Three main characteristics of successful procedures should be stressed in detail:

a) Clarification of the task

The often discussed 'first idea' actually does not have an important role within the procedures of either experts or novices. Rather they first of all scan their knowledge in order to start an analogous reasoning [16]. The procedures of successful experts differ from those of less successful novices with respect to a refined clarification of the task, a subtle definition of goals concerning subfunctions and a thorough extraction of those aspects that are important for product functions – not simply for its shape.

Consequently, going into detail, the characteristics of a successful design procedure are:

- a thorough clarification of the task
- concentration on components that determine the functions of the objects
- an exhaustive analysis of the main functions
- the weighted integration of the goal characteristics to be implemented
- the written or sketched fixation of the solution principles of the goal
- a goal-oriented but nevertheless flexible method of processing.

In an experimental analysis, a significant relationship was found between the number of goal characteristics analysed and the quality of the resulting design solution [16].

A second group of characteristics of successful procedures concerns

b) The type of search for a solution principle
Engineering design is more than design problem-solving. In most cases there is no well-defined problem given, which just needs to be solved. Instead, first of all the key problem itself must be defined. Therefore several variants of a solution principle need to be developed, then the most promising of these needs to be selected in order, finally, to solve just that problem. However, this procedure is not only rule-governed deduction but also creative induction.

Going into some detail again, the characteristics of a successful procedure here are [9, 13, 15, 20–22]:

- A procedure generating new solutions instead of only modifying already given ones.
- Conceiving a few different principal solutions before starting to work out one selected solution principle. This so-called *divergent* search phase in successful design must be followed by
- A *convergent* restriction of the variety of solution principles (the 'search space') on as few variants – which are worked out – as possible.
- A well-scheduled phase of detailed design, corresponding with the proposals of design methodologies: to develop and logically follow the ordered subgoals [23]. However, in spite of this goal-directed processing, the expert procedures are not obviously straightforward. There are anticipations of later steps and regressions to already processed ones, although less frequently than within less successful procedures.

The third group of characteristics of successful procedures concerns

c) The evaluation of the solution procedures
In successful procedures, more feedback is processed which guides further operations. In detail:

- The process and its outcome are more often evaluated in terms of goal characteristics, and
- The evaluation is made for both levels, the global solution principle and the detailed solution steps worked out.

This macro-procedure of engineering design discussed so far may be broken down into two components of problem-solving, the development of mental representations and the operations on these representations. What about the differences between design procedures in terms of these components?

4. Characteristics of Solution-supporting Mental Representations

Problem-solving requires the development of correct and complete mental representations as the materials of internal processing operations. The mental representations are the search space; no solution will be better than the search space will allow: the final solution will depend on whether the relevant solution possibilities were represented or stored mentally. Consequently the designer ideally should imagine all relevant solutions through an exhaustive combination of their features in order to select an optimal feature pattern from all possible ones. Just this, however, is often impossible because of the limitations of human mental capacity, often labelled 'capacity of consciousness' or 'working memory'. Thus the designer is forced to concentrate on a selection of features only. However, he will restrict not because he could not imagine further feature patterns, but because he cannot hold all of them within consciousness. Nevertheless the selection should contain just the optimal solutions of design. Thus the critical issue turns out to be: designers should have in mind the optimal feature patterns of a required solution which, however, in most cases they cannot know, since they should have been selected from the total of possible combinations which they cannot hold in consciousness.

Obviously the bottleneck of creative design in the first place is not the variety of designer's solution ideas but the limitation of working memory. Actually we could show experimentally with simulated computer-based design tasks that the lower the working memory span of a subject, the more superfluous steps are considered, the more inefficient solutions paths are tested repeatedly, and the higher the time consumption will be.

Thus, a possible way to support design problem-solving is to relieve working memory load. Just this was one of the above mentioned characteristics of successful procedures: experts tend to fix their ideas by writing and sketching. A labelled sketch as an external store offers a way out of the conflict discussed.

5. Solution-supporting Design Procedures: Designing as Interaction of Internal and External Processing

Now consider a designer who has developed a search space in mind, and who must operate on its representations. He should compare the characteristics of the solution variants, recombine characteristics, introduce new characteristics and reject irrelevant ones. Here the procedures of human beings strictly differ from those of computers. Designers in most cases will not start from the details and proceed to a complete solution. Rather they will start from a vague impression of a complete solution and proceed to details. To quote a description by Bach, the designer frequently will start 'not from individual elements of a solution, but from a holistic solution. It will contain already all details which will be unfolded within the process of design. Thus designing is a stepwise decomposition and clarification of partial functions ... It is a holistic-analytical procedure ... The designer creatively works on two levels: On the one hand he more or less unconsciously applies abstract, complete solution structures, on the other hand he consciously sketches specific combinations of elements ...' [24, p. 4].

This human design procedure needs external, sensorimotor support because human beings are 'thinking with their hands'. There is an analogy with 'the gradual development of an idea while speaking' [25]: often solution ideas are developed gradually within the process of sketching by hand. Solution ideas are fixed as an abstracting sketch of the solution principle and are specified afterwards within feedback-circles of thinking, sketching and critical inspection of the refined sketch. Thus, quoting Görners observations '... the sketch not only offers the results of the designer's thinking processes but mainly serves him as a working means' [16, p. 240]. In an interview study he analysed 74 experienced designers. One of the questions was, whether they use – in order to develop a solution principle – 'mainly thinking about it or mainly sketching'. 69.3 % of the designers were 'mainly sketching', 3.8% 'mainly thinking' and 26.9% reported a balanced combination of thinking and sketching (Figure 2).

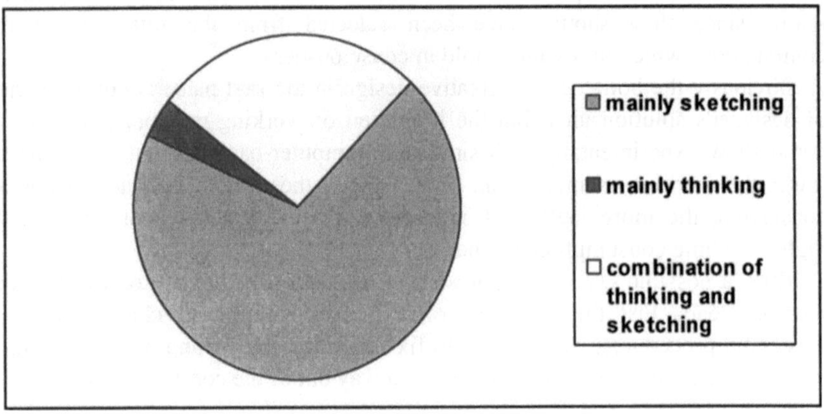

Figure 2: Interview study ($N = 74$ experienced designers).

A further question dealt with the reasons for sketching. The respondents here could apply more than one category: 61.5 % of them applied sketches in order to clarify an idea, 44.4% stressed the role of a sketch as memory aid, and 30.8% stressed the role of communication (Figure 3).

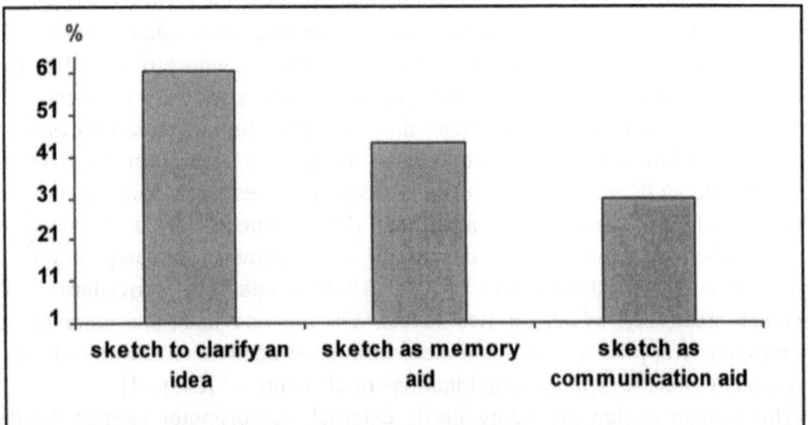

Figure 3: Interview study ($N = 74$ experienced designers).

To overcome the shortcomings of an interview study we carried out the following experiment. Two matched groups of students by means of a computer program dealt with three design tasks of different complexity. One group was asked to sketch their ideas first of all, the other was not. With the most complex task a significant difference was revealed. In the case of previous sketching, the number of operations in task accomplishment was lower, fewer operations needed to be corrected, the perceived difficulty of the task was lower, and – in spite of the additional sketching phase – the total working time was lower, each in comparison with the group not using sketching [26].

In addition to sketching, a couple of still more 'materialising' kinds of operations may support design problem-solving. Even in the area of CAD and digital prototyping, many simple physical prototypes and models are applied. Designers use pencils, rods, paper, cardboard, wire or plasticine models or elements of building kits, in addition to the more sophisticated computation, CAD or stereolithographic models. An interview study with a further 40 experienced engineering designers revealed the following main functions of the concrete simple prototypes: they serve as a means of problem analysis, an aid to create new ideas, an instrument of evaluation, especially of functions, a means of storage relieving working memory, and a communication support [27].

Consequently – as was shown recently also by Ehrlenspiel, Bernard & Günther [21] – the simple concrete prototypes are very useful, especially within the early phases of design problem-finding and solving. Therefore the application of this 'early low-cost rapid prototyping' should be facilitated along with the application of the more extensive high-tech types of modelling and prototyping, and it should be integrated into design education.

6. Design Procedures Continued: Knowledge-based Opportunistic Solution Procedures

Design problem-solving involves more than just thinking. To a great extent it is a knowledge-based development of solutions. Thus, it integrates the reactivation of already known solutions and the development of actual new ones.

Through this integration the design process becomes an opportunistic procedure: it does not just proceed systematically towards the goal without any jumping between subfunctions or modules. Rather the procedure will start at some isolated islands of given knowledge and try to rearrange and to combine them stepwise by thinking. For that reason, backward jumps to already processed steps and forward ones to future steps of processing are inevitable. A systematic goal-directed procedure will only become possible when the problem analysis has found a suitable way of breaking down the total problem into subproblems. Thus, empirical results contradict the assumption of a systematic problem-solving procedure, as was assumed for instance with the waterfall metaphor of software design [28]. Instead of a systematic procedure from task clarification, searching for solution principles, and detailed development of a solution, another procedure prevails. It is based on the detection of solution-relevant knowledge while working. This opportunistic procedure shows the following characteristics:

- It switches seemingly irregular between mental and external (sketching, modelling) processing of subproblems.
- It accomplishes no complete clarification and decomposition of the total task before starting the design. Thus, its initial problem definition is global and incomplete.
- It switches between subproblems and levels of problem description, caused by the experience-based identification of applicable knowledge.
- The transferred knowledge leads to redefinitions of the problem and to changes of developed action programmes (plans).

Again an important consequence arises: the broadly announced methodologies of design may support engineering design only if they do not force a strictly systematic procedure following theoretically optimal phases. Rather they should offer optional guidelines which in particular will not restrict the naturally opportunistic initial steps of design problem identification and solving.

7. General Consequences

The initial phases of design problem-solving are task clarification with the identification and decomposition of the problem and the selection or development of a solution principle. These phases are of essential impact for the result and its costs. A couple of teachable procedures may be recommended to improve engineering design education concerning these phases.

Because design combines knowledge activation with problem-solving, the initial phases cannot proceed systematically towards a final result until the problem is clarified and decomposed into its subproblems. Therefore a useful design methodology should offer flexible, heuristic guidelines instead of strict algorithmic rules.

The bottleneck of creative design is less a deficit of ideas, but rather a limitation of conscious information processing, measurable by the working memory span. This limitation can be diminished by sketching and the development of external models.

Moreover, sketching and modelling will support the development, selection and correction of ideas. Therefore only in the initial phases of design should these aids be applied, and their application should be supported.

Acknowledgement

This work was sponsered by a research grant within the 'Sonderforschungsbereich 374, Rapid Prototyping' and the project 'Effectivity and Learnability of Design Methodology' (HA 2249/2-1) by the German Federal Science Foundation (DFG).

References

1. Steuer K. 1968: *Theorie des Konstruierens in der Ingenieurausbildung.* Fachbuchverlag, Leipzig.
2. Carroll J.M., Thomas J.C., Malhotra A. 1980: Presentation and Representation in Design Problem Solving. *British Journal of Psychology* 71, 143-153.
3. Pahl G., Beitz W. 1997: *Konstruktionslehre. Handbuch für Studium und Praxis.* 4th edn. Springer, Berlin.
4. Ehrlenspiel K. 1995: *Integrierte Produktentwicklung. Methoden für Prozeß-organisation, Produkterstellung und Konstruktion.* Hanser, München.
5. VDI-Richtlinie 2221, 1993: *Methodik zum Entwickeln und Konstruieren technischer Systeme und Produkte.* VDI-Verlag, Düsseldorf.
6. Hansen F. 1965: *Konstruktionssystematik.* Verlag Technik, Berlin.
7. Pahl G. (ed.) 1994: *Psychologische und pädagogische Fragen beim Konstruieren.* TÜV Rheinland, Köln.
8. Dörner D. 1976: *Problemlösen als Informationsverarbeitung.* Kohlhammer, Stuttgart.
9. Fricke G. 1993: *Konstruieren als flexibler Problemlöseprozeß – Empirische Untersuchung über erfolgreiche Strategien und methodische Vorgehensweisen beim Konstruieren.* VDI-Verlag, Düsseldorf. (VDI-Forschungsberichte, Reihe 1: Konstruktionstechnik/Maschinenelemente).
10. Müller J. 1990: *Arbeitsmethoden der Technikwissenschaften. Systematik, Heuristik, Kreativität.* Springer, Berlin.
11. Langner T. 1991: *Analyse von Einflußfaktoren beim rechnergestützten Konstruieren.* Thesis TU Berlin, Berlin. (In: W. Beitz (ed.) *Schriftenreihe Konstruktionstechnik,* Bd. 20).
12. Dörner D., Schaub H., Stäudel T., Strohschneider S. 1988: Ein System zur Handlungsregulation. Oder: Die Integration von Emotion, Kognition und Motivation *Sprache & Kognition* 7, 217–232.
13. Ehrlenspiel K. 1993: Denkfehler bei der Maschinenkonstruktion: Beispiele, Gründe und Hintergründe. In: S. Strohschneider, R. von der Weth (eds.) *Ja, mach nur einen Plan.* Huber, Bern, pp. 196–207.
14. Ehrlenspiel K., Dylla N. 1991: Untersuchungen des individuellen Vorgehens beim Konstruieren. *Konstruktion* 43.
15. Fricke G. 1994: Erfolgreiches individuelles Vorgehen beim Konstruieren, *Konstruktion* 46, 183–187.
16. Görner R. 1994: Zur psychologischen Analyse von Konstrukteur- und Entwurfstätigkeiten. In: B. Bergmann & P. Richter (eds.) *Die Handlungsregulations-theorie. Von der Praxis einer Theorie*. pp. 233–241.
17. Günther J., Frankenberger E., Auer P. 1996: Investigation of Individual and Team Design Processes. In: N.G. Cross, H.H.C.M. Christiaans & C.H. Dorst (eds) *Analysing Design Activity.* Wiley, London, pp. 117–132.
18. von der Weth R. 1994: Konstruieren: Heuristische Kompetenz, Erfahrung und individuelles Vorgehen. *Zeitschrift für Arbeits- und Organisationspsychologie* 38, 102–111.
19. von der Weth R., Frankenberger E. 1995: Strategies, Competence and Style – Problem Solving in Engineering Design. *Learning and Instruction* 5, 357–383.
20. Dylla N. 1991: *Denk- und Handlungsabläufe beim Konstruieren.* Hanser, München (Konstruktionstechnik, vol. 5).

21. Ehrlenspiel K., Bernard R., Günther J. 1995: Unterstützung des Konstruktionsprozesses durch Modelle. *Bericht über das Werkstattgespräch 'Bild und Begriff 3' 1995 in Seußlitz.* TU-Eigenverlag, Dresden.
22. Rückert C., Schroda F., Gaedeke O. 1997: Wirksamkeit und Erlernbarkeit der Konstruktionsmethodik. *Konstruktion* 49, 26–31.
23. Carroll J.M., Miller L.A., Thomas J.C., Friedman H.P. 1980: Aspects of Solution Structure in Design Problem Solving. *American Journal of Psychology* 95, 269–284.
24. Bach K. 1973: Denkvorgänge beim Konstruieren. *Konstruktion* 25, 2–5.
25. Kleist H. von 1925: Die allmähliche Verfertigung der Gedanken beim Reden. *Sämtliche Werke*. Volksverlag, Weimar.
26. Sachse P., Hacker W. 1997: Unterstützung des Denkens und Handelns beim Konstruieren durch Prototyping. *Konstruktion* 49, 12–16.
27. Sachse P., Hacker W. 1995: Early Low-cost Prototyping: Zur Funktion von Modellen im konstruktiven Entwicklungsprozeß. *Forschungsberichte 19, Institut für Allgemeine Psychologie und Methoden der Psychologie.* Technische Universität, Eigenverlag, Dresden.
28. Guindon R. 1989. The Process of Knowledge Discovery in System Design. In: G. Salvendy & M.J. Smith (eds) *Designing and Using Computer Interfaces and Knowledge Based Systems*. Elsevier, Amsterdam, pp. 2, 727–734.
29. Schroda F., Leinert S., Sachse F. 1996: Anforderungsstruktur (task structure) und Problemraum (problem space) – theoretische und empirische Problemanalyse. *Forschungsberichte 31, Institut für Allgemeine Psychologie und Methoden der Psychologie*. Technische Universität Eigenverlag, Dresden.
30. Beitz W., Hacker W., Timpe K.-P. 1997: Design Methodology: Learnability and Effectiveness. *Forschungsberichte 41, Institut für Allgemeine Psychologie und Methoden der Psychologie*. Technische Universität Eigenverlag, Dresden.
31. Beitz W., Timpe K.-P., Hacker W., Rückert C., Gaedeke O., Schroda F. 1997: *Konstruktionsarbeit studentischer Übungsgruppen. Empfehlungen für die konstruktionsmethodische Ausbildung an Technischen Universitäten.* Schriftenreihe Konstruktionstechnik 40, Berlin.

Event Scales and Social Dimensions in Design Practice

David F. Radcliffe

Abstract

One of the difficulties in comparing results across empirical design studies is the disparity in duration of the events studied and different levels of social complexity involved. This paper proposes a schema that characterises a piece of design work according to the event scale of interest (the duration of the event that you are paying attention to) plus the corresponding social dimension (the degree of socio-cultural complexity of the work setting). A Social Dimension – Event Scale framework is then constructed and used to map a diverse range of design studies conducted in natural settings. The same schema is used to highlight the use of physical objects including prototypes as carriers of design knowledge. Three types of prototype are described; impromptu, proof-of-concept and embodiment. The schema is also used to describe the nature of tacit knowledge. Three domains of tacit design knowledge – technical, commercial and cultural – are identified.

1. Introduction

Over the past decade, design researchers have conducted an increasing number of empirical studies investigating the way in which engineering and other designers practise their art. Some studies have been conducted in natural settings with designers doing their everyday work, e.g. [1-4]. These formal investigations using participant observation, ethnographic observation and video analysis of group work are complemented by anecdotal accounts of practice, e.g. [5-7].

Given the problematic nature of observing design *in situ* it was inevitable that many studies of designers are conducted in laboratory or otherwise contrived work settings [3, 8–11]. These studies usually involve students doing staged exercises of one or two hours' duration. It is usually difficult to compare results between these studies. The Delft Protocol Study [12] attempted to overcome some of these shortcomings by bringing design researchers from many countries and research traditions together to study a single data set, based on contrived exercises done by a team of professional product designers.

This flourishing of empirical research in design complements other scholarly, historical and autobiographical accounts of what it means to design [13–17]. Such accounts do not feature prominently in discourses on design research. This is unfortunate as many insights and much wisdom about the design process is lost. With notable exceptions, such as Nevil Shute's *No Highway,* there have been few fictional accounts of engineering designer – more is the pity.

One of the difficulties in comparing results across different design studies is the disparity between the settings, the nature of the work and the perspective and methodology of the researchers. For example, how do we compare the results from a multi-year study of a large industrial design project in a commercial setting with those from a study based on a few graduate students doing a two-hour protocol? Apparent design performance also depends upon who perceives it, a participant or an external observer, the stance they take, for example positivism or naturalism, and the observational tradition from which they come. We are left asking – how do we generalise findings or conclusions about the design process? How do we make sense of what may seem contradictory accounts?

Frankenburger and Auer [18] collated and characterised numerous empirical studies of design by grouping them according to two determinants: (1) the approach taken (either design-scientific or design scientific + psychological) and (2) the scale of the work (either a single designer or a group). The approach taken was divided into two sub-types; laboratory studies and those conducted in industrial practice. Their schema provides one basis for discriminating between design studies of different types.

This paper proposes an alternative schema for characterising engineering design practice and hence empirical accounts of design work. It is based on the notion of event scales and social dimension as an orthogonal pair that span the entirety of design practice. This schema can be used to map the type of design practice we are concerned with, how we perceive it and the internal and external influences on it. The concept is illustrated with examples from the Engineering Design Practice Group at the University of Queensland.

2. Design Characteristics

Design work depends not only on the capabilities of individual designers, how they work as a team, the tools they employ and their work environment, but also on the scale of the product development activities that we are considering. At its simplest, a piece of design practice can be characterised by the event scale of interest and its governing social dimension. This is illustrated in Figure 1. An analogy may be drawn with turbulence phenomena in fluid flow. What you see depends the event scale – 'the characteristic length' – you choose to focus on.

Both scales are highly non-linear, i.e. the social dimension is foreshortened as you go up the scale and the event scale is compressed as you move to the right. Both scales are qualitative. Each is defined in the following sections.

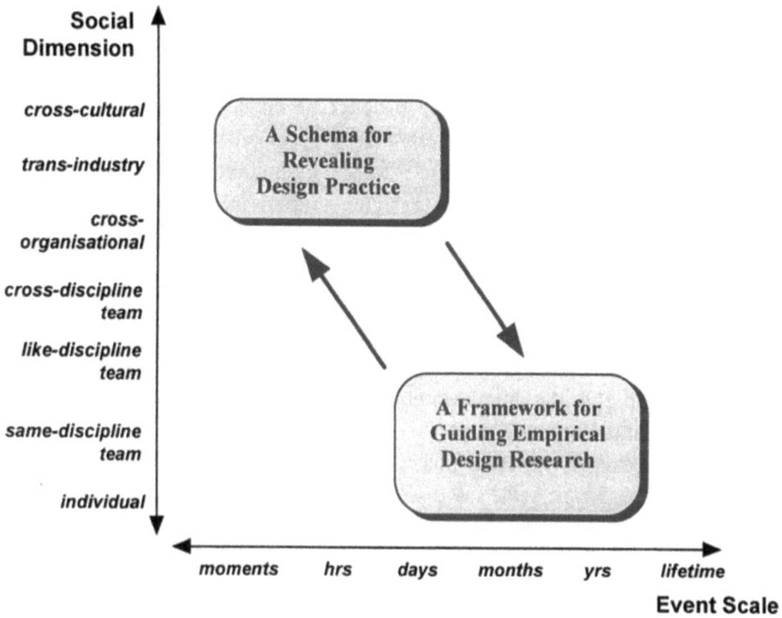

Figure 1: Mapping design characteristics.

2.1 Event Scales

The event scale is the characteristic time scale of the activity of interest to the design practitioner or the design researcher. These are different from the overall duration of the design project or product development task. For instance, an observer or participant may be concerned about moment-to-moment events such as a discussion over a prototype which is just part of a task that lasts several weeks or months. Conversely, we may be interested in the stages of personal and professional development of a designer over a career. While the moment-to-moment experiences of design accumulate to create large-scale events, these larger event scales have meaning beyond the simple summation of everything.

Models of design tend to focus on event scales of the order of hours, days or possibly weeks. Many everyday design tasks fit into this category. While projects may last months or even years, we often work on intermediate planning horizons of the order of days and weeks. In the process, many moment-to-moment processes occur outside the conscious attention of the designer. The sheer volumes of data have forced design researchers to concentrate on the moment-to-moment while bypassing the mid-scale events that characterise much of design practice.

Longer scale events, stretching out to a lifetime, have not been studied as part of design research. These more gradual trends go largely unnoticed, except at periodic times of reflection or at dramatic points of life change.

2.2 Social Dimensions

The social dimension is concerned with the relationships that exist between the designers, the other stakeholders in the product being developed and with the intermediate and ultimate work products. The level used reflects the degree of complexity of these relationships. At one extreme we have the individual designer working alone. The individual artist or designer-craftsperson is perhaps the only authentic example of this extreme; a single person who creates an artefact as a solitary act. While many excellent designers are portrayed as rugged individuals, e.g. [19], and exhibit a fierce independence, their creative efforts take place in a socially constructed work environment. Product development is an interdependent activity. The immediate work group may only comprise two or three people, but many others are pivotal to the development of a new product: the client, suppliers, the venture capitalist, the manufacturer.

At the other extreme, as we move to engineering design and product development on a global scale, the social aspects assume a cross-cultural dimension. Cultural here refers to indigenous culture, including language and customs as they influence the workings of a team or a whole project that is distributed around the world, located in several countries and conducted across different time zones.

Between these extremes there is a range of states that are of intermediate social complexity. These are represented on the vertical axis in Figure 1. The following descriptive titles and accompanying examples illustrate the continuum in the social dimension.

2.3 Observational Perspectives

Our appreciation of the social dimension is coloured by the observational perspective we use. We can view the social dimension from many different perspectives: from an engineering perspective, from several social science perspectives or from the personal perspective of a practitioner. No single perspective has a monopoly on the truth; it is through a combination of them that design can be fully appreciated.

Prescriptive models of design are the most enduring accounts of design from an engineering perspective. This technical view of product development as information processing fails to take account of the social nature of design. Engineering perspectives of design have tended to be limited to event scales of hours or days and usually assume either individual or same-discipline to like-discipline teams. Information system-based accounts of product development that deal with longer event scales are beginning to appear in the concurrent engineering literature. They usually ignore the social issues.

The social sciences offer many different perspectives. As you move away from the simple social dimensions, the social sciences contribute most to our understanding of issues. At the individual level, cognitive psychology can tell us something of the designer as information processor. On the other hand, for teams of various levels of complexity, behaviourists or psycho-dynamacists may shed light on the workings of the team [20]. Social psychologists and linguistics also have an obvious role here. While cultural anthropology [21] and specifically ethnography [22] offer insightful approaches to understanding design work and engineering culture, they have particular relevance to cross-discipline teams, cross-organisational and cross-cultural issues. Historical and sociological accounts offer broader, contextual perspectives in the trans-industry and cross-cultural domains. A range of case study techniques can be deployed here [23].

While we usually understand design research in terms of an observer (possibly a participant observer) we should not overlook the personal accounts of practitioners. These insider perspectives may be revealed through reflexive journals or personal diaries, enhanced by interviews or focus groups and gathered up in autobiographies. These complement the other external accounts of design.

3. A Framework for Empirical Design Research

The Social Dimension – Event Scale framework has been used in the Engineering Practice Research Group at the University of Queensland since 1992 as a means of mapping our diverse research projects. These projects have ranged through the exploration of the philosophical foundations of engineering practice [24–25] and the transition to professional work, to design education and empirical studies of prototying and DFA usage (Figure 2). All are concerned with engineering design or more generally, engineering practice. This mapping framework unites the work by presenting relationships between the different studies. It also helps identify gaps or research opportunities.

3.1 Applying the Framework

Each project is represented by a ellipse located at the intersection of the appropriate event scale and social dimension. The aspect ratio of the ellipse conveniently conveys the sense that, while projects typically span a range of event scales, they have a primary focus at the one aligned to the minor axis of the ellipse. In contrast, the social dimension is more narrowly defined in most cases. Many had multiple foci in terms of the design process of interest and all drew on multiple sources of data.

These studies were carried out principally with one enterprise in each industry over many years with a developmental perspective. One of the advantages of conducting multiple projects in close collaboration with a limited number of workplaces is the synergistic opportunities that arise. A subtle advantage of a protracted engagement is the level of trust and understanding that accrues over time. You can get close to an industry partner and begin to appreciate a totality of issues in a way that is simply not possible in short duration projects.

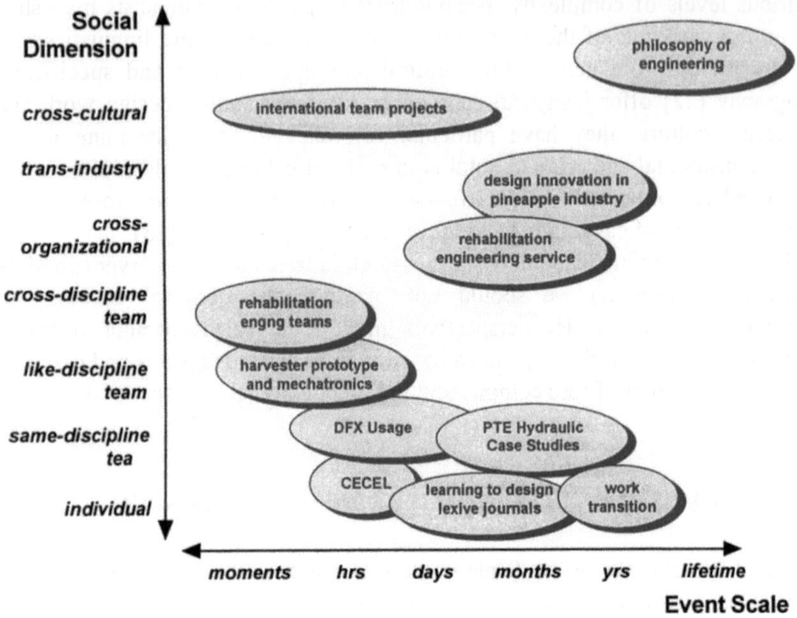

Figure 2: Engineering practice research group.

3.2 Illustrative Examples

A series of design studies was conducted with the pineapple industry. These involved the development of proof-of-concept prototypes for planting pineapples and for de-topping them at harvest. The design experience gained by two graduate students via these projects lead to two contrasting theses. One was on the systems level impact across an industry like the pineapple industry of the introduction of a new technology like a detopper [26]. The second thesis focused on the moment-to-moment actions of designers as they create and work with prototypes [27].

For several years, we have worked with the Rehabilitation Engineering Centre at the Royal Brisbane Hospital examining how their cross-discipline team designs and makes special seating inserts for disabled children [4]. This includes fine-grained studies of the moment-to-moment work of the team, specifically the relationship between talk and the use of objects and gesture [28].

A third industry research partner is a small manufacturing company that makes hydraulic cylinders. PTE Hydraulics is typical of many small manufacturing companies in Australia. Projects have included a study of planning for advanced manufacturing technology, a fundamental study of flexibility in manufacturing, a major new product development cycle, introduction of FEA analysis and QFD, and a study of information integration with a focus on the role of tacit knowledge. We have also observed the transition into such a company of one of our graduates.

One of the techniques used to capture this transition is the reflexive journal. We have been trialing its use with undergraduate students in design subjects as a means of developing their self-awareness as designers. This is part of a wider range of initiates in design education including the Contextual Experiences in Concurrent Engineering Learning (CECEL) project [29]. We have embarked on several international student design projects with the focus on the cross-cultural dimension of these teams and the development of relationships at a distance.

3.3 The Big Picture

These projects illustrate the utility of the Social Dimension – Event Scale mapping to distinguish between projects. It graphically represents the relationship between projects that might otherwise seem to be of a similar type and provide a 'big picture' summation of a diverse range of design research. This mapping contrasts with others used for characterising design or similar work practices. With the growing interest in computer supported cooperative work (CSCW), time and place are often used to identify four domains of team work, co-located and synchronous, co-located and asynchronous, non-co-located and synchronous, and non-co-located and asynchronous. These can be accommodated within the Social Dimension – Event Scale paradigm, while preserving time and place as additional variables.

4. Mapping Aspects of Practice

The Social Dimension – Event Scale schema can also be used to display understandings of engineering practice that arise from observations in natural design settings. The studies outlined above reveal a variety of ways in which engineering design knowledge is manifested in practice. In simple terms it can be embodied or embedded. In the former it is an explicit part of the design environment, potentially available to all parties. In the later it is concealed within the means of doing design. Knowledge is a contested term. It is possible to have endless definitional debates about what we mean by the word. The focus here is on knowledge as actionable information in the hands and minds of designers.

4.1 Embodied and Embedded Design Knowledge

Knowledge is embodied symbolically in formal and informal documents, in text, in sketches and in drawings, physically in products, components, catalogues and prototypes, virtually via 3D solid models on computer, and abstractly in, for example, models of material performance. It is preserved in articulated design principles. Design knowledge is embedded collectively in standards, conventions and in unarticulated design principles. It is embedded personally in tacit understandings, and expressed in individual appreciations and talk during design. It is told in our stories of design [30]. Examples of design knowledge embodied in physical artefacts and embedded as tacit understandings are discussed below.

4.2 Physical Objects in Design Practice

The importance of sketching in conjunction with thinking is described by Hacker and colleagues [31]. They also point to the importance of physical representations such as models and prototypes to aid design thinking. From our studies it is possible to distinguish at least three types of prototypes, *impromptu*, *proof-of-concept* and *embodiment*, and two classes of physical artefact, *familiar* and *unfamiliar*. The regions of significance for each in engineering practice is mapped in Social Dimension – Event Scale space in Figure 3.

4.2.1 Impromptu Prototyping

Impromptu prototyping refers to the use of things-at-hand to express physically a design idea, embodiment or detail. It includes the use of human gesture to augment available materials as a means of doing this. Talk over, with and through physical objects is an integral part of most design conversations. Designers and other team members do it all the time. It may take place in a scheduled design meeting where an existing component or whole product is used as a surrogate to convey to the other team member a new proposal. In the field, impromptu prototyping is sometimes used with quite large artefacts to achieve a similar end. This is often associated with the clarification of issues about the context in which a new device is to be installed and operated.

It happens on the spot, in chance conversations, in the corridor or by the workstation. The materials at hand may be everyday work pieces, the stuff on the desk, a ruler, a disk box, a water bottle. In episodes of solitary work, individuals are seen to use impromptu prototyping, presumably as a means of clarifying or testing their own thoughts.

Impromptu prototyping happens in the moment. It is usually short lived. As the technical diversity of the people involved increases, i.e. the social dimension of design increases, the corresponding event scale is longer. Ideas cannot be expressed as easily in say a cross-discipline team as is the case with a like- or same-discipline team. When it seems clear that others are not understanding an idea expressed by impromptu prototyping, variants are developed. Sometimes different artefacts are recruited into the process, mimicry is used, talk – especially the use of metaphor – is used to complement the physical construction. In a cross-discipline team people may return to a variant of the impromptu prototype some time later. Nevertheless they remain essentially ethereal objects. They are conduits for design conversations, not fixtures. While aiding mutual understanding and shared appreciations they also invite ambiguity, thus ensuring potentiality in the device being developed.

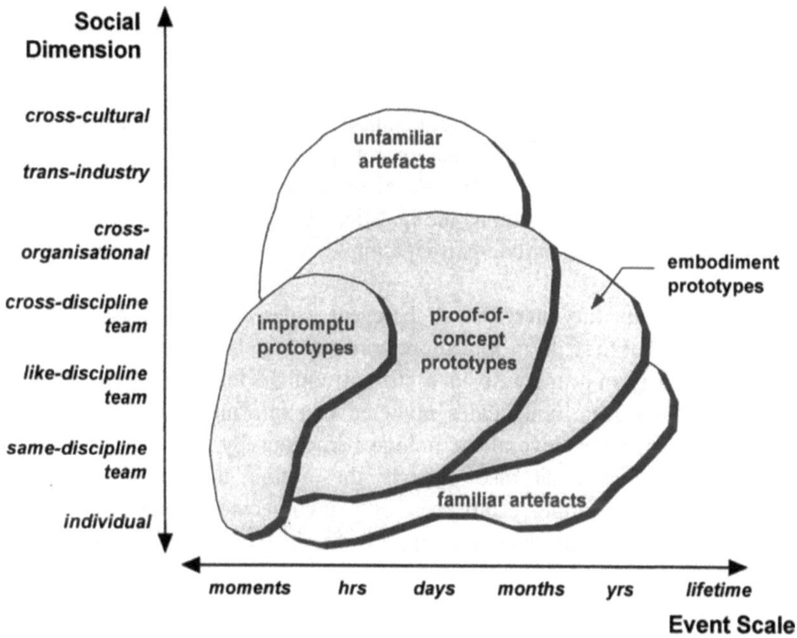

Figure 3: Physically embodied knowledge in design practice.

4.2.2 Proof-of-concept Prototypes

Proof-of-concept prototyping is about expressing a design concept in sufficient detail to enable the idea to be tested in hardware. However, these prototypes accomplish much more. Based on his work with the pineapple planter and harvester prototypes, Glen Horton [27] draws the distinction between a prototype as a representation and as a device-in-action. He points out that one advantage of a physical, device-in-action prototype as means to test a concept is that it will behave in the physical environment as it must, i.e. it is subject to the Laws of Nature. This distinguishes it from a virtual prototype which only responds to the properties it has been given. Thus prototypes reveal much more than we may have planned for when we conceived them as a means to prove a concept. The trick is to distinguish which of the unexpected behaviours they exhibit are an artefact of the prototype embodiment, and hence of no significance in the final device, and which are telling of the performance of the final device.

Device-in-action prototypes also present themselves as accessible sites for impromptu prototyping, especially suited to team members who are not enculturated in the language and symbolic representations used by engineers (e.g. sketches and drawing conventions). Thus they can be modified in the mind and modified in their embodiment.

In mapping this aspect of practice, it became clear that proof-of-concept is limited to social dimensions around the cross-discipline and cross-organisational level. Beyond this level it is unrealistic to think that the participants are going to be concerned with that sort of technical detail. Proof-of-concept demands a more immediate level of engagement. Such prototypes may be used as 'proof-of-competence' prototypes at the higher social dimensions, the trans-industry and cross-cultural. Here it is more about the question 'Can your team or firm do the job?', not so much about active participation in the creative potential of the prototype.

As the social complexity increases so the event scales become larger, for much the same reasons that they do in impromptu prototyping. In addition, the level of technical complexity may increase in a cross-discipline team, for example in a mechatronics device. The time-scales involved in exploring a proof-of-concept prototype that incorporates several technologies are naturally longer.

The mapping (Figure 3) also reflects the notion that proof-of-concept prototypes, unlike impromptu prototypes, only involve individuals in so far as they are a member of a team working on part of a device that will be tested in a team context.

4.2.3 Embodiment Prototypes

As the name implies, an embodiment prototype contains a substantial amount of the structure, functionality, interfaces and finish of the production device. Like the device-in-action prototype, such prototypes are subject to all aspects of the natural environment and respond accordingly. Being a more extensive embodiment of intent they are a significant repository of design knowledge. Accordingly they are associated with a wide range of event scales from hours and days, extending out to months or sometimes years.

In the case of like-discipline or cross-discipline teams, e.g. mechatronics design, embodiment prototypes are useful over the widest range of event scales. They are semi-permanent objects. Their relative completeness and resolved ambiguity compensate for the lack of familiarity of the different members with embodied technology outside their home discipline. This is especially so for cross-discipline teams. It makes them valuable in event scales that might be months in duration, for instance 'next model' discussions and product planning exercises. They may also be helpful starting points for impromptu or proof-of-concept prototyping explorations of next generation products.

In devices developed by same-discipline teams, much of the knowledge carried in an embodiment prototype is also embedded in their practices or readily available through company design resources. Accordingly the event scale of such prototypes is smaller for these teams. Its utility is limited to its immediate role as, say, a pre-production or field test prototype.

4.2.4 Familiar and Unfamiliar Artefacts

Prototyping activity takes place within the natural and built environments, a world filled with physical objects. These naturally occurring and human fashioned artefacts inform and ground prototyping activity. This is especially so for impromptu prototyping. From the point of view of participants in the design process, these influences can be viewed as either familiar or unfamiliar artefacts. Our ability to prototype depends on the extent of our familiarity with a wide range of artefactual resources. Physical artefacts are taken to include existing commercial and in-house components, products and systems.

The term familiar is defined generously. It is not limited to those things with which someone has had first-hand experience. Rather it is all those artefacts that are within the cultural, social and professional perview of the person; things that you are readily capable of knowing even if you have not seen one. An artefact is familiar if when you first come across it your background and experience enable you to inquire of it and reason about it and quickly come to an appreciation if its essential workings. Unfamiliar artefacts are those for which you have not been prepared. They do not fit into your world of understandings. When explained they may make partial sense to you; nevertheless they do not sit comfortably with your world.

Familiarity is contextual and situated. A pineapple may be familiar in the sense of a 'piece of fruit' to both a pineapple farmer and to an engineer designing a pineapple harvester. As an artefact that is to be de-topped (i.e. to have the leaves and top of the stem broken off the top of the body of the fruit) the farmer who has years of experience of doing this by hand is very familiar with the pineapple while the engineer is unfamiliar with it in that sense.

A vital part of the transition to practice is the expansion of the domain of familiar things, in a design sense. Over the years, familiarity grows. It is more than the simple sum of experiences but is deeply internalised by the Gestalt's of appreciation and cognitive integration that occur periodically. Its event scale encompasses significant episodes that occur in the moment and over a lifetime. Almost by definition an individual working in a same-discipline environment will be exposed predominately to familiar artefacts. With widening exposure and a willingness to extend one's comfort zone, the limits of familiarity are pushed back into high levels in the social dimension. This growth into more elaborate social dimensions often goes unnoticed, or at least unreflected upon.

Unfamiliar artefacts are prominent at high levels of social complexity. They are a source of uncertainty, potential misunderstandings and even irritation in cross-discipline teams and in more complex design settings. They are also a potential source of design ideas – the stimulus of the unexpected. Their event scales are bounded as the impact of their unfamiliarity is finite in duration. Either an understanding is developed to the point where they become familiar and part of the design fabric or they become an impediment to progress. In this case their unfamiliarity is of little further design value.

4.3 Patterns of Tacit Knowledge Usage

Tacit knowledge is a slippery concept. By definition something that is tacit is not able to be articulated, written down or codified. Polanyi [32] defined tacit knowledge as 'knowing more than we can tell'. The late Donald Schön [33] expressed it thus – 'Our knowing is ordinarily tacit, implicit in our pattern of action and in our feel for the stuff with which we are dealing'. As such it cannot really be isolated. Nevertheless we know from experience that there are things like *know-how* that enable design work to happen, even when we cannot describe them explicitly. We can only describe in general terms the nature and impact of tacit knowledge in a design context.

Tacit understanding can be thought of as a substrate that supports design activity. It makes design work possible; a necessary but not sufficient condition for it. It is integral to the use of prototypes as part of design. The whole notion of familiar and unfamiliar artefacts is predicated on the tacit knowledge of the design participant. This knowledge is embedded in the individual and collective wisdom of designers. It is diffuse.

For these reasons tacit knowledge is mapped onto the Social Dimension – Event Scale schema in a very general way, as depicted in Figure 4. Three groups of tacit knowledge are identified, regions on the map that contain related types of knowledge. These overlapping domains of tacit knowledge are referred to here as 'technical', 'commercial' and 'cultural'. Each domain contains a group of elements representative of the myriad of abilities that might together describe the domain. Each is mapped at its characteristic location using the schema. In reality, many of these facets apply across large tracts of the map.

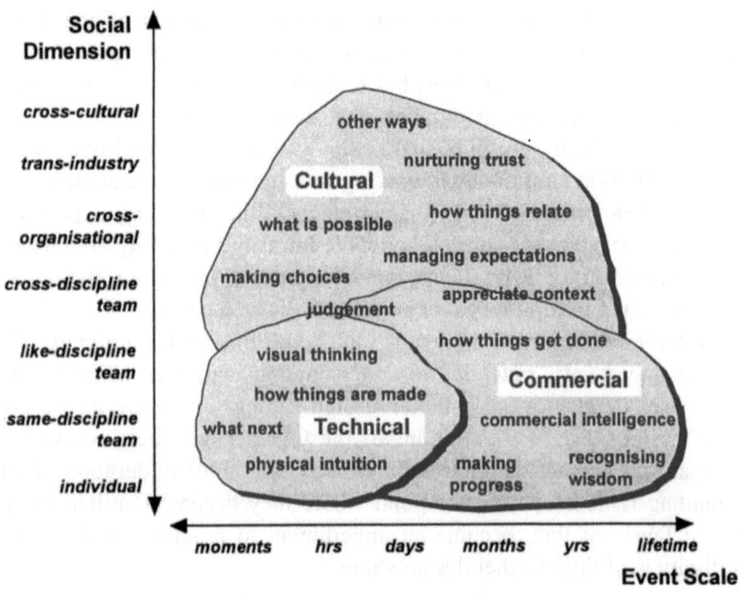

Figure 4: Tacit knowledge in design practice.

Tacit 'technical' design knowledge falls into the lower left-hand end corner of the map. It is this form of tacit knowledge that makes the transition from university to engineering practice smooth or difficult. Some is embedded in later year subjects, like design projects. Some is made explicit in management subjects, in as far as this is possible. Nevertheless it often comes as a shock to the novice designer. 'How things are made' exemplifies the grey area between what can be made explicit and what remains tacit. Students can read about or see unit manufacturing processes. Through these they learn something explicit about the processes. Through hands-on experience they may even acquire some tacit knowledge about operating machine tools.

However they do not acquire the tacit knowledge that is necessary to 'get things made'. This knowledge goes beyond the mechanics of the process, an understanding of tolerances or even an appreciation of manufacturing operations. It is those intangible things which in a real work environment make production possible. These are a blend of technical and management issues in a bounded social context and they have relatively short event scales. Numerous examples of tacit 'technical' knowledge appeared in the projects studied.

In contrast, tacit 'commercial' knowledge, located in the lower right-hand region, is less localised and deals with learning 'how things get done' in a system of work and production. They have larger event scales and are slightly more complex in the social dimension. It involves appreciating the wider context of design work. 'Commercial intelligence' is about developing an appreciation for the commercial imperatives. One indicator of this is the shift from an 'ego' design orientation (technical elegance primarily for the benefit of the designer) to a 'market' design orientation (satisfising designs that meet customer expectations and make a profit).

In the upper part of the map, the region of relatively high social dimension, is tacit 'cultural' knowledge. It is about achieving possibilities against a background of heightened uncertainty due to contrasting ways of looking at the world. Trust is the vital social glue that binds design. It is often taken for granted in a small organisation where relationships have been built up unnoticed over many years. In transitory cross-discipline teams or a cross-organisational project, the relationship is to be formed and it may be much more delicate and fragile. Managing expectations and knowing what is possible, given the possible fragility of relationships, relies on tacit design knowledge. As the social complexity increases to that of a trans-industry or trans-national consortium, cultural sensitivities, other ways of doing business, special matters of commercial confidence and intellectual property must be attended to.

The three domains of tacit knowledge are interdependent in that technical knowledge occurs within the commercial sphere and both are informed by the individual's cultural tacit knowledge. Each is being continuously updated. This gives rise to design judgement and professional wisdom.

5. Discussion

Barkan and Iansiti [34] described prototyping as a learning tool for product development. They classified prototypes according to the stage in the product development process at which they were used and the 'degree of representativeness' or prototype integrity. In sequence, commencing with project initiation, these different prototypes were labelled respectively as mock-ups, computer simulations, sub-systems prototypes, mechanical prototypes, engineering prototypes and production prototypes. These normative labels are based on the form or fabric of the prototype. They do not tell us anything about the relationship between the prototype and the design process and how it influences it.

Our earlier papers [4] and subsequent studies by others [35] identified and described the use of physical objects – not prototypes *per se* – in design work. This paper develops these observations into the notion of impromptu prototypes. The term impromptu was chosen in part because of its connotation of opportunity taking, of seizing the moment. In contrast, proof-of-concept prototypes involve more commitment, and hence longer event scales. Embodied prototypes, being the most committed of the three, have the longest event scales.

There is a growing recognition of the important, if often unseen, role that tacit knowledge plays in engineering practice including design and product development. Wagner [36] proposed three categories of tacit knowledge on which 'practically intelligent behaviour in professional and management careers' depend in part: tacit knowledge about managing self, about managing others and about managing career. There is a degree of similarity between these three and the three domains of the tacit knowledge mapped here (Figure 4). The schema presented has the advantage that this mapping is based on two independent variables that are common to each category of tacit knowledge. This enables the relationships between them to be explored.

Designers and other stakeholders in the product development process bring with them personality traits, a personal history and a set of values – all tacit. They operate in a world containing current design practices (some tacit, some explicit), within a pre-existing work culture, coloured by professional traditions and values. They have both peer relationships and personal expectations. Tacit knowledge takes shape amidst this social milieu. An individual's tacit knowledge is unique. Gathering evidence of its existence (for it cannot be measured directly) is always going to be problematic. While the empirical basis is of necessity limited, the potential of the schema is clear.

Underpinning the paper is the whole concept of measuring design practice in terms of its characteristic social dimension and the event scale of interest. Neither of these metrics is easy to define precisely. Nevertheless, in combination they create a useful framework or schema for mapping aspects of design practice and research. The ecology of engineering design and new product development is a complex one. The Social Dimension – Event Scale offers a new approach to making sense of this reality. It allows empirical design studies to be delineated on the basis what setting the designers are in and what things they are attending to.

6. Closure

The potential of the Social Dimension – Event Scale schema to map and hence make sense of quite diverse pieces of design research is demonstrated. The type of prototype used depends on the prevailing event dimension and social scale. By their nature, impromptu prototypes arise in the instant. Their impact depends upon the range of materials at hand. Proof-of-concept and embodied prototypes have impact over much longer event scales but they are less flexible. Tacit technical knowledge has most impact in short event, low social complexity tasks. Tacit commercial and cultural knowledge become more influential as the design task involves more stakeholders and the consequences of the design become more critical. As an engineer's career develops, tacit commercial and cultural knowledge displace tacit technical knowledge as key enablers.

Acknowledgements

I acknowledge the contribution and support of my colleague Tim Holt. The ideas developed in this paper have their origins in many stimulating discussions with members of the Engineering Practice Research Group (EPRG) including Margot Brereton, Bob Gammie, Brian Mathesien, Sandra Munoz and Fiona Solomon. I am especially indebted to stimulating thesis discussions with Louise Hearder, Glen Horton, Gil Logan, Michael Lucas, Jinglu Qi, Peter Slattery, Philip Teakle, Brett Watson and Paul Wong.

References

1. Hales C. 1987: *Analysis of the Engineering Design Process in an Industrial Context.* Grant Hill Pub., Eastleigh.
2. Bucciarelli L.L. 1988: An Ethnographic Perspective on Engineering Design. *Design Studies* 9, 159–168.
3. Minneman S.L. 1991: *A Social Construction of a Technical Reality.* PhD Thesis, Stanford University.
4. Radcliffe D.F., Slattery P. 1992: Emergence, Learning and Interaction in a Cross-Discipline Design Team. *DETC 92, ASME DE* Vol 42, 123–130.
5. Cuff D. 1991: *The Story of Practice.* MIT Press, Cambridge, Mass.
6. Radcliffe D.F., Harrison P. 1994: Transforming Design Practice in a Small Manufacturing Enterprise. *DETC 94, ASME* Vol DE-68, 91–98.
7. Cross N., Cross A.C. 1996: Winning by Design: the Methods of Gordon Murray, Racing Car Designer, *Design Studies* 17, 91–107.
8. Ullman D.G., Dietterich T.G., Stauffer, L. 1988: A Model of the Mechanical Engineering Design Process Based on Empirical Data. *AIEDM* 2, 33–52.
9. Tang J., Leifer L.J. 1991: An Observational Methodology for Studying Group Design Activity. *Res. Eng. Des.* 2, 209–219.
10. Radcliffe D.F., Lee T.Y. 1990: Models of Visual Thinking in Novice Designers. *Design Theory and Methodology, ASME*, Vol. DE-27, 145–152.

11. Christiaans H., Dorst K. 1992: An Empirical Study into Design Thinking. In: N. Cross et al. (ed.) *Research in Design Thinking.* Pub. Delft University Press, pp. 119–126.
12. Cross N., Christiaans H., Dorst K. 1996: *Analysing Design Activity.* John Wiley and Sons, New York.
13. Schon D.A. 1983: *The Reflective Practitioner, How Professionals Think in Action.* Basic Books, New York.
14. Petroski H. 1985: *To Engineer is Human: The Role of Failure in Successful Design*, Macmillan, London.
15. Vincenti W.G. 1990: *What Engineers Knows and How they Know It: Analytical Studies from Aeronautical History.* The Johns Hopkins University Press, Baltimore.
16. Ferguson E.S. 1992: *Engineering and the Mind's Eye.* MIT Press, Cambridge, Mass.
17. Shute N. 1954: *Slide Rule, Autobiography of an Engineer.* Heinemann, Melbourne.
18. Frankenberger E., Auer P. 1997: Standardized Observation of Team-work in Design, *Research in Engineering Design* 9, 1–9.
19. Barker R., Harding A. 1992: *Automobile Design: Twelve Great Designers and their Work.* SAE, Dearborn, Mich.
20. Smith R.E. 1993: *Psychology.* West Publishing Co., St Paul, Minn.
21. Kessing R.M. 1981: *Cultural Anthropology: A Contemporary Prespective*, CBS Publishing, New York.
22. Hammersley M., Atkinson P. 1983: *Ethonography: Principles and Practice*, Tavistock Pub., London.
23. Yin R.K. 1994: *Case Study Research: Design and Methods.* 2nd edn. Sage Pub., Thousand Oaks, Calif.
24. Solomon F.L. 1997: *Towards a Philosophy of Engineering.* PhD thesis, The University of Queensland, Australia.
25. Holt J.E. 1996: On the Nature of Mechanical Engineering – an Engineering Ethos, *Int. J. Eng. Ed.*, 12, 333–338.
26. Buchanan M. 1995: Approach to Engineering Innovation in Agriculture. *Trans. Mech. Eng.*, 20, 293–298.
27. Horton G.I. 1997: *Prototyping and Mechanical Engineering.* PhD Thesis, The University of Queensland, Australia.
28. Logan G.D., Radcliffe D.F. 1997: Role of Artefact in Rehabilitation Engineering Design. *Proc RESNA 97.*
29. Radcliffe D.F. 1996: Exploring the Principles of Concurrent Engineering. In: *Manufacturing Education for the 21st Century.* SME, Vol III, pp. 233–236.
30. Lloyd P. 1998: *This volume*, pp.113–123.
31. Hacker W., Sachse P., Schroda F. 1998: *This volume,* pp. 205–216.
32. Polanyi M. 1966: *The Tacit Knowing.* Routledge & Kegan Paul Ltd, London.
33. Schön D.A. 1983: *The Reflective Practitioner.* Basic Books, New York.
34. Barkan P., Iansiti M. 1993: Prototyping: A Tool for Rapid Learning in Product Development. *Concurrent Engineering: Research and Applications* 1, 125–134.
35. Harrison S., Minnemen S.L. 1994: A Bike in Hand: a Study of 3D Objects in Design. In: N.Cross, H. Christiaans & K. Dorst (eds) *Analysing Design Activity.* John Wiley & Sons, New York, pp. 417–436.
36. Wagner R.K. 1987: Tacit knowledge in Everyday Intelligent Behaviour. *J Business and Psychology* 1, 301–312.

Information Processing and Storage during the Design Process: The Use of a Flexible Information System[1]

Henry H. C. M. Christiaans & Jost van Andel

1. Introduction

One of the characteristics of problem solving within the domain of industrial design engineering is the typical strategy that designers use in their information gathering. First, every designer will (consciously or unconsciously) build up a personal archive of his own, partly stored in long-term memory and partly based on external sources. When the designer has some knowledge on the topic he often tends to rely on what he already knows and/or simply reacts on already existing solutions. He understands incoming information to the extent that it conforms to his knowledge structure or ways of knowing. Information is filtered and modified by the existing knowledge structure so that it fits with someone's understanding of things. Sometimes, the designer can even be blocked by past experience. Second, in making use of external sources designers mainly look for technical information, while user-related information is often difficult to find or is given low priority. And if there is any user information at all, it is often confined to physical data, see [1]. Third, much of the external information is not very accessible, such as catalogues, directives, database software programs, picture collections etc. These sources often have a static, invariable and biased character and through this nature might be an impediment for use.

The consequence is that designers are constrained in their solution-seeking behaviour. Therefore, it seems fruitful to look for tools and media through which the designer can look for and process information in the most efficient way, a way to overcome the limitations of working memory – a topic discussed by Hacker *et al.* in this volume.

The computer becomes more and more a tool for designers. Mostly it is used as a drawing aid and as a text editor, but since the rise of Internet facilities the computer is an important information source as well. A number of studies show how the processing capabilities of the computer influence the mental

[1] The authors wish to thank Gijs Verrest and Els Fonteyne for their help with the database, the experiments and statistical matters.

representations and cognitive processes of learners [2]. But the effectiveness of the computer as an information source will depend on the characteristics of the external environment. The design process, like the learning process, is sensitive to the availability of specific information at a given moment, the duration of that availability, the way the information is structured, the ease with which it can be searched, etc. Research on the influence of these environmental variables is still incomplete. Therefore, the study described in this paper looks for the effects of factors that influence the use of a computerised information source. The factors that in our opinion play an important role in the context of industrial design engineering will be described in the next section.

2. The Use of a Computerised Information System

2.1 Interface between the User and the Computer

A good information system needs an interface between the user and the computer that allows for an optimal focus on the output function: the search for information. The software and the data must be applicable to the task at hand. It has been suggested that well-designed systems promote an effective mental model for the user of how the system operates [3]. Mental model refers here to both the way in which the information is structured and operational issues such as the menu structure. The user adopts a conceptual model of computer operations that may relate abstract ideas to concrete issues. The menu structure should be congruent with the user's cognitive structure. Another cognitive factor has to do with the mental effort by the user of the system while performing a design task. The system should reduce the time it takes for the user to perform the task. Our goal was to allow the user to concentrate on the relevant aspects of the design task and not on system operation. Hence, the menu selection should be such that response input was straightforward, minimising subjective operating complexity, and that the user would not be forced to recall commands from memory.

2.2 The Structure of the Information System

The structure of the database generally dictates the way in which users may search for information. Because of the growing popularity of Internet and WWW, hypertext is used more and more as a flexible way to link pieces of information that in one way or another are related to each other. From the literature it is known that hypertext structures allow a topic to be explored in multiple ways using a number of different concepts or themes [4]. Recent studies reported by De Vries [5] and De Vries & De Jong [6] showed that in the domain of architectural design an information system with a hypertext structure was mainly advantageous for divergent tasks such as browsing for new ideas. For convergent tasks, such as searching for specific information, a hypertext did not appear to have specific advantages.

2.3 Verbal and Visual Material

An important question in the lay-out of an information system is the use of text and pictures. Research into the effect of the use of two modalities is mostly performed within the educational setting. However, even in this field no firm theory has been developed yet. But research supports the assumption that student learning is affected positively by presenting text and illustrations together. This contiguity effect yields that students are better able to transfer what they have learned from a science book when text and illustrations are presented next to each other on the page rather then when they are isolated from each other. It might also facilitate retention by all readers. As an extension of the 'dual-coding theory' [7] Mayer & Sims [8] describe how visually and verbally presented material might be integrated within the learner's working memory during learning. Together with verbal encoding (building a verbal representational connection between an external and an internal representation) and visual encoding, referential connections between the two mental representations are built. For example, with an explanation in an information system of how a specific part or tool works, the designer can build referential connections between visual and verbal representations of the essential parts [8]. Other studies show that readers use the pictures to evoke a schema that serves as a preliminary mental model of the situation. Subsequently it seems that the text carries the primary message while the pictures are used to map this information onto this preliminary mental model [2]. A study by Van Andel [9] on the effect of different types of information on an architectural design task showed that the combined presentation of numerical and visual information had a positive effect on the amount of detail and the elaborateness of students' intake of information.

2.4 The User as an Active Participant

In educational literature on learning and teaching strategies, the issue of active involvement of the learner stands for an approach known as 'constructionism'. In brief, constructionism holds that students acquire knowledge by individually building their own knowledge structures rather than by passively receiving them from some knowledge source, such as a teacher [10]. This active processing of information into knowledge has proven to be a much more effective way of learning. The process of integrating prior knowledge and new knowledge involves elaborating information. Elaboration is said to serve several purposes: it will augment meaning making by adding examples, details and analogies to the existing information, and it will facilitate comprehension by creating a context for constructing meaning [11].

We assume that there is a similarity between a design process and a learning process, because it is an active constructive process whereby the designer strategically manages the available cognitive resources to create new knowledge. This is done by extracting information from the environment and integrating it with knowledge already stored in memory. However, it is also known that not all information is processed and stored. A study by Christiaans [12] showed that information gathered during the design process was hardly processed or even used in solving the design problem. Partly it was a consequence of the subject's problem-solving approach, but another reason could be the way in which this information was offered to the designer. Subsequently an information system should allow for this active operating of the user.

Research shows that information ordered by means of hypertext facilitates cognitive flexibility because the user can choose his or her own way through the information [4]. But even then the information can be presented in different ways. One option is to present the information system as a passive database, structured by the authors with freedom to navigate through it in whatever way the user may want. Another option is to allow the designer to add his or her own information and construct his/her own relationships. As Kozma [2] states, such systems can be made to correspond to the processes that learners use when constructing interrelationships among concepts in real memory. They provide an explicit model of information representation that, under certain conditions, learners may come to use as mental models of their thinking.

If the user is allowed to adjust the information system, user-friendliness and effectiveness are important factors. This involves the acquisition of knowledge about the system, skill in using the system and the ability to accommodate changes [13].

2.5 Teamwork

Teamwork in design is common practice and is fundamentally a social process. So far, most design studies on supporting design tools focus primarily on individual activity. As Minneman [14] states, tools for group design activity are virtually non-existent. Hence this study also aims at investigating the use of an information system by a team. It is important then to observe how a design team will work with the tool we offer, especially from a group dynamics point of view: What role does the information from the system play while the designers negotiate about their opinions? Does the information facilitates communication? What about the designers' prior knowledge? Designers in a team have different knowledge bases and may come up with arguments that will not always be understood by other team members in the way it is meant [15]. Therefore, Minneman [14] argues that the social exchange in the group should work to establish a 'shared understanding'.

3. Research Questions

The first question of the research project is: What is the effect of a personal information system on the use of both technical and user-related data in the design process and in the design itself?

A second question is: Is there any difference in effect of information storage on the basis of richness of the data? This question relates to the use of pictures together with text. Both questions will be studied in an explorative way.

Third, is there any relationship between the adjustability of the information system, according to individual needs and wishes of the designer, and the extent to which the information is used and integrated into the design? Adjustability concerns the database structure and the information itself. According to our theoretical notions we expect that adjustability will have a positive influence on the processing of information and thus on designing.

Fourth, is there any relationship between the characteristics of the team and the use of the information system? And in reverse, has the (use of the) information system any influence on the team behaviour?

Besides these general questions, we are interested in the way in which the information system is used. For example, are users dictated by the structure of the system? Do they scan menus in an idiosyncratic manner, glancing across the list of alternatives, hoping to light upon the desired alternative? Are they scanning in a systematic fashion, and if yes, how?

A small-scale study is conducted in which this broad range of aspects will be looked upon mostly in an explorative way.

4. Experiment

4.1 Subjects

Third-year and fourth-year students from the faculty of Industrial Design Engineering at Delft University were e-mailed and asked to participate in a contest organised by a Dutch bicycle manufacturer. They were encouraged to register as a team. In order to get six teams of three students each, 18 students were selected from a total of 27 applicants. Half of them actually registered as a team (three teams in total). For reasons of uniformity only fourth-year students were selected. Because of loss of part of the information from the three experimental groups, two months later a new selection was made after the same kind of e-mail message. Another 9 students, divided into three groups, enrolled in the experiment. Six students registered as members of one of two teams. The members of one team were third-year students.

All students were rewarded with one credit point (40 hours) and with the prospect of winning the contest.

4.2 Design Assignment

In order to tune the information system to a design problem the first step in the study was to define a design assignment, which would force the use of external information and come close to the core activity of industrial design (engineering): bringing together the intentions of the designer, the users' needs and experiences, and the system image of the product to be designed. In order to increase the reality of the design brief and the motivation of the participants, the project was presented as a contest organised by the bicycle manufacturer SPARTA. A bicycle for people of 55 years and older (55+), a topical subject in The Netherlands, was chosen as the design assignment. Both technical and functional information, and specific knowledge on the users, are essential in the solution of this design problem. Details of the assignment are available from the first author.

4.3 Information System

An Internet hypertext information system was developed with information on several aspects regarding the assignment. The Internet WWW browser at Delft University is Netscape.

The structure chosen in our information system is called a 'connected graph menu structure': acyclical with cyclical subgraphs (see Figure 1). Higher levels in the hierarchy pertain to divisions that are more global in nature. Lower levels move toward a stepwise refinement. Nested classifications follow this plan.

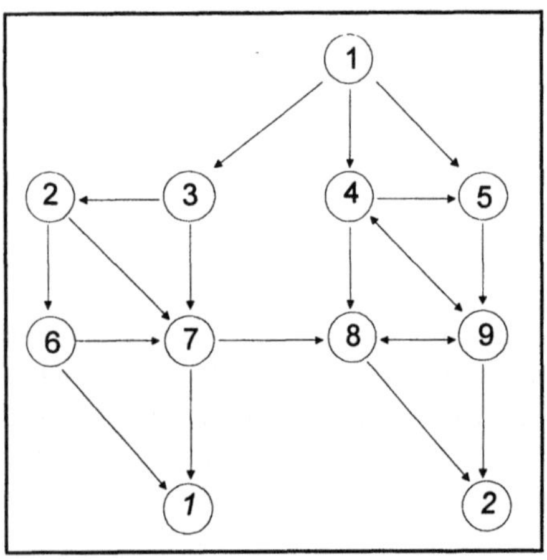

Figure 1: Asymmetric connected graph menu structure: acyclical with cyclical subgraphs [16].

According to expectations about how designers should look for information, the main menu started with three general categories: (1) information on the manufacturer and the market, (2) information on bicycles and (3) information on the user. A second level of navigation pages referred to either bicycle-related or user-related topics. The paths leading to the general issues (manufacturer, bicycle, user) differ in length. There are some multiple paths between two nodes also. Moreover, two special function keys, 'main menu' and 'dictionary', were superimposed on the connected graph menu system and could be activated at any point in the system. Links to other pages are blue-coloured and placed at the bottom of each page, preceded by: 'See also'. Underlined, blue-coloured words in the text refer to the dictionary. In using the information system the standard facilities of Netscape could also be used by the subjects, such as the buttons 'back', 'forward' and 'bookmark'.

In the information system both text and pictures are used. Except for the navigation pages, 48% of the pages have text together with a picture which can take the form of a symbolic representation, such as a graph or a real world image (in colours). None of the 'navigation' pages had an illustration included.

4.4 Measurements

Several instruments were used to measure the different aspects regarding the design process, the use of the information system, the designers and the team:

- Logbook. A logbook was prepared by the research team using three categories: (a) phase (problem analysis, synthesis, evaluation etc.), (b) activities (sketching, consulting the information system, writing the report etc.) and (c) topics (divided into three main headings: design characteristics, internal and external characteristics) [17]. Every quarter of an hour the team had to fill in the logbook by just marking one or more words from these three categories. In this way they could also indicate activities or topics dealt with simultaneously.
- Log-files. The server with which the computers were connected kept record of the 'pages' of the information system that were consulted by each team. These log-files were used to analyse the amount and kind of information asked for, the sequence and the time spent on each page.
- A Competence Questionnaire [18]. This questionnaire is already used in studies on design activities [19], and measures the factors: heuristic competence, emotional load, problem-solving performance, regression and resignation. This questionnaire was filled in during the instruction at the start of the two-day period.
- Group Dynamics Questionnaire. In order to control for differences in team composition, a questionnaire was developed regarding the topics: leadership, group climate, conflict handling, co-operation and efficiency. This questionnaire was filled in at the end of the two-day period.
- Evaluation Questionnaire. In this questionnaire subjects were asked their opinion about the information system as a design tool and a 'team tool', and to indicate the most important information they had used. In order to test the influence of illustrations on remembering the information of the information system, subjects had to write down all 'pages' of the database that they could remember. Finally, they were asked to report their learning experiences.

4.5 Procedure

The teams worked for two days (09.00 till 22.00) on the design assignment. The meeting started with a general instruction on the design project, the use of the information system and the contest requirements. The edit facilities were explained to the experimental teams. Next, subjects filled in the Competence Questionnaire.

Each group had its own studio in which a telephone and a computer were available (McIntosch Performa 630; screen Iiyama Vision Master pro21) with the information system installed. Sketching materials were also available. On request, the teams were allowed to make use of a printer. In order to prevent the printing of too many data from the information system, only one printer was available at a central place in the building. During the design process, teams kept the logbook every quarter of an hour.

At the end of the second day the teams presented their results. This presentation was meant as a preparation for the official jury presentation.

About 10 days after the design session, the teams came together for (1) an evaluation of the project, and (2) the judgement of the design contest by a jury. The subjects started with the Evaluation Questionnaire. Next, all teams presented their results. Finally, the jury, presided over by the manufacturer, presented the results of the assessment and offered every participant a bicycle-related present.

5. Results

In the following paragraphs, descriptive results of the first data analysis are presented.

5.1 Activities during the Process

Through the logbook filled in by every team, a detailed report was obtained about the phases and activities within phases of the design process. In general there are differences between teams in both duration and sequence of each phase, and in kind and number of activities per phase. First, some activities are connected to certain phases. Consulting the information system (bicycle database) was mostly done in the analysis phase on the first day. But the time spent on this activity varies from 8% (team 8) to 30% (team 9) of the total time. A number of activities were performed simultaneously, depending on the team and on the phase in the process. Teams differed in the way they worked together. Seven out of nine teams started working on the same activity without dividing their labour, while in two teams the members immediately worked separately on mainly information gathering and sketching. In general, on the second day the team members worked more in parallel, while they were busy with sketching, writing a report, checking the information and preparing the presentation. Looking at the kind of activities that were performed simultaneously the following were mentioned most frequently: sketching, gathering information (either through the information system or through other sources), writing a report and preparing the presentation. Often these activities were mentioned in combination with 'group discussion'.

Two out of three experimental teams did spend some time on editing the information system through adding own pages, changing existing information or marking words or sentences in some of the pages.

5.2 Individual and Team Characteristics

Differences in the output of the teams might be influenced by the particular composition of the teams. Moreover, it is interesting to see if teams that registered as a team differ from the other teams. Conclusions are based on an average team score. For 'heuristic competence' this score varies from 44.0 (team VIII) to 52.3 (team VI); 'emotional load' varies from 44.3 (II) to 67.6 (VIII); problem-solving competence from 79.6 (VIII) to 101.3 (IX); 'regression' from 44.7 (III) to 75.0 (VIII); 'resignation' from 22.3 (VI) to 27.3 (III).

Significant differences between teams are also found on the Group Dynamics Questionnaire, such as 'leadership', 'conflict handling' and 'efficiency'. It might be expected that teams that applied as a team, with members who know each other quite well, would have higher scores on group climate, conflict-handling and co-operation. But no such relationship can be found.

Next, the correlation will be defined between these differences and the design results.

5.3 Use of the Information System

5.3.1 Consulting the Information System

The first question regarding the use of the information system is how many 'pages' of the information system are consulted by the teams and, second, what kind of information is asked for. Data were available for 6 teams. Table 1 shows the number of pages consulted per day, while the 'navigation' pages are not included. Moreover, the number of pages consulted for the first time are distinguished from the re-consulted ones.

While most of the teams asked for more than 50% of the total content of the system, team VIII is an exception, with only 37%. There is no explanation possible based on the experimental condition. The expertise of the subjects seems more likely. As mentioned before, the students of team VIII are from a younger generation than the other participants. Previous research showed that novice design students (2nd-year) show less information-seeking behaviour than senior design students [12].

On the second day, the proportion of re-consulted pages was much higher than on the first day, except for team VIII.

Table 1: Number of topics (pages) consulted by each team (navigation pages excluded). Total number of pages = 169. *New* = pages consulted for the first time. *2nd* = pages consulted for the second (or 3rd, 4th) time. % of pages used is put in brackets.

Condition	Team	day 1 new	day 1 2nd	day 2 new	day 2 2nd	Total new	Total 2nd
	I	69	11	23	17	92 (*0.54*)	28
control	II	76	6	19	26	95 (*0.56*)	32
	II	68	12	24	12	92 (*0.54*)	24
experimental	VII	48	1	45	23	93 (*0.55*)	24
	VIII	39	5	23	1	62 (*0.37*)	6
	IX	69	10	12	5	81 (*0.48*)	15

5.3.2 Type of Information Consulted

In order to give an overall view of the kind of information consulted, the information is summarised in 11 headings, 171–173 dealing with bicycle-related topics and 174–177 with user-related topics. As an example, in Figures 2 and 3 the type of information asked for is presented for team I and team VII. In Figure 2 the headings are explained.

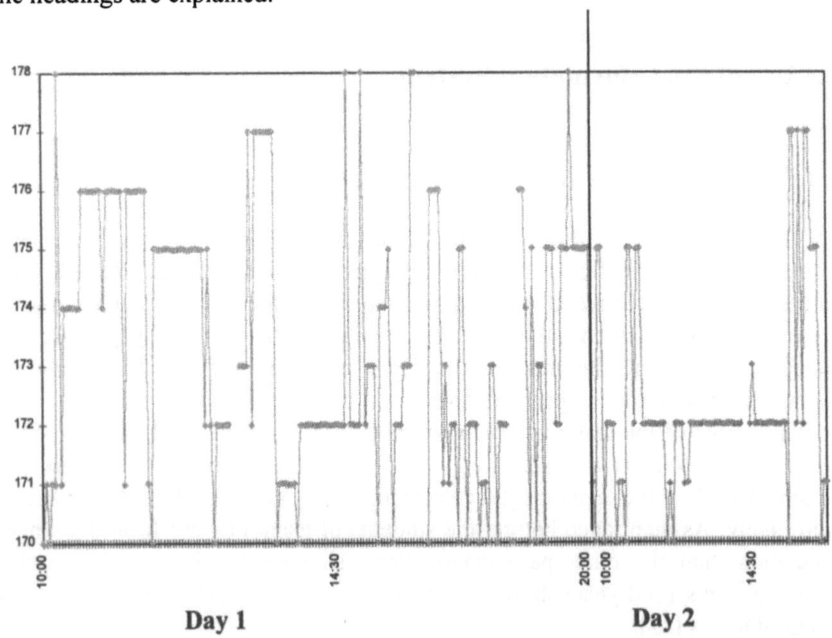

Figure 2: Information asked for by team I. Y-axis: 170=Main menu, 171 to 173=Bicycle aspects, 174 to 177=User aspects, 178=Dictionary, 180=pages edited by team.

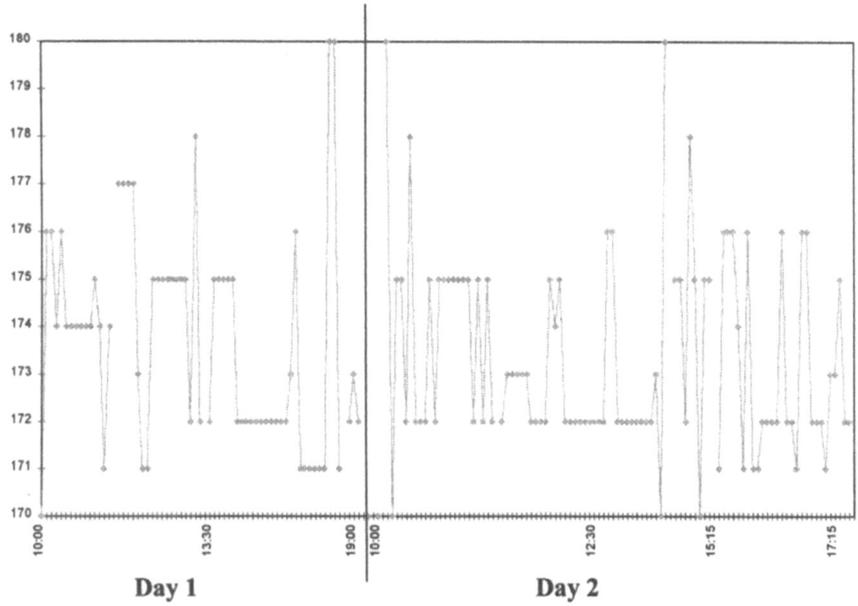

Figure 3: All information asked for by team VII. Values on the Y-axis: see Figure 2.

Each team shows its own specific graph, representing the way it goes through the information system and also indicating what information is most important for the team. As can be seen from the graphs, the two teams differ not only in the amount of information consulted on both days, but also in the choice of topics. On the other hand, teams also show some similar behaviour. The following conclusions can be made:

- Four out of six groups started with information about the target group of users.
- Team VIII alone hardly gathered user information.
- When subjects chose one of the topics on a general level, for example 'bicycle parts' (see Figures 2 and 3), they tended to stick to the topic, following the links belonging to this topic. At the end they went back to the main menu, chose another topic and went through the same procedure.
- The type of information asked for on the second day seems very close to the phases of the process that the teams were involved in. These phases can be defined as: conceptualisation, decision making and evaluation. Two sources of information were looked for then: (1) technical data, such as the geometry of the bicycle, materials and production methods, and (2) legal norms and standards. While on the first day teams looked for user information, this was hardly the case on the second day.

Subjects were asked to evaluate the system on the criteria mentioned in Table 2. A 7-point scale was used, from very unsuitable (1) to very suitable (7). Most of the scores are on the positive side except for the 'adjustability' criterion in the case of the control teams. This could be expected because no editing facilities were provided. But even the experimental teams were not very positive about the adjustability of the system. We will come back to this later.

Regarding the content of the information, 37% of the subjects were negative about the completeness. Usability and relevance on the other hand were very positively valued, for three main reasons: the system provides the necessary basic knowledge, which is very helpful in the first phase of the design process and which is a very efficient and time-saving method of knowledge transfer.

Table 2: Evaluation of the information system – average scores for control and experimental teams.

Evaluation criterion	Control	Experimental
Usefulness	5.57	5.33
Completeness	4.43	4.13
Relevance	6.29	5.60
Format (Internet, pages)	5.14	5.27
Structure (hierarchy, links)	4.14	4.40
Use of pictures	4.71	4.67
Adjustability	3.71	4.20

Another way to test the relevance of the information is to consider the design specifications, as reported by the teams, and to look for data that are derived directly from the information system. Table 3 gives an overview of the system-related specs together with the assessment of the jury regarding the bicycle design. The application of data from the system is evident. The table also reflects the use of the information system in that the little use by team VIII is confirmed (see also Table 1).

Table 3: Information about the system used in describing design specifications.

Condition	Team	System-related specs/ Total specs	Design assessment
	I	12 / 16 (0.75)	6
Experimental	II	9 / 14 (0.64)	5
	III	8 / 10 (0.80)	1
	IV	15 / 22 (0.68)	4
	V	6 / 8 (0.75)	2
Control	VI	14 / 24 (0.58)	3
	VII	11 / 13 (0.85)	9
	VIII	5 / 15 (0.33)	8
	IX	6 / 10 (0.60)	7

5.3.3 Structure of the Information System

Does the structure of the information system dictate the way in which users search for information? Analysis of the way in which subjects go through the database shows that:

shows that:

- Regarding the content at the general level of the system, subjects are free to choose the topic they want, be it information about users, bicycles or manufacturing and market.
- At a deeper level of the system, all subjects mostly follow the links provided on every page. Deviating from the advised links is mostly seen after subjects have worked for a couple of hours with the system and have learned to make 'bookmarks' or to use the 'back' and 'forward' buttons.
- All pages were accessible through the navigation pages. The data show that only a few pages are ignored by all teams. These pages had to do with: muscular strength related to cycling, some of the legal rules, and some accessories such as helmets and bags.

In their Evaluation Questionnaires some subjects are critical about the structure. In their opinion, an overview of all pages in the system is missing. Moreover, there are some problems with understanding the hierarchical structure. Other remarks deal with the facilities for editing and printing.

5.3.4 Richness of Data

It was hypothesised that information with both text and illustrations would have a positive effect on information processing. In our study, 48% of the pages of the information system included a picture: coloured photographs, graphs, tables, etc. Because one of the hypotheses is that providing information through two modalities facilitates retention, subjects were tested 10 days after the design meeting. They were asked to write down all topics that they remembered from the information system. On average, 60% of the pages remembered have both pictures and text.

The conclusion is that there is an overall retention effect due to the way the information is provided. There are individual differences: for 8 subjects, no effect can be demonstrated.

5.3.5 The User as an Active Participant

Our hypothesis was that the possibility to adjust the database to one's own wishes and preferences should have a positive influence on the processing of knowledge from the information system and thus on the influence of this knowledge in designing. However, as the results show, this possibility is hardly used by the experimental teams. Hence, this hypothesis cannot be tested in this study. There are at least three explanations why teams did not use the editing facilities:

- The editing facilities offered were not user-friendly (the Netscape Gold version was used). Subjects complained about it. Moreover, they also complained about the speed of the computer. Both facts discouraged activities of adjusting the information or the structure.

- The time available for the design project was very short; teams had to give priority to the design activity itself.
- The information system was positively judged by all teams, so the necessity to adjust the system was hardly present.

5.3.6 Teamwork

Two aspects of this subject will be briefly considered. First, the use of the information system will be related to the work of the team, and, second, the team characteristics will be considered with regard to the results of the design process.

In evaluating the suitability of the information system as a 'team tool', two opinions could be heard. First, a minority of the subjects (3) thought that the system hampered team activities because only one person could operate the computer. Most subjects, however, have the opposite opinion. They argued that the use of the system forced the team to divide the work among the team members. The most interesting argument was that the information gave direction to the team discussions, providing the team with a common knowledge base. As a positive effect, long discussions were avoided because team members talked 'at the same wavelength'.

The second question is related to the performance of the teams in connection with the individual and group-related characteristics of the teams. So far, most of the analysis has still to be done. Some conclusions can be tentatively drawn.

Relating the data from the Competence Questionnaire and the Group Dynamics Questionnaire to the design assessment by the jury (see Table 3) shows some surprising results:

- The teams with a high score on 'problem solving competence' have only a low rating in the contest.
- Regarding the Group Dynamics Questionnaire, the score on 'conflict handling' does not seem to predict the success or failure of the team.

6. Discussion

In this study the main issue was to investigate the way in which designers use information with the help of an information system. In developing a tool for designers we agree with Meerkamm (elsewhere in this volume) that the kind of information management that is aimed at determines the specifications of the tool. Hence, in order to explore our typical information system, the quality of both the system structure and the content of the information were studied. But, at the same time, assumptions were explored regarding the active involvement of the designer in using the system. A connected consideration was the possible profit for a designer to develop a computerised personal archive. The question then should be what the necessary conditions are to build such a system. The assumptions, mostly derived from learning theories, were tested through an experimental design in which both control groups and experimental groups were involved. Senior design students from the School of Industrial Design Engineering, Delft University of Technology, were selected to participate in the experiment. They worked for two days in teams of three persons each on an assignment to design a bicycle for people of 55+.

It is important to realise that the situation as created in this study deviates from daily practice because there is hardly any topic for industrial designers that has all specific information gathered in one database. Nevertheless, the bicycle company involved was very interested in the information system because this company considers the system to be a valuable tool, even for experienced designers. In spite of the rather unrealistic situation, the study could provide relevant information as to how an information system can be used and to what extent it is used.

The first conclusion of the study is that the information system played a very relevant role during the design process. All teams, except for one, made extensively use of the system and were very positive about its usability and efficiency.

As the reports of their work show, the data from the system are used mostly in the analysis phase and in defining the design specifications. There are also some critical remarks regarding the completeness of the information and the adjustability of the system.

Although there are differences in the way the teams go through the information, most teams start with user-related topics. On the second day of the project, the type of information asked for is related to the typical phases in the design process: conceptualising, evaluation, and preparing the report and the presentation. The 'logbook' in which teams kept record of activities appeared to be a practical instrument to obtain detailed information about the progress in design.

The effect of the adjustability of the information system could not be tested. The experimental teams hardly made use of the editing facilities, partly because the facilities were not user-friendly, and partly because the satisfaction with the system did not elicit any need to adjust the system.

References

1. Christiaans H.H.C.M., Andel J. van 1993: The Effects of Examples on the Use of Knowledge in a Student Design Activity: The Case of the 'Flying Dutchman'. *Design Studies* 14, 58–74.
2. Kozma R.B. 1991: Learning with Media. *Review of Educational Research* 61, 179–211.
3. Norman D.A. 1989: *The Psychology of Everyday Things*. Basic Books, New York.
4. Spiro R., Jehng J. 1990: Cognitive Flexibility and Hypertext: Theory and Technology for the Nonlinear and Multidimensional Traversal of Complex Subject Matter. In: D. Nix & R. Spiro (eds) *Cognition, Education and Media*. Lawrence Erlbaum Associates, Hillsdale, NJ, pp. 163–206.
5. Vries E. de 1994: *Structuring Information for Design Problem Solving*. Dissertation, University of Technology, Eindhoven.

6. Vries E. de, Jong T. de 1997: Using Information Systems while Performing Complex Tasks: An Example from Architectural Design. *Journal of Human–Computer Studies* 46, 31–54.
7. Paivio A. 1986: *Mental Representations: A Dual Coding Approach.* Oxford University Press, Oxford.
8. Mayer R.E., Sims V.K. 1994: For Whom is a Picture Worth a Thousand Words? Extensions of a Dual-coding Theory of Multimedia Learning. *Journal of Educational Psychology* 86, 389–401.
9. Andel J. van 1994: Behaviour Mapping and Urban Design: Graphic Versus Non-graphic Information about Environment-behaviour Relations. In: S.J. Neary, M.S. Symes & F.E. Brown (eds): The Urban Experience. A People-Environment Perspective. *Proceedings IAPS 13.* E & FN Spon, London, pp. 507–517.
10. Self J. 1997: From Constructionism to Deconstructionism: Anticipating Trends in Educational Styles. *European Journal of Engineering Education* 22, 295–307.
11. Greene S., Ackerman J.M. 1995: Expanding the Constructivist Metaphor: A Rhetorical Perspective on Literacy Research and Practice. *Review of Educational Research* 65, 383–420.
12. Christiaans H.H.C.M. 1992: *Creativity in Design.* Lemma, Utrecht.
13. Rumelhart D.E., Norman D.A. 1978: Accretion, Tuning, and Restructuring: Three Modes of Learning. In: J.W. Cotton & R.L. Katsky (eds) *Semantic Factors in Cognition.* Lawrence Erlbaum Associates, Hillsdale, NJ.
14. Minneman S.L. 1991: *The Social Construction of a Technical Reality: Empirical Studies of Group Engineering Design Practice.* Dissertation Xerox Parc, Palo Alto SSL-91-22.
15. Trousse B., Christiaans H.H.C.M. 1996: Design as a Discursive Activity. In: N.G. Cross, H.H.C.M. Christiaans & C.H. Dorst (eds) *Analysing Design Activity*, Wiley, London, pp. 365–388.
16. Norman K.L. 1991: *The Psychology of Menu Selection: Designing Cognitive Control at the Human/Computer Interface.* Ablex Publishing Corporation, Norwood, NJ.
17. Roozenburg N.F.M., Eekels J. 1995: *Product Design: Fundamentals and Methods.* John Wiley & Sons, New York.
18. Stäudel T. 1986: Der Kompetenzfragenbogen. *Memorandum no. 41, Lehrstuhl Psychologie II.* Universität Bamberg.
19. Frankenberger E., Badke-Schaub P. 1997: Influences on Design Productivity – Empirical Investigations of Group Design Processes in Industry. In: A. Duffy (ed.) *The Design Productivity Debate.* Springer, London.

Information Management in the Design Process – Problems, Approaches and Solutions

Harald Meerkamm

Abstract

Design work is information management and processing. In order to fulfil their tasks during the complete design process, designers need a large amount of information. The character and size of this information depend on several parameters and are in direct relation to the stage of the design process. In order to find out the best solution the engineer will use various sources of knowledge data and information.

It is essential that the designer gets the correct information at the right time, in an appropriate quantity and state. As he cannot have all the necessary knowledge in memory, it is important to provide a quick and comfortable means of access to the different sources and kinds of information. Conventional ways can be used as well as the current and effective tools of information technology. For both approaches we need an appropriate methodological base, i.e. this complex quantity of knowledge must be structured in the right way and form. Otherwise quick access and comfortable use cannot be guaranteed.

This paper discusses the quantity and complexity of the required knowledge, and provides some approaches for information management by structuring this knowledge and using different tools of information technology. An Engineering Workbench with an integrated XPS (knowledge based system) is discussed as well as the potential of Tele-engineering. There is no universal and standard approach however, since the process and set of tools depend on the problem/product/company/user.

Some examples are presented of the requirements of a prototype realisation:

- context-sensitive presentation of solution elements via Internet
- presentation of different calculation tools via internet
- simulation tools in Tele-engineering
- tolerance analysis based on an XPS
- presentation of recycling information via an assistance system.

1. Introduction

The design process is driven by information. In order to fulfil his task, the design engineer must obtain a lot of different information of varying complexity, work with it and then provide some information to the other people involved in the process of product realisation [1]. This process of gaining information and managing it is most important for the result of the design process (Figure 1). The quality of the designer's decision and therefore the quality of the solution obtained depend on the quality of information that the design engineer obtains.

It is essential that the engineer gets information

- at the right time
- in the right quantity
- in the right form, and
- at the right place.

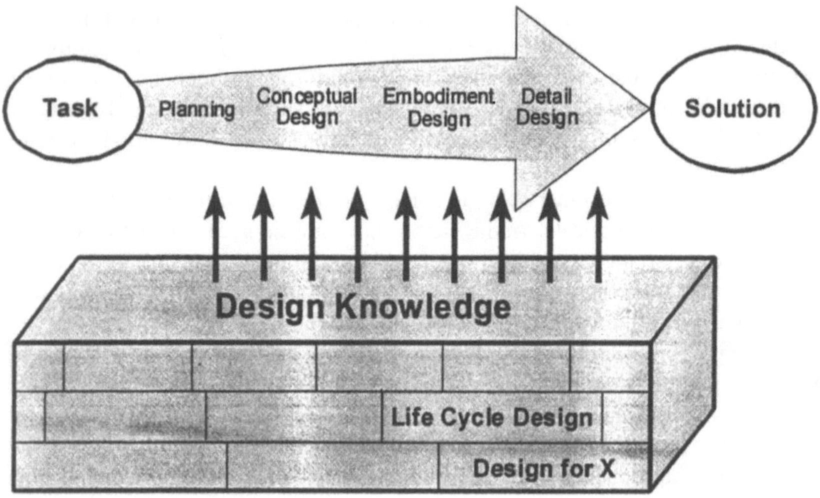

Figure 1: Information management in the design process.

1.1 Problem of Information Quantity and Complexity

To realise the necessary information turnover is not easy for the designer. Getting the right information takes a lot of time during the design process. Statistical investigations with designers in practice report that it takes approximately 25–60% of their working time.

Designers have not only to handle a great quantity of design knowledge from various fields, but its nature varies in accordance with the particular 'design for X' and the fact that total life cycle alters according to the stage of the design process that is actually being worked on. The various sources of knowledge are thus many for designers. Both the quantity and complexity of information and new restrictions are ever increasing (e.g. design for environment).

1.2 Requirements/Aims

As the designer alone is not able to handle and manage all the information with the desired quality and promptness, adapted support should be available. This must be based on the approved methodological approach and should include the availability of the powerful tools of information technology. These modern software and hardware tools with their huge variety and potential, can be helpful and effective. There is no universal approach, however, because there are no single tools available that can deal with all the kinds of information needed, for all companies and for all designers.

In order to guarantee quick and optimal access to the various kinds of information needed by a designer, it is therefore necessary to structure this huge field of information in a manner that allows all the various tools of information technology to play their part. This could be one reason for shortening the product development process and improving its quality.

2. Approach

A solution to the problem is just described could be sought using the following two strategies:

- optimal structuring of the design knowledge
- selecting the best tools of information technology that are available.

2.1 Methodological Structuring

The basic information is unstructured knowledge (see Figure 2) [3, 7–9], and the aim is to find the best and most practical structure in which to put it.

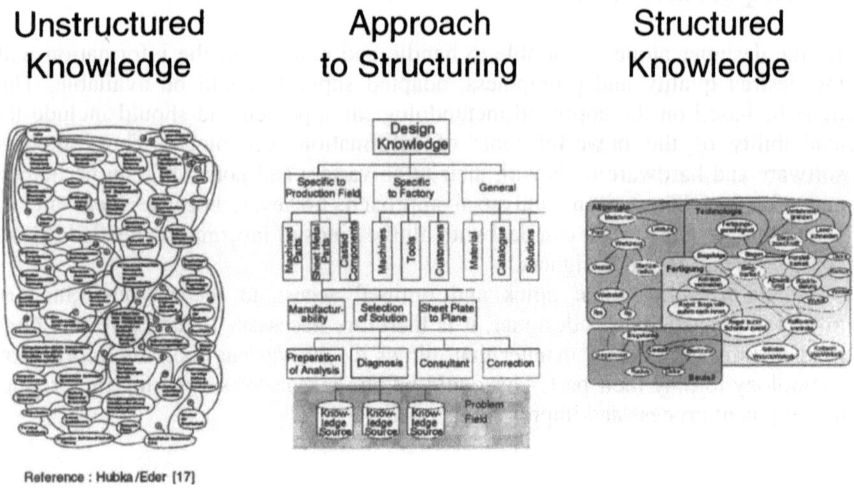

Figure 2: Knowledge structuring.

There are many possible ways of structuring design knowledge. In [2], the following categories are proposed:

- formal and informal knowledge
- knowledge resources
- generic and specific knowledge
- knowledge for the several stages of the design process
- structural, descriptive and procedural knowledge.

The basic framework of the structuring can be kept general, however some details must be structured specifically for the user, the company or the actual situation. The most important factor is that the method of structuring allows the different views of design knowledge to be incorporated (see Figure 3). Each application, each user etc. has a specific requirement. The whole range and complexity of knowledge need not be available to the designer at all times but it is essential, however, that each specialist has immediate access to the information he needs where unusual design problems are concerned. Each specialist has a specific view of the knowledge base. The stress calculation specialist needs different information from that required by the production specialist. Both need just their special range of information from the complex knowledge base, so they want their specific requirements built in.

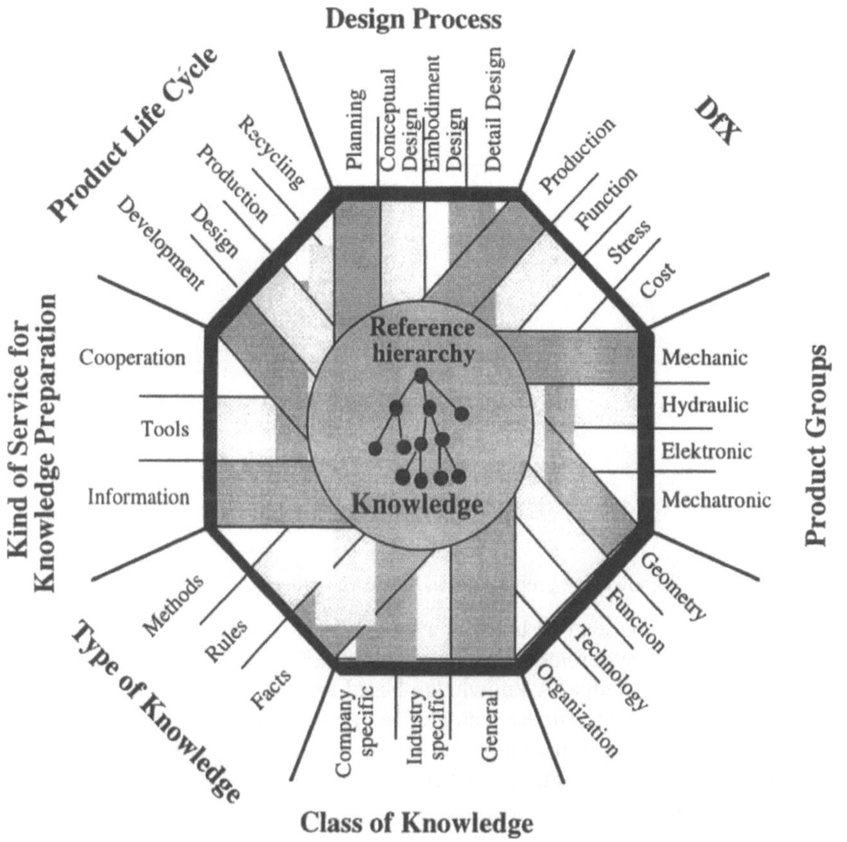

Figure 3: Views at design knowledge.

2.2 Tools of Information Technology

Methodological structuring of knowledge is one essential precondition for effective information management, the selection and use of the right tool is the other. Depending on the form of knowledge presentation and the range of information transfer, there are many possibilities for knowledge preparation (see Figure 4): conventional (i.e. without using a computer) and/or using the various tools of information technology. There are a wide range of tools available starting with databases, expert systems, all the way up to Hypermedia, World Wide Web, Multimedia Conferencing etc.

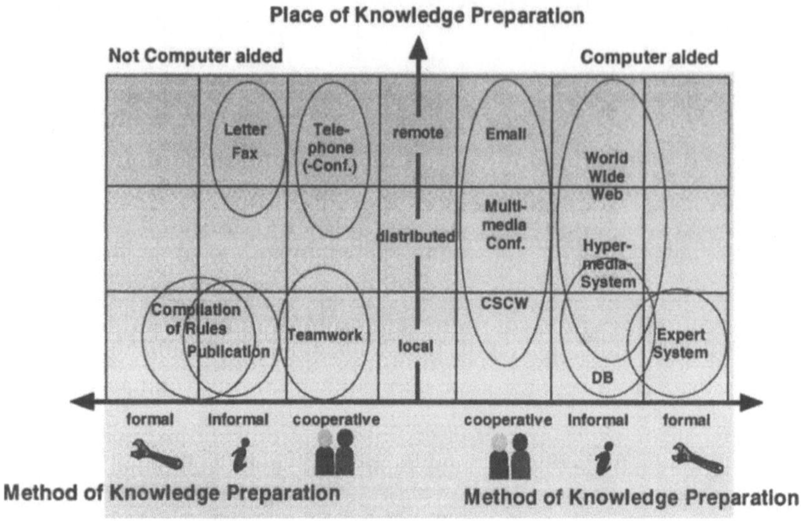

Figure 4: Types of knowledge preparation in the design process.

No tool should be seen or used as a universal aid. Each tool has its specific capability, and the requirements for application are as different as the need for formalising the knowledge. A knowledge-based system, for example, needs a lot of formalisation and knowledge acquisition. A collection of data needs a low degree of formalisation, but acquisition of the data normally takes a lot of time. With hypermedia systems, the needed for formalisation and the work required to obtain the information are reduced.

Which tool for knowledge preparation is the best for preparing knowledge of a specific task depends on [4]:

- frequency of use
- importance of the information
- required degree of confidence in the results
- time available for product development
- time available for preparing the knowledge tool
- possibilities of formalising the knowledge
- cost.

3. Results/Findings

At the Institute for Design Technique (KTmfk) we have been investigating the field of information management for some years, with regard both to effective computer tools and the various basic methods of structuring the work (see Table 1). In the following, some results of this work are described.

Table 1: Tools and methods for information management.

Tools	Methods
• Database • Knowledge-based system • World Wide Web (WWW) distributed client/server systems Internet/Intranet/Extranet • Hypermedia Systems • Video/Audio Conferencing Shared Application/WhiteBoard Joint X . .	• Context sensitive knowledge preparation • ISO 13584 PLIB (Part Library) • CSCW Computer supported cooperative work Tele engineering . .

3.1 Context Sensitive Knowledge Preparation

The aim of this approach is to support the designer during the design process by providing him with the necessary information in the right context. It is essential that the engineer gets the information (what), at the right time (when), at the right place (where) and in a well- prepared form (how). However, he should be overloaded with information, so that he is drowned by information but thirsting for knowledge.

In our approach to context sensitive knowledge preparation, we aim to deliver the right information in the context of the current problem and the existing stage of design (see Figure 5).

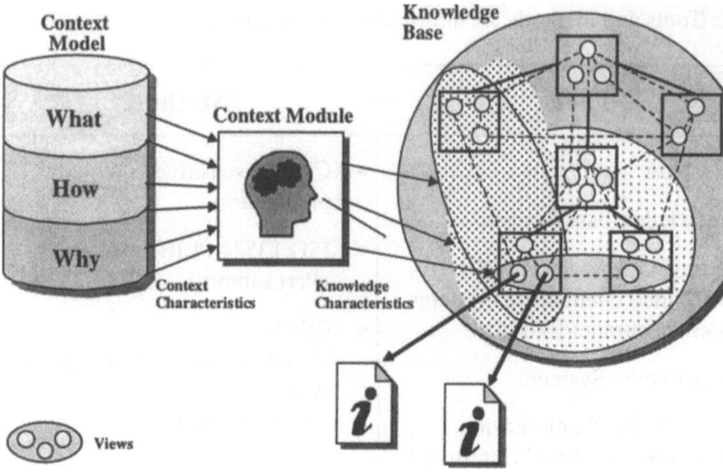

Figure 5: Context sensitive knowledge preparation.

The designer describes his problem in a language that can be understood by the computer, i.e. by design features. The context module accesses the structured knowledge base by comparing the design features/elements of the context model with the special features of the information system / knowledge base. The more knowledge characteristics can be identified by the context module, the more the knowledge is considered relevant to the context within the intersection set (context characteristic versus knowledge characteristics via the context module) [4].

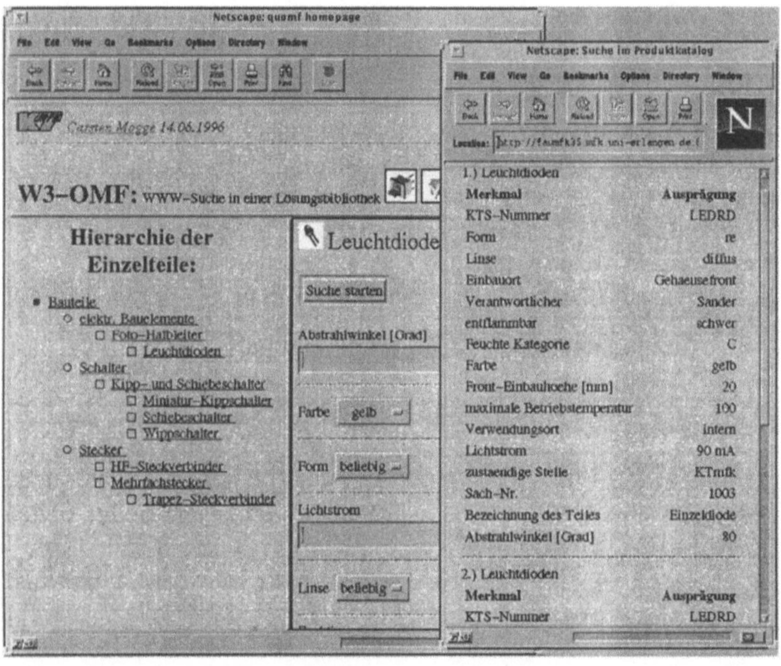

Figure 6: Access to a part library via WWW.

Figure 6 shows an application of the context sensitive search for solution elements. The aim of this system is support to the designer when searching for solutions/parts/elements that fulfil the requirements of the product. In this case the system allows access to a part library which can be company specific and/or contain information concerning subsuppliers.

In the context model, the requirements (general, specific) are specified. The system searches for suitable solutions that fit with the hierarchy of demanded parts and the required characteristics, and that define the range of possible solutions.

This part library was implemented according to a product data management system. An interface to WWW was also made available. This access via an engineering network is advantageous both for the user and supplier.

For the user, the WWW browser presents a cheap user interface, which is easy to handle. Information is always current, as the vendor has to keep the information up to date.

For the browser-based application, efficient development tools are also available. This is an advantage for the supplier as well as for control of access to the library. Costs are lower than those that would be incurred using product information via CD-ROM.

If the part library also provides access to the catalogues of the subsuppliers, application of the concept and format of ISO 13584 (Part Library) is useful [10].

3.2 Part Library ISO 13584

If a company wants to use part libraries from other subsuppliers, this often causes a problem because of the different data formats and different nomenclatures. Integration of several external subsuppliers is made easier if one uses a standard format for the realisation of the part libraries. ISO 13584 PLIB offers an interesting approach for solving this problem (see Figure 7) [10, 11, 12].

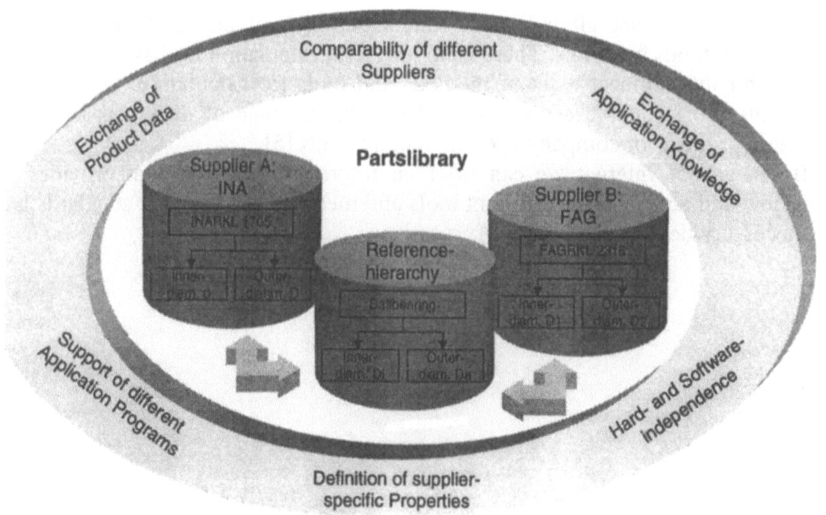

Figure 7: Concepts of ISO 13584 PLIB.

The aim of ISO 13584 is the exchange, processing and use of digital part libraries. It allows a comparison of various subsuppliers. Normally the designation and description of different suppliers are not uniform, therefore the computer is not able to compare the products of different suppliers and find alternatives. A reference hierarchy gives a neutral description for defined categories of products. The products of all those suppliers who refer to ISO 13584 can be compared with respect to their characteristics by the computer. The computer can then find all the alternatives available from the subsuppliers for a defined product needed by the user.

Furthermore, ISO 13584 PLIB allows the exchange of application knowledge. In this way the customer can obtain knowledge about, for example, conditions of use, possibilities of variation (dimensions, material etc.) and compatibility with other products. This knowledge can be exchanged in the form of programs (e.g. for calculation), algorithms or rules.

An important advantage is that the users can define optional features/characteristics,. therefore they are not restricted in describing their products.

Different user programs can be supported by PLIB. Therefore CAD systems, PPS systems etc. can all obtain access to the part library and get the data they need.

ISO 13584 PLIB is an international standard and is compatible with the STEP standard, so a part library based on it is independent of both the hardware and the software.

3.3 Information for Stress Calculation via Internet

During the design process – especially for complex products – the engineer must do different types of calculations. Stress calculations to determine the right dimension/size of a part, calculations to find out the most suitable component/part, calculations based on FEM or BEM to optimise structure and/or form of components etc. can all be static, kinematic, thermal etc. and are based on different methods and tools. There is no universal calculation tool available, so we must offer the engineer a set of methods and tools for calculation, and he must chose the most suitable according to the problem, stage of design process and probable product- or company-specific requirements [5].

By means of Internet we can offer an information system that provides an overview and access to the different tools and methods, the relevant standards and the necessary details (see Figure 8).

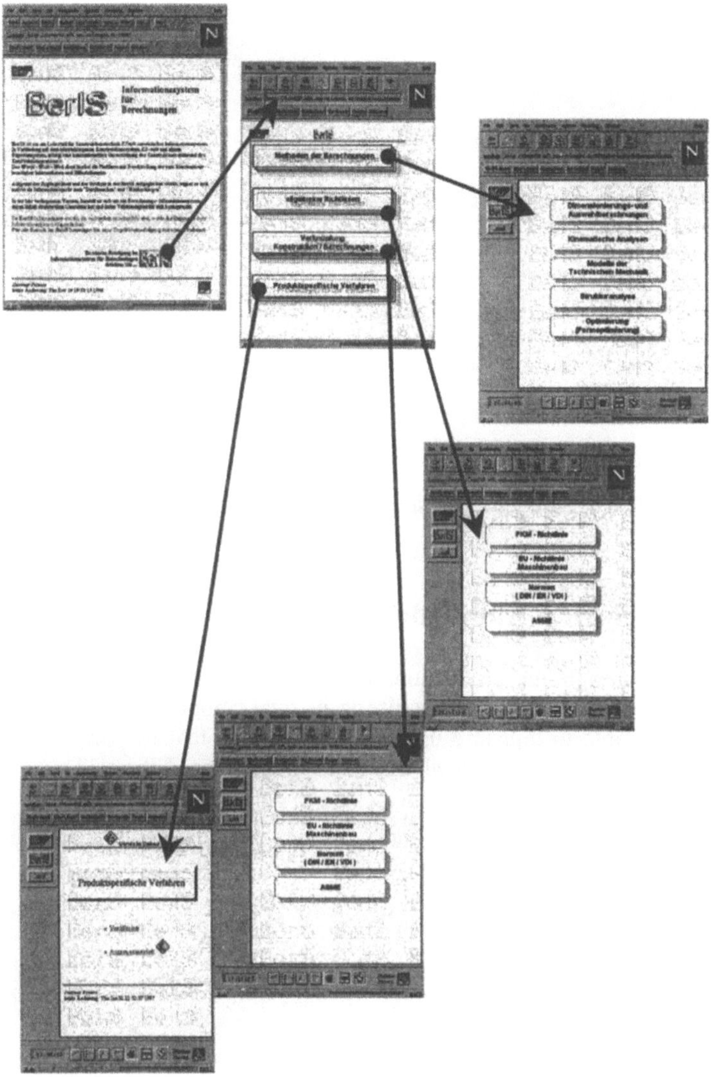

Figure 8: Information system for stress calculation based on Internet.

Furthermore, Internet provides a quick and special service from the supplier for the customer. Suppose a customer needs access to a special stress and bending calculation for shafts, which is offered by a supplier for roller bearings. Up to the present this service has been provided by mail or fax, but via Internet it can be assessed more quickly and effectively. The customer simply describes the calculation problem via a 'CAD-function' and sends a calculation order to the supplier (see Figure 9) [13].

The calculation is initiated by the order of the customer. Nobody in the organisation of the supplier is involved with it, because it is effected automatically. Using Internet, the customer is automatically supplied with the result (see Figure 10).

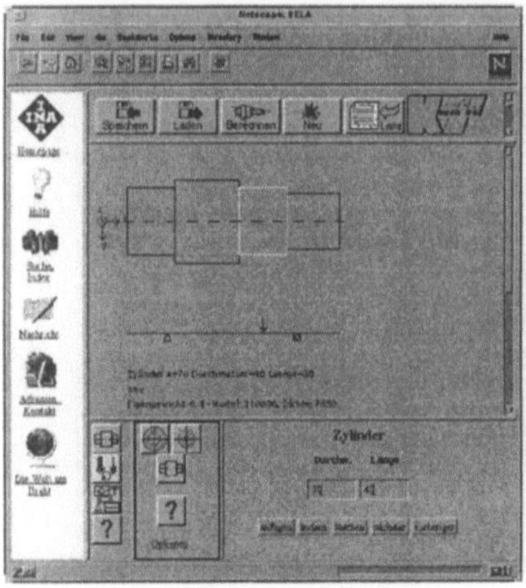

Figure 9: Problem description via Internet.

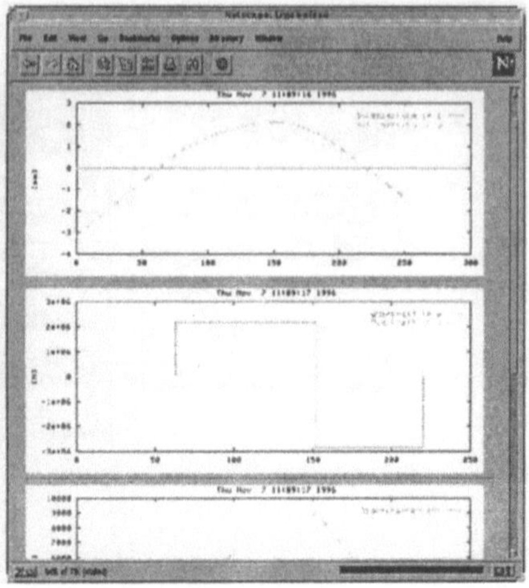

Figure 10: Result of calculation via Internet.

3.4 Technological Knowledge for Design for Production

In order to design for production, the engineer needs a lot of technological knowledge, which he may not require in its total complexity. Therefore we must make it available to him in a user-friendly way, so that he can make decisions based on the technological knowledge that is offered (see Figure 11).

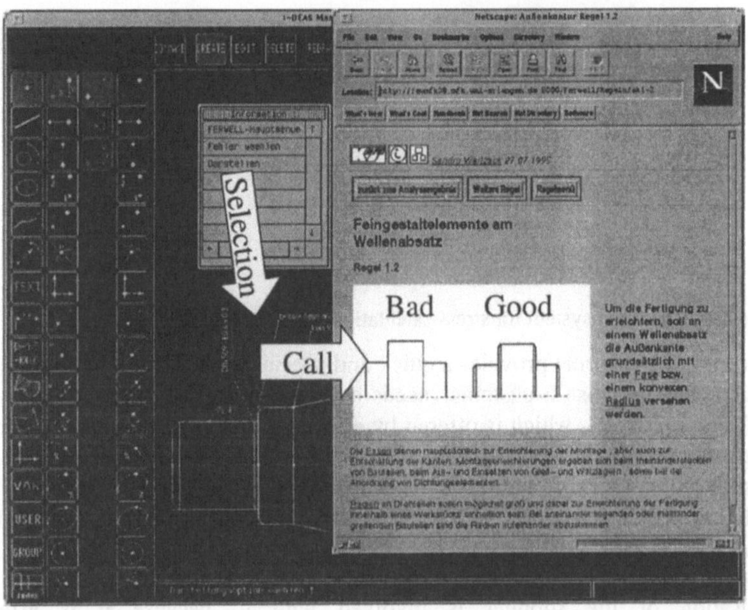

Figure 11: Information system 'Design for Production'.

When working with the CAD system the designer can ask for information in the field of 'Design for Production'. The necessary information is offered by an information system that allows an analysis of the design solution and provides additional consultancy possibilities [4].

3.5 Tele-engineering

Using this powerful network it is possible to realise a cooperation between different experts working simultaneously at separate places (see Figure 12).

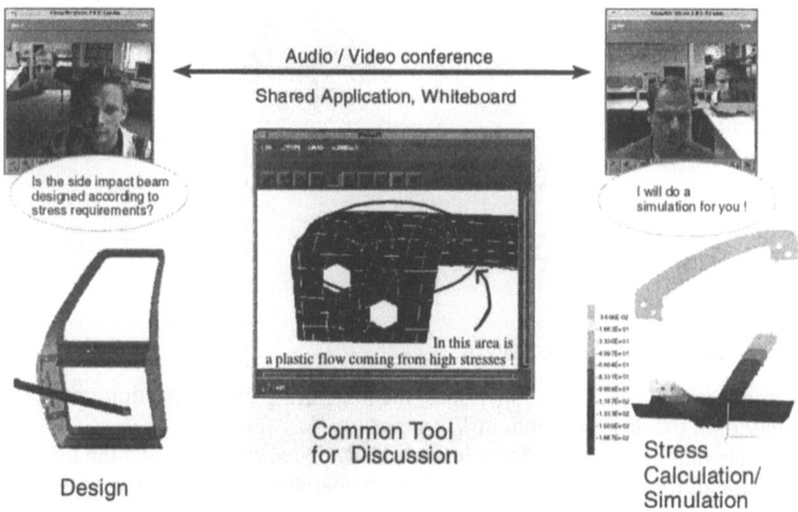

Figure 12: Tele-engineering.

Via this network, the designing engineer can contact for example the calculation/simulation expert. They intercommunicate by video and audio, a whiteboard and shared application software guarantees that they both have the same design solution on the screen of their CAD system. The designer can describe the current problem and might ask for a stress calculation and simulation. The result, created by the calculation expert, is transferred to the designer via the network and they both have the possibility of discussing it immediately [14].

3.6 Tolerance Analysis

Selecting the correct tolerances (dimensional, form, position) is one of the important tasks of the designer. Very often he cannot foresee the consequences of his choice on function, production process, inspection and cost. Therefore he needs assistance, for example from an integrated tolerance analysis based on the design system mfk [15] (see Figure 13).

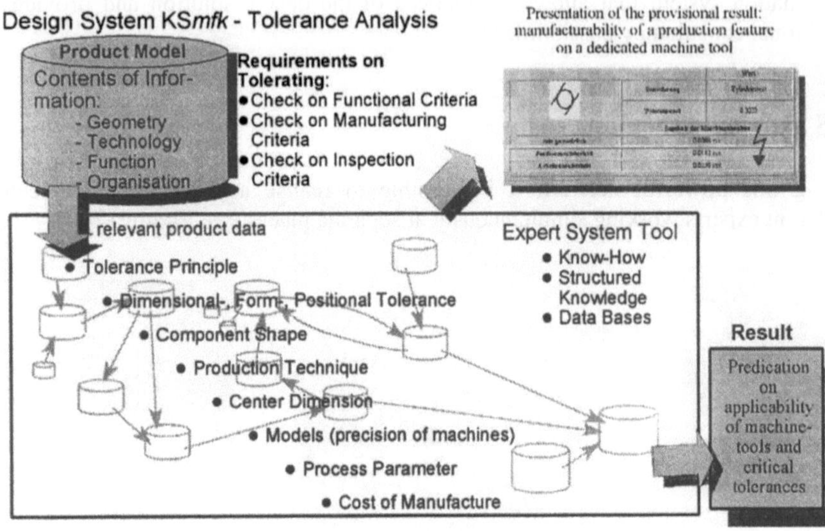

Figure 13: Integrated tolerance analysis.

The necessary checks on functional, manufacturing and inspection criteria are also based on the design system mfk.

The relevant product data for tolerance analysis are extracted and the tolerance (dimensional, form, position) is defined. In cooperation with an expert system tool (Nexpert Object) the design solution is related to the different options of the production processes and machines. The manufacturability is then checked and the costs of manufacturing are calculated. Critical tolerances are brought to the attention of the designer [16].

3.7 Assistance System for Design for Environment

Design for Environment is a field where the engineer needs the maximum of information from several knowledge fields. He needs information about various materials, harmful substances, the best way of manufacturing, assembly and disassembly, ecological balances etc. This great quantity of complex information is better handled using the right methodological approach and a computer-based assistance system (see Figure 14).

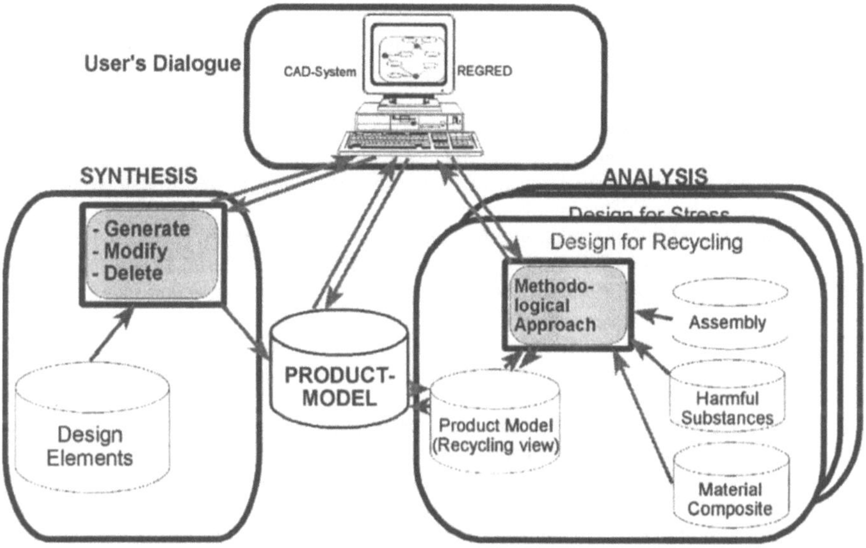

Figure 14: Assistance system design for environment.

The engineer is assisted by the system in the process of synthesis as well as in analysing the solution with regard to its environmental aspects. It is possible to create a recycling graph automatically, showing the structured group, the different materials, the recycling strategies etc. Integration of this assistance tool at an engineering workbench guarantees that the designer gets the right information at the right time, in an adapted quantity and well prepared for the task he is actually working on [6].

4. Conclusions and Recommendations

Information management and processing in and for the design process is a most demanding task. To solve the problem of information quantity and complexity, and in order to guarantee that the engineer gets information at the right time, in the right size and form and at the right place, we need a methodological approach in combination with computer assistance.

The most important precondition is structuring the knowledge and preparing it for the use of computers. Then it will be possible to use the different and powerful tools of information technology. There is no universal tool, however, that is applicable to all cases. Therefore we must determine, which tool is the best according to the actual task, the specific company and the product.

References

1. Pahl G. 1989: Anforderungen an die Wissensverarbeitung aus der Sicht der Konstruktion. In: *VDI Bericht Nr. 775,* VDI-Verlag, Düsseldorf, pp. 1–24.
2. Abeln O. (ed). 1997: Produktentwicklung im Wandel – das CAD Referenzmodell in der Praxis (in press December 1997).
3. Rösch S. 1996: *Konstruktionssystem für ein rechnerunterstütztes Konzipieren und Entwerfen komplexer Blechteile.* PhD thesis, Universität Erlangen-Nürnberg.
4. Storath E. 1996: *Kontextsensitive Wissensbereitstellung in der Konstruktion.* PhD thesis, Universität Erlangen-Nürnberg.
5. Löffel C. 1997: *Integration von Berechnungswerkzeugen in den rechnerunterstützten Konstruktionsprozeß.* PhD thesis, Universität Erlangen-Nürnberg.
6. Weber J. 1997: *Konzept eines rechnerunterstützten Assistenzsystems für die Entwicklung umweltgerechter Produkte.* PhD thesis, Universität Erlangen-Nürnberg.
7. Büttner K. 1997: *Rechnerunterstütztes Konfigurieren von Baukastenprodukten.* PhD thesis, TH Darmstadt (also published in *VDI-Fortschrittsberichte Reihe 20 Nr. 246,* VDI-Verlag, Düsseldorf).
8. Hubka V. 1976: *Theorie der Konstruktionsprozesse – Analyse der Konstruktionstätigkeiten.* Springer-Verlag, Berlin
9. Pahl G., Beitz W. 1997: *Konstruktionslehre.* Springer-Verlag, Berlin
10. Sander S. 1996: KONSUL – Ein System zur kontextsensitiven Suche anforderungsgerechter Lösungen. In: H. Meerkamm (ed.) 1997. *Fertigungsgerechtes Konstruieren – Beiträge zum 7. Symposium 1996.* Universität Erlangen-Nürnberg, pp. 118–123.
11. Meerkamm H., Sander S., Mogge C. 1997: Das Internet als Medium zur kontextsensitiven Bereitstellung von Konstruktionswissen auf Basis der ISO 13584. In: VDI-Tagung Karlsruhe *Der Ingenieur im Internet.* VDI-Verlag, Düsseldorf.
12. Sander S. 1997: Die Modellierung von Baugruppen auf der Basis der ISO 13584 PLIB im Konstruktionssystem mfk. In: *Proceedings 3. Magdeburger Maschinenbau-Tage – Entwicklungsmethoden und Entwicklungsprozesse im Maschinenbau.* Logos-Verlag, Berlin, pp. 159–168.
13. Meerkamm H., Storath E. 1996: Produktentwicklung mit Tele Engineering – Potentiale von verteilten und integrierten Netzwerkdiensten. In: *VDI Bericht Nr. 1302,* VDI-Verlag, Düsseldorf, pp. 155–173.
14. Wartzack S. 1998: Optimieren der Prozeßkette durch Auswählen der bestgeeigneten Baustruktur und fertigungsgerechtes Gestalten der Bauteile. *Arbeits- und Ergebnisbericht 1996-1998 DFG - Sonderforschungsbereich 396, Teilprojekt A1.* (in preparation).
15. Meerkamm H. 1995: Engineering Workbench – ein Schlüssel zur Lösung komplexer Konstruktionsprobleme. In: *ICED 95 Vol. 4.* Heurista, Zürich, pp. 1261–1268.
16. Hochmuth R. 1997: Konstruktionsbegleitende Analyse auf Fertigbarkeit von Toleranzfestlegungen auf Basis des Konstruktionssystems mfk. In: H. Meerkamm (ed.) *Fertigungsgerechtes Konstruieren – Beiträge zum 8. Symposium.* Universität Erlangen-Nürnberg, p. 49.
17. Hubka V., Eder W. E. 1996: *Design Science.* Springer, London.

Résumé of Section V

There are many methods in systematic design with the emphasis on highlighting the significant design steps and, thus, supporting the whole design process. But surprisingly, there are only very few approaches that aim to characterise the design tasks to make the designers more aware of a problem's specific requirements. Such a distinction of problem requirements seems to be important, not only for a successful design strategy (e.g. in terms of the suitable use of methods), but also for the comparison of results and processes in design research.

Starting with the analysis of the design task, designing can be described as the aggregation and structuring of information. Consequently, the central question for supporting design work is how to structure information so that it can easily be understood and processed. Such design support is mainly done via design methods and tools. These tools have to be quick and easy to use, i.e. they have to allow flexible changes of resolution level – abstract/concrete, total/part – and different methods of representation.

These requirements can often be partly very easily met by traditional but effective tools such as sketching (and models). A main advantage of sketches is their back-up to mental capacity, which is limited by the designer's working memory span.

However, the enormous amount of relevant information and the trend to globalisation in the development of complex products also necessitate support by modern tools, such as tele-working, Internet online services, video conferences, etc. The rapid evolution of information technology is a great chance to evolve user-friendly design support. The new tools are used in various situations with different goals for specific individuals and/or groups, but with little knowledge about the question of efficient application. Additionally, the design process is not only supported but also *changed* by the use of modern tools.

In summary, the use of basic and high-tech tools depends on several factors, such as the individual preferences of the users, which are often neglected by the technical-oriented focus on design tools. For example, designers may differ in regard to preferences for a more textual or more graphically based arrangement of information. Obviously, further research is necessary in order to improve design tools by considering the prerequisites of the designers and the specific influences in different design situations.

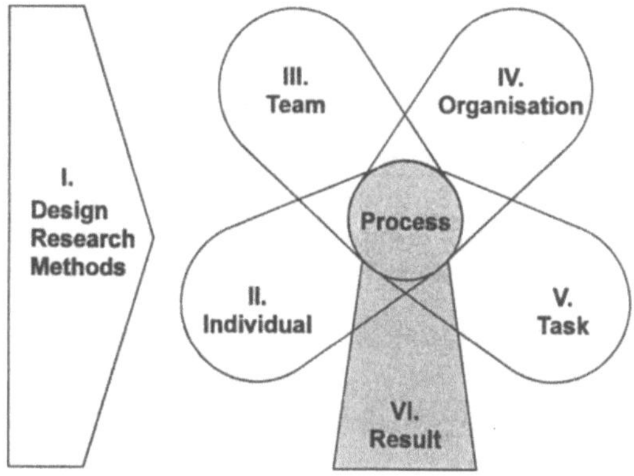

Section VI:

Evaluation of Design Work and its Result

The evaluation of design work is a key problem in design research and practice. This question concerns the evaluation of single steps as well as connected activities in the design process and especially the evaluation of the result. The course of this section considers the questions:

What theoretical considerations can we use to *model the development of information* in design steps in order to assess design performance?

How can we *evaluate* a design activity, process or product?

How reliable are those evaluations and what *influences* on their result do we perceive?

A popular approach to design productivity is the overall evaluation of design as a business activity with 'inputs' and 'outputs'. Thus, this section starts with the introduction of a *pre-model* for assessing product development performance, based on theoretical considerations. Next, the real difficulties in evaluating design concepts in empirical investigations are presented. The section closes with an overall analysis of the factors that influence design success.

- *Duffy & O'Donnell* approach the question of design performance from a business perspective. They introduce the concept of evaluating design activities via a black-box approach based on systems theory (the sum total of knowledge/information/data in–out) and define efficiency and effectiveness as characteristics to evaluate in later empirical studies.

- The problems of evaluating design results are shown by *Wiegers, van Breemen, Knoop & Vergeest*. They illustrate differences in rankings based on four concepts, giving reasons for their conclusions. In addition, they present interesting observations on individual factors that influence how information is dealt with in the problem-solving process.

- From the point of view of a general manager, *Fricke* discusses various influences on success. He stresses the different dimensions of complexity in industrial product development in terms of goals, product, process and team.

A Model of Product Development Performance

Alex H.B. Duffy & Frank J. O'Donnell

Abstract

The overriding aim of much product development research is to improve the performance of the Product Development Process (PDP). Yet the concept of performance as applied to this area is currently without underlying theory and no consensus exists on its definition or the most appropriate metrics to use in evaluation. This paper presents a review of research in PDP performance to illustrate the various issues that have received attention in this area. A definition and model of performance are proposed which are directed at developing a fundamental understanding of the phenomenon. The model illustrates the main elements of performance – efficiency and effectiveness – and provides a basis upon which to identify, formalise and analyse suitable performance metrics at any level or range in the product development process. Factors influencing performance have been assessed by the authors and sample results are presented. It is concluded that, although the model determines the main elements of performance, further research is necessary to truly formalise performance and specify a suitable approach for its measurement.

1. Introduction

In the continuous struggle to maintain or increase market share in current competitive environments, organisations are tending to focus many of their improvement efforts on their Product Development Process (PDP). Having defined and implemented the PDP within the company, the interest often moves towards the performance of that process and ways in which performance can be monitored and increased.

Current research efforts in areas such as Concurrent Engineering, Team Engineering and Design Management are aligned with this focus on performance improvement. It is anticipated that the adoption of results from these research areas by industry, through process improvement initiatives, will lead to increased performance. The ultimate success of such research efforts may then become apparent through systematic measurement of performance, e.g. on a 'before-and-after' basis or as part of an internal/external benchmarking initiative. Yet there are many uncertainties surrounding the practice of performance measurement in product development. A great deal of work has already been carried out in measuring the performance of the manufacturing process [1] but there are inherent differences in the nature of development compared with manufacturing [2] which make performance measurement more difficult [3]. Indeed the whole concept of performance as applied to this area is currently without underlying theory and little consensus exists on its definition or the appropriate metrics that may be used. Some of the issues that have attracted research interest include investigations of performance at different levels, such as at project or programme level, and the relationships between performance at these levels [4–9], the comparison of product development performance across industries [10–13] and the factors that affect performance [6, 14–17].

The aim of this paper is to investigate the issue of how to formalise product development performance and to develop a better understanding of the phenomenon as a basis for performance measurement. A model of performance is developed, based on the concepts of efficiency and effectiveness, and related to other views in the area to illustrate generic elements. Results from assessing the effectiveness of various product development aspects in meeting a company's goals are presented as part of the current work. Finally, research areas requiring further investigation are identified.

2. Performance in Product Design and Development

The task of determining relevant values for performance of the product design and development process is a complex one. The definition of a process model such as that described in [18] provides a starting point in understanding the process. However further issues must be resolved in order to fully understand performance aspects. The following discusses some of the research in this area to illustrate these issues and describe some of the approaches adopted in attempts to address them.

2.1 Definition of Performance

A concise definition of performance as applied to product development is not found in the literature, although in one of the more extensive works in this area a model is proposed (Figure 1). Clark and Fujimoto [7] suggest that performance is composed of three parameters i.e. lead time, productivity and total product quality. This model forms the basis of their study of the automobile industry in which they collect data from 25 projects in 20 companies. The authors investigated whether performance along the three parameters is ultimately reflected in the firm's competitive position. The study is heavily based on data analysis from completed projects and aimed at developing correlation between different aspects of product development and performance. Consequently the work is focused more on aspects influencing performance than the nature of performance itself. The proposed model appears to have little theoretical foundation and consequently does not contribute significantly towards a generic understanding of performance.

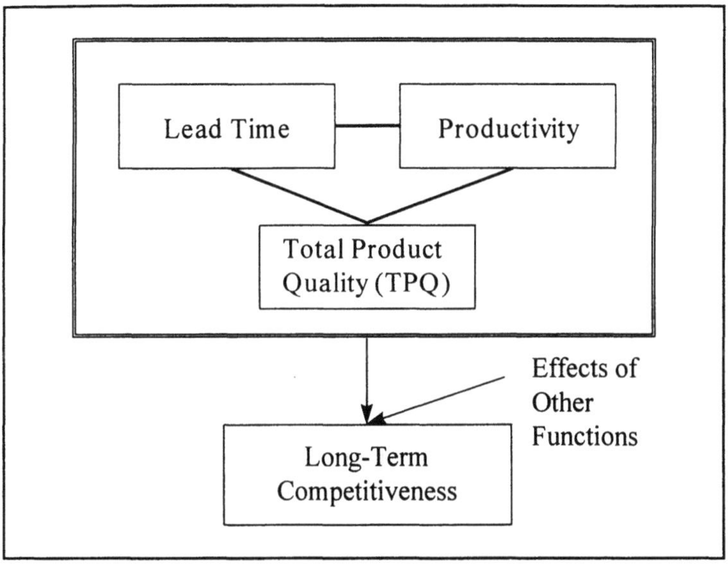

Figure 1: Parameters of performance [7].

There are various aspects of the PDP that relate to performance, such as efficiency, effectiveness, capacity, productivity, capability, etc., yet we do not understand how these relate or indeed their true meaning when used to describe product development. The most common terms to appear in work aimed at evaluating the PDP have included *efficiency* and *effectiveness* [6, 14, 15, 19], although in only a small number of cases have there been attempts to explicitly define these terms, e.g. [18, 20]. In both [19] and [20] efficiency is seen to relate to the process and could be described in terms of the level of resources used, such as time, man hours, materials etc., whereas effectiveness is determined from the result of the process, e.g. through analysis of the final product characteristics, such as quality and functionality with respect to the initial requirements and goals.

The use of effectiveness in describing performance supports the fact that different companies will have different sets of goals and importance weighting on these goals arising from differences in programme and/or project strategy. For example, a company that wishes to break into a new market with a completely novel product may specify goals for the PDP that relate to the degree of innovation of the final product, with little concern for the resources used. In contrast a company, such as a car manufacturer, which wishes to launch a revised 'model' on an already established market, may define goals that reflect a desire to focus on reducing the cost of the process. In addition to these explicit goals there are often implicit goals that should be represented, e.g. learning may be an implicit goal of all projects but, unless explicitly stated and the result measured in these terms, it is unlikely that this goal will be achieved adequately or inefficiencies highlighted. Performance should therefore be measured with respect to specific goals so that the values obtained are directly relevant to the company.

2.2 Performance Metrics

One of the initial concerns in measuring performance is establishing the correct metrics with which to obtain performance values. Although the general aspects of performance, such as time, cost and quality, are evident in most studies, there is little similarity between the specific metrics used for measurement and in some cases no recognition of the different nature of each company, i.e. there is no effort to adjust metrics to allow for product complexity before comparison between different projects or companies. It is widely recognised that performance involves a complex set of variables and viewing it in terms of time, cost and quality does not give sufficient insight and understanding of performance [3]. The metrics used in studies range from those directly related to the product, such as quality and performance, to process metrics and more corporate metrics related to the impact of the product on the market. It is evident from the extensive range presented in the literature that the difficulty lies not in defining performance metrics but in narrowing down the resultant list to a more concise and relevant set.

The most common metric quoted in the literature is *lead time* which generally reflects the time taken from the beginning of concept development to the introduction of the product to the market place, although it may also be used in relation to specific project phases such as detailed design. Taken in isolation this measure may mean little and indeed it has been shown that the link between speed of development and eventual success is a tenuous one [21]. Lead time often forms the basis for many related metrics such as the number of *man-hours*, *break-even-time* and *time-to-profitability*. More complicated metrics that estimate the productivity of the process are also time related, such as *elapsed time per component per unit* [12] which is used in an effort to reflect product complexity and allow more accurate comparison of results across industries. In addition, the metrics used will depend on the performance viewpoint that is being taken, i.e. the analysis may be at programme level, investigating the combined performance of all product development projects, or it may be focused on the performance of a specific project phase.

Given the range of metrics and the complexity of their links, it is necessary not only to define a set of performance metrics, but also to classify them in order to formalise the relationships between them. The application of specific metrics, without understanding how they influence each other, can result in an inaccurate and incomplete picture of performance. Some effort has been applied to classify these metrics. For example, Griffin & Page [22] have used a project strategy topology to assist in selecting metrics. However, further work is needed to derive a more detailed classification with defined relationships. This must be based on a better understanding of the nature of product development performance.

2.3 Scope of analysis

Within an organisation the performance of product development can often mean different things to different people. In general, product development performance may be viewed in terms of different *levels,* from the overall product development programme to the implementation of that programme within individual projects and eventually to specific activities. In addition, the performance analysis will be applied to product development over a particular *range,* e.g. a phase in a project or a series of individual activities within a phase. Therefore the subject of analysis can exist within a particular range and level (Figure 2) and it is necessary to define the scope clearly to ensure accurate interpretation of any results.

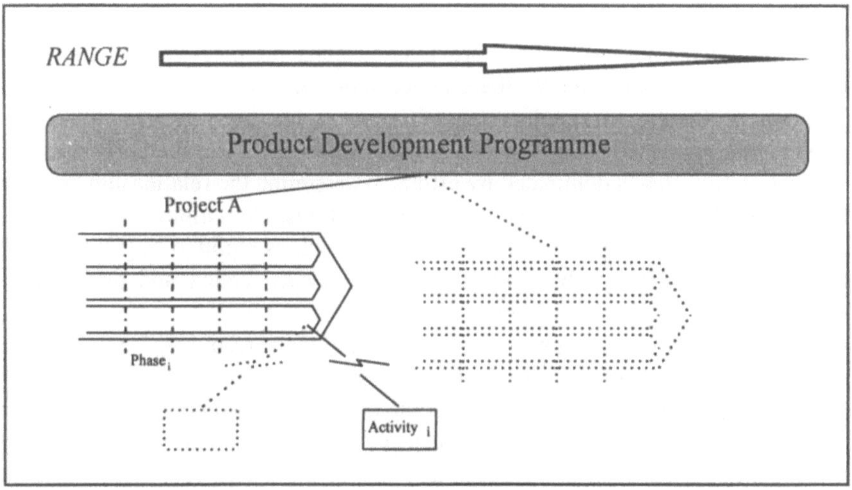

Figure 2: Scope of performance analysis.

Some of the studies of product development performance reported in the literature have been focused on a company, or macro, level viewpoint, such as that reported in [10], although most have investigated project level aspects, e.g. [19]. At the company level the issue of performance is generally related to the totality of the product development programme, encompassing a number of projects at different stages, and their combined impact on company level issues. Often this level of analysis is centred on the success of the product [23] and is therefore based on the effectiveness of the product development process. That is, it investigates how well the product meets the high-level programme goals relating to issues such as market share, corporate image and technical success. In an assessment of project performance the focus is likely to move closer to the process and encompass aspects of both efficiency and effectiveness. Typically these studies investigate issues such as lead time and productivity, in addition to properties of the process output (the product) in terms of quality, functional performance, degree of innovation, etc.

It is clear that there are links between different levels of performance [4, 7], and effective management and coordination of these links is required to ensure that successful projects do not exist in isolation but contribute fully to the overall programme success at company level [3]. In addition, formalising these links will help to ensure that performance metrics employed throughout a company are properly aligned with strategic goals and without conflicts [24]. This prevents the situation where a project team's performance may be analysed using unaligned metrics, resulting in high performance values, while in reality contributing little to the overall company goals.

2.4 Factors Influencing Performance

Perhaps of most interest to researchers in this area is the question of what factors affect performance of product development and whether it is possible to measure this effect in some way. There are many claims on how various initiatives improve performance, such as the claimed reduction in cycle time through the adoption of Concurrent Engineering (CE), yet it has been proven difficult to provide hard evidence in support of such claims [25, 26]. The literature reporting research in this area is dominated by efforts to determine the relationship between the organisational structure, practices, methods, tools, etc. employed in the PDP and the level of development success [8, 11, 13] but, as reported in [6], the result is often an empirically observed correlation between factors and performance values with limited theoretical understanding.

Currently many organisations are spreading their resources too thinly in seeking to improve various aspects of their business without focusing on the most relevant areas. It has been observed that the more successful companies are often quite weak in some aspects of their business but lead the field where it matters [27]. If the link between factors such as tools, methods and approaches, and the aspects of performance that they influence can be determined, this would provide a basis for focusing process improvement efforts. Companies could use this understanding to focus their investment on those areas that have the greatest impact on achieving their performance goals.

3. A Performance Model

From the material reviewed in this paper it has been concluded that there is a need for a model of product development performance that contributes towards a more fundamental understanding of the phenomenon. In addition it is considered here that:

- efficiency and effectiveness are the central aspects of product development performance;
- appropriate performance metrics are dependent on individual cases and their definition requires a better understanding of performance;
- company specific goals must be addressed in measuring performance to ensure relevant results;

- the scope of the performance analysis must be represented and the links between levels/ranges maintained;
- identifying the factors that influence performance provides the key to successful process improvement.

We next outline a model of performance based on these observations, and the relationship between the model and existing views in this area is discussed. Work aimed at identifying factors that influence performance, based on the understanding provided in the model, is also presented.

3.1 Model Overview

In order to represent and discuss the nature of performance, the IDEF0[1] [28] formalism has been adapted to illustrate the concepts of efficiency and effectiveness within product development (Figure 3). The IDEF0 box is used to represent the product development function. Essentially this function is a boundary that encompasses a product development activity or number of activities across a specific range and at a particular level in the organisation. The standard IDEF0 formalism includes controls and mechanisms that are represented here as the goals/constraints and resources respectively. The input to the activity is represented as *K*nowledge, *I*nformation and/or *D*ata (KID_i) which is transformed by an activity/activities within the function resulting in an increase in one or more of these elements when delivered as output (KID_o). This transformation is achieved at a cost in terms of resources such as people, methods, tools, materials, etc., while being directed and controlled through the input of goals and constraints.

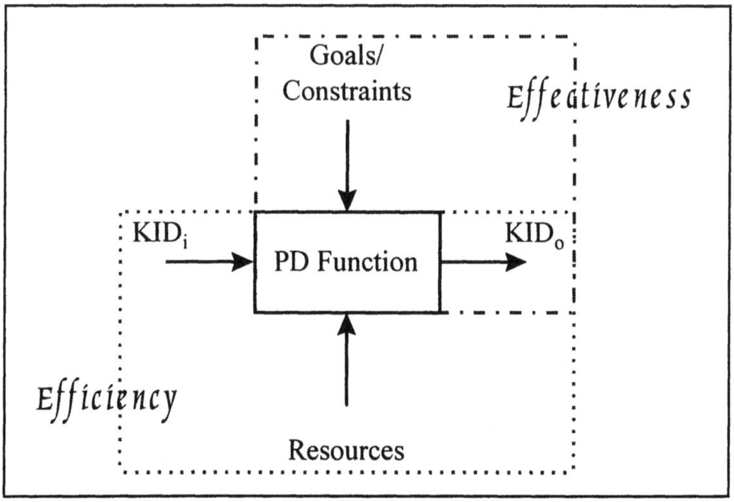

Figure 3: Product development performance.

---

[1] Integrated Computer-Aided Manufacturing DEFinition.

The efficiency of the Product Development (PD) function is presented here as a measure of how economically the output is generated, i.e. efficiency is represented as the ratio between the change in knowledge/information/data ($KID_o - KID_i$) and the cost incurred:

$$Efficiency \equiv KID_o - KID_i / (Cost\ of)\ Resources$$

It is proposed here that effectiveness is a measure of how well the output of the function, i.e. KID_o, meets the goals and satisfies the constraints. Thus, it may be measured through comparison of these two elements:

$$Effectiveness \equiv Comparison\ of\ KID_o\ with\ Goals/Constraints$$

The overall performance of the activity is a combination of its efficiency and effectiveness. Therefore:

$$Performance \equiv Combined\ measure\ of\ Efficiency\ and\ Effectiveness$$

It is believed that these definitions form the basis from which appropriate performance metrics may be derived. Thus, metrics should support the measurement of inputs, e.g. past design information, resources used, etc., and the outputs obtained in a way that allows the calculation of efficiency. In addition it is necessary to devise metrics that allow the comparison of the PD output with goals, resulting in values for effectiveness.

From the definition it can be seen that it is not feasible to measure the performance of the PD function directly but only the activities that are contained within it, i.e. at its most fundamental level, performance is an aspect of, and is used to describe, behaviour. This principle also applies when analysing the performance of a particular product. For example, the fuel consumption of a car while driving at a constant speed is considered to represent an aspect of its performance. This value is essentially an output of the activity, created for test purposes by car manufacturers, describing the car in constant motion. The efficiency and effectiveness of the car are illustrated in the same way as that related to the PD function above. Thus, metrics for efficiency should be derived which relate:

- the input, i.e. current speed at a particular position (datum);
- the output, i.e. new speed and distance travelled; and
- the resources used, i.e. the fuel consumed, which is represented in KID_o after measurement.

Similarly, effectiveness metrics must allow the difference between the actual fuel consumed and the design goal for consumption to be determined.

The scope of analysis may also be clearly identified using the above formalism. This modelling approach supports the decomposition of activities into sub-activities, goals into sub-goals, etc., while maintaining consistency between the different levels of the model (Figure 4). For example, the higher level goals and constraints associated with the product development programme may be decomposed into goals and constraints that relate directly to a project, project phase, etc. The approach therefore supports the development and maintenance of links between different levels of performance analyses as discussed in 3.3. Both the efficiency and effectiveness of activities may be analysed at different levels, provided the inputs and output, i.e. KID_i, goals/constraints, cost of resources and KID_o, are known.

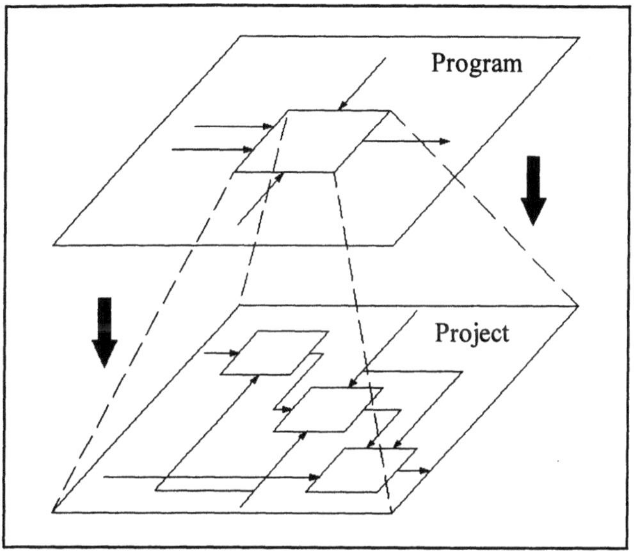

Figure 4: Decomposition of performance model.

3.2 Relation to Existing Views

The model presented above is aimed at developing a more fundamental understanding of performance than much of the previous work in this area and therefore is different in nature from previous efforts. However, in the following section the model is compared against a range of views on aspects of performance in order to further illustrate its key elements.

Andreasen and Hein [18] address efficiency, one of the key elements of performance, and define it as:

$$\text{Efficiency} = \frac{\Delta\text{clarification} + \Delta\text{risk reduction} + \Delta\text{detail} + \Delta\text{documentation}}{\Delta\text{costs}}$$

It may be seen that this definition of efficiency relates closely to the one for efficiency previously given in this paper, i.e. the change in KID relative to the costs incurred. While the meaning of documentation is not explicit in the Andreasen's & Hein's definition, the degree of clarification, risk reduction and detailing all relate directly to an increase in knowledge, information and/or data. Having said this, the scientific basis for these four parameters is unclear and it would seem that they are elements of, but do not completely describe, the total change in knowledge information and data.

The three parameters of performance described by Clark and Fujimoto [7] in relation to project level performance may be related directly to the representation of performance provided here. The lead time is seen as the time taken to transform the input to the output and is viewed here as a resource.[2] These parameters define the productivity in terms of the level of resources used in the process, which is again directly reflected in the model here. Finally the concept of total product quality is seen as a measure of how well the output of the PD function meets the goals and satisfies the constraints. Thus the parameters in the model proposed by Clark and Fujimoto [7] are seen to represent elements of the formalism presented here, i.e. lead time and productivity relate to efficiency while total product quality is an aspect of effectiveness. In contrast, the parameters have been applied in a study without explicitly addressing their relationship to company specific goals.

At a corporate level of analysis, Marks & Reilly suggest that there are three primary business objectives in any organisation [29], i.e. to increase:

a) market size – e.g. through innovation and provision of new products
b) market share – e.g. through providing preferred products
c) margin on sales – e.g. through reducing costs.

Although there is no clear demarcation between the three objectives, it may be seen that the first and second objectives require a focus on effectiveness in order to ensure the customer requirements are met by the product(s). However the last objective may be achieved with a greater focus on efficiency of the PDP and specifically the reduction of costs.

In addition, Marks & Reilly argue that all performance goals must be defined in terms that contribute towards achieving one or more of the three business objectives, and that any measured increase in performance must translate into an increase in market size, market share and/or margin on sales. For example, if we reduce the lead time of a project and identify this as an increase in performance, then we must be able to determine the relationship between lead time and market size, market share or margin on sales in order to justify our conclusions. This type of relationship has not been formalised in existing research and further work is required in this area to ensure alignment between performance strategy, goals and metrics. The performance model presented here provides a starting point for this task in that the strategic objectives stated above may be applied to the business activity and decomposed to relevant goals and activities in product development while maintaining consistency.

---

[2] While time per se is not truly a resource, the time allocated to a project, phase, etc. may be viewed as such.

3.3 Assessing Influencing Factors

Recognising the two main elements of performance as efficiency and effectiveness, the authors have been developing a technique for measuring the effectiveness of different Product Development aspects. This involves assessing the impact of various tools, methods and approaches (typically CAD, QFD, CE, etc.) on achieving the specific PD goals of the company.

The method of obtaining values for effectiveness is similar to Quality Function Deployment (QFD) in that it is a team-based exercise employing matrices to determine the relationships between the tools/methods/approaches and the overall PD goals of a company. Thus, it ensures that any initiatives aimed at enhancing the PD performance are directly aligned with overall company strategy. A form of internal benchmarking [30] is employed where both the current and ideal effectiveness of the PD aspects are estimated. The exercise is demanding and requires a dedicated team of personnel to provide honest and well thought out input to the process. The data are analysed and presented in a number of different forms, e.g. Figure 5, to give both a comprehensive picture of current and ideal effectiveness and the level of investment required for improvement. This supports the need for an organisation to identify particular product development aspects where improvement initiatives should be focused to best achieve specific and/or overall goals. That is, it allows the organisation to become more effective in the use of their key tools, methods and/or approaches.

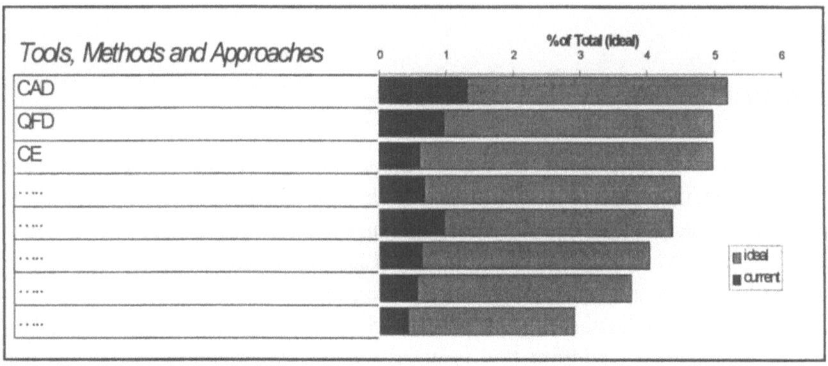

Figure 5: Typical data analysis (effectiveness of PD aspects).

4. Discussion

Research in the area of performance analysis in product development has been found to be empirically based and descriptive in nature. That is, current work tends to report the results of surveys on product development performance and the degree of success of particular projects. Little effort has been made to draw theoretical or prescriptive conclusions from the majority of this work. In addition, such research is not based on a common model of the product development process and/or performance, leading to ambiguity in the scope and nature of the analysis.

The model of product development performance presented here provides a better understanding of the phenomenon and a basis from which an approach to performance measurement may be developed using metrics derived from the model. It relates closely to other research in this field and supports elements of other work. The formalism used allows the model to be applied at any level in an organisation and across any range. However, obtaining values for performance is in itself worth little if the factors that influence these values cannot be identified. Without a better understanding of these factors it is difficult to fully justify some of the research efforts in the field of process improvement. The technique discussed in 4.3 provides a starting point based on an analysis of the effectiveness of product development aspects with respect to individual and collective goals.

Although a model of performance, based on efficiency and effectiveness, has been developed, further research is required to determine exactly how efficiency and effectiveness can be measured, the exact nature of their relationship to each other and how this will translate into an overall value of performance. In addition, the technique presented in 4.3 allows a degree of effectiveness to be established between aspects and goals. However, the technique is subjective and does not involve an analysis of the relationships to determine why they exist. Further work is currently under way, including industrial case studies, in an attempt to evaluate and validate the model and technique presented.

5. Conclusion

This paper has attempted to clarify some of the central issues related to performance in product development which seem to have received insufficient attention in research to date. A study of current work in the area was based on four key aspects of product development (PD) performance: definition, metrics, scope of analysis and influencing factors. From this study it can be concluded that a fundamental understanding of product development performance is lacking within the field. To address this issue a generic performance model has been proposed and related to other work in the area. This model provides a basis upon which to identify, formalise and analyse suitable performance metrics, while ensuring that performance measurement is carried out in order to directly enhance the achievement of company goals. Consequently the model supports specified objectives at any level or range in the product development process. Based on an improved understanding from the model, a technique has been developed to assess the effectiveness of product development tools, methods and approaches in realising identified PD goals. This technique is currently undergoing evaluation within an industrial context.

Particular issues that this paper has addressed include:

- *What are the central issues in product development performance?* The central issues include the definition of performance, the metrics for performance measurement, the relations between different scopes of analyses and the identification of factors that influence performance.
- *How may performance be defined and measured?* Performance has been defined in relation to its two key elements, i.e. efficiency and effectiveness. It is unlikely that a complete set of metrics can be created for use in all performance measurement exercises. Specific metrics must be defined based on a fundamental understanding of performance and these should relate to company specific goals in order to ensure relevance in the results.
- *How can we relate different levels/ranges of performance?* The model presented here supports decomposition and may be used to represent performance at any level and across any range while maintaining consistency. Thus it provides a basis upon which to ensure consistency in the approach to performance measurement from a strategic level down to the level of individual activities. In addition it is recognised, and addressed in the model, that in order to optimise performance it is necessary that the measurement approach is directly aligned with the specified goals at all levels.
- *How can we identify the factors that influence performance?* A technique has been presented, based on the definition of effectiveness in the performance model and on the method of Quality Function Deployment, which allows the key factors that influence performance to be identified. This technique has been applied in an industrial context, resulting in the definition and implementation of performance improvement initiatives which are currently ongoing.

The answers provided here are a starting point towards developing a more fundamental and generic understanding of the phenomenon of product development performance. There remains a considerable task in truly formalising performance and specifying an approach for its measurement. The achievement of this task would provide companies with a basis upon which to target scarce resources in order to maximise performance and profit.

References

1. Neely A., Gregory M., Platts K. 1995: Performance Measurement System Design: A Literature Review and Research Agenda. *International Journal of Production and Operations Management* 15, 80–116.
2. Gilbreath R.D. 1988: Working with Pulses Not Streams: Using Projects to Capture Opportunity. In: D.I. Cleland & W.R. King (eds) *Project Management Handbook.* Van Nostrand Reinhold, New York, pp. 3–15.
3. Cusumano M.A., Nobeoka K. 1996: Strategy, Structure, and Performance in Product Development: Observations from the Auto Industry. In: T. Nishiguchi (ed.) *Managing Product Development.* Oxford University Press, Inc., pp. 75–120.

4. Nobeoka K., Cusumano M.A. 1995: Multiproject Strategy, Design Transfer, and Project Performance: A Survey of Automobile Development Projects in the US and Japan. *IEEE Transactions on Engineering Management* 42, 397–409.
5. Loch C., Stein L., Terwiesch C. 1996: Measuring Development Performance in the Electronics Industry. *Journal of Product Innovation Management* 13, 2–20.
6. Brown S.L., Eisenhardt K.L. 1995: Product Development: Past Research, Present Findings, and Future Directions. *Academy of Management Review*, 20, 343-378.
7. Clark K.B., Fujimoto T. 1991: *Product Development Performance: Strategy, Organisation and Management in the World Auto Industry*. Harvard Business School Press.
8. Cantamessa M. 1997: Design Best Practices at Work: An Empirical Research upon the Effectiveness of Design Support Tools. In: A. Riitahuhta (ed.) *Proceedings of the 11th International Conference on Engineering Design (ICED'97)*, Tampere University of Technology, Tampere, Finland.
9. Cordero R. 1989: The Measurement of Innovation Performance in the Firm: An Overview. *Research Policy* 19, 185–192.
10. Cooper R.G., Kleinschmidt E.J. 1995: Benchmarking the Firm's Critical Success Factors in New Product Development. *Journal of Product Innovation Management* 12, 374–391.
11. Griffin A. 1997: *PDMA Research on New Product Development Practices: Updating Trends and Benchmarking Best Practices*. Institute for the Study of Business Markets, The Pennsylvania State University, University Park, PA.
12. Oliver N., Gardiner G., Mills J. 1996: *Design and Development Benchmarking in the UK Electronics Industry*. Judge Institute of Management Studies, University of Cambridge, Cambridge, UK.
13. Schrijver R., Graaf R.d. 1996: Development of a Tool for Benchmarking for Concurrent Engineering. *Concurrent Engineering: Research and Applications* 4, 159–170.
14. Griffin A. 1996: *The Impact of Engineering Design Tools on New Product Development Efficiency and Effectiveness*. Institute for the Study of Business Markets, The Pennsylvania State University, University Park, PA.
15. McDonough E.R., Griffin A. 1996: *The Impact of Organisational Tools on New Product Development Efficiency and Effectiveness.* Institute for the Study of Business Markets, The Pennsylvania State University, University Park, PA.
16. Hull F.M., Collins P.D., Liker J.K. 1996: Composite Forms of Organization as a Strategy for Concurrent Engineering Effectiveness. *IEEE Transactions on Engineering Management* 43, 133–142.
17. Kahn K.B. 1996: Interdepartmental Integration: A Definition with Implications for Product Development Performance. *Journal of Product Innovation Management* 13, 137–151.
18. Andreasen M.M., L. Hein 1987: *Integrated Product Development.* IFS (Publications) Ltd and Springer, London.
19. Hauptman O., Hirji K.K. 1996: The Influence of Process Concurrency on Project Outcomes in Product Development: An Empirical Study of Cross-Functional Teams. *IEEE Transactions on Engineering Management* 43, 153–164.
20. Duffy A.H.B. 1998: Design Productivity. In: A.H.B. Duffy (ed.) *The Design Productivity Debate.* Springer, London.
21. Meyer M.H., Utterback J.M. 1995: Product Development Cycle Time and Commercial Success. *IEEE Transactions on Engineering Management* 42, 297–304.

22. Griffin A., Page A.L. 1996: PDMA Success Measurement Project: Recommended Measures for Product Development Success and Failure. *Journal of Product Innovation Management* 13, 478–496.
23. Cooper R.G. 1994: New Products: The Factors that Drive Success. *International Marketing Review* 11, 60–76.
24. O'Donnell F.J., Duffy A.H.B. 1997: Understanding Product Development Performance. *Internal Report, CAD Centre, University of Strathclyde. CADC/97-22/R/04.*
25. Lawson M., Karandikar H.M. 1994: A Survey of Concurrent Engineering. *Concurrent Engineering: Research and Applications* 2, 1–6.
26. Tomiyama T. 1998: Concurrent Engineering: A Successful Example for Engineering Design Research. In: A.H.B. Duffy (ed.) *The Design Productivity Debate*. Springer-Verlag, London pp. 175–186.
27. Coulson-Thomas C. 1997: *The Future of the Organisation: Achieving Excellence through Business Transformation*. Kogan Page, London.
28. *IDEF0 Forms and Procedures Guide*, 1979, SofTech Inc., MA, USA.
29. Marks P., Reilly K. 1995: *Aligning Technology for Best Business Results.*
30. Rolstadas, A. (ed.) 1995: *Benchmarking – Theory and Practice*. Chapman and Hall, London.

Valuation of Design Concepts

Tjamme Wiegers, Ernest van Breemen, Willem Knoop & Joris Vergeest

1. Introduction

1.1 Assigning a Value to a Design Concept

In the early stages of design, the designer handles large amounts of information, of many kinds. Furthermore, between the moment when the customer requirements are recorded until the first embodiment proposal is finished, many uncertainties exist. Examples are dimensions that are not yet known, weights and shapes, but also product structures and even product functions may be unknown beforehand, especially in innovative projects. In this situation the designer has to assess the importance of different kinds of information, and how these influence each other. The designer has to reach an understanding of the design problem as a whole, and at the same time to incorporate more detailed information of the individual aspects. This requires complex and intensive information handling. It is still a major research issue as to which method would be the most effective to support the designer in this task. In particular, there is little being written about measuring the effectiveness of a new method or tool in a given type of design process [1]. Valuation of design concepts is used as a measure, a reference, for the impact of design support.

1.2 Measuring Influences on the Value of a Design Concept

In order to know what the influence of design support is, we have to compare a design process in which a certain means of support is incorporated, with a reference design process. We need to make the design processes measurable. Then we can make statements about whether or not the proposed influence leads to an improved result [1, 2].

1.3 Structure of this Paper

After discussing some valuation methods, we will describe the results of the application of these methods to our data. The differences between the methods are discussed, and we conclude with some advantages and disadvantages of the applied methods.

2 Some Valuation Methods

2.1 Introduction

Here we will show that there is no one-and-only method to evaluate a design process. It is not meant as an in-depth study or a complete overview. The following methods are discussed:

- The designer's opinion
- The expert's opinion
- Fulfilment of the requirements
- Internal process parameters.

The different methods mentioned here all have their own strengths and limitations. Because of that, often a combination of several methods is used. For example, Christiaans [3] studies internal process parameters and relates them to experts' opinions. Blessing [1] uses both internal process parameters and designer's opinions for the verification of her hypotheses. Fricke [4] studies internal process parameters and relates them to the fulfilment of the requirements and to the technical validity of the designs. Furthermore, he considers the overall drawing quality, as a condition for proper validation. Van Elsas [5] considers both the designer's opinion and the designer's performance to evaluate the support tool that he developed.

2.2 The Designer's Opinion

One method to evaluate a design process is very simple: just ask the designer's opinion. The designer is biased, but if you want to know the impact of a change in the design process, the designer may give you important information. The designer is thrown into the design process and experiences it from the inside, without explicitly recording all his experiences. Therefore, asking the designer's opinion may lead to inside information that cannot be retrieved from other data capturing methods. This method is often used in addition to other, less biased methods, e.g. [1]. Some methods use the quality of the design concept as a measure of the quality of the design process (see 2.3 and 2.4). This assumes that a better design process results in a better product, but ignores the effects of other characteristics of the design process, such as the time devoted to it. In a laboratory setting we can overcome this by performing design experiments of equal duration.

2.3 The Expert's Opinion

To ask for the opinion of one or more experts is a realistic method. Realistically, in an industrial situation, a design concept will often be judged by an expert, or by a group of experts. With this method, the judgement does not depend on the opinion of the designer. But it is still biased. Individual experts may, for example, use their own criteria, or they may deal differently with compromises between conflicting criteria. An advantage of this method is that it can be used in domains for which there are no simple, unbiased quality criteria. An example of this is the work of Christiaans [3] in which he used experts' opinions to rate the creativity of design concepts.

2.4 Fulfilment of the Requirements

Valuation based on the fulfilment of requirements [6, 7] seems an obvious method if we want to be independent of personal judgement. This method uses an inventory of the essential requirements. If the requirements differ considerably in importance, weighting factors can be assigned to them. General technical and economic characteristics can also be taken into account. For each of the requirements or characteristics, a value is assigned to the design. The value of the design is obtained by the summation of these individual values (and possibly other calculations). The use of a predetermined scheme reduces bias, but the assignment of values still depends more or less on the judging expert. Fricke [4] showed that a great consensus can be achieved by the application of this method. However, the domain of the method is restricted. Requirements with regard to the appearance of a product, for example, are hard to judge in an unbiased way. Another problem is that this method considers the requirements to be beyond discussion. In real life, however, requirements are often re-interpreted or changed during the course of a project [8, 9].

2.5 Internal Process Parameters

The designed concept is not the only base for rating a design process. The quality of the design process can also be determined from internal process parameters. For example, the kinds of activities that occur, and the time spent on those kinds of activities; or the numbers and types of drawings that are made. To capture these parameters, the design process is observed and its events are recorded in protocols. When the values of such parameters are used to rate the process, this implies assumptions about the optimal values of these parameters. In other words, it requires certain hypotheses about the ideal design process. Blessing [1], for example, formulates characteristics of design before she analyses process parameters. The observation of process parameters can also serve other goals. If another method is used for evaluating the design process, we can look for relations between the process parameters and the value of the process [3]. This may result in empirically based considerations of the process parameters. Finally, this method can be used to test a hypothetical model of the design process [10].

3 Valuation Methods Applied in this Research

3.1 Our Design Case

The goal of our research is to gain more knowledge about the conceptual design process, especially about the information that is handled by designers. This knowledge will be translated into requirements for tools or methods to improve the process. We gain our knowledge through empirical research. Our research contains the following steps:

1. Definition of a design assignment that can be performed in a laboratory setting.
2. Observation of a designer who performs the design assignment, followed by analysis.
3. Formulating requirements for tools or methods to improve the process.
4. Proposal of such a tool or method.
5. Repeating the design assignment, this time however with a change: the use of the proposed tool or method.
6. Comparison of the design process with and without the proposed change.

We have gathered data from ten similar design sessions in which a subject (an industrial design engineer) had to generate a concept for a new lighting system for a bicycle. The assignment that the subjects had to fulfil is a typical industrial design engineering project, it is multi-disciplinary (mechanical and electrical) and it is a re-design problem. We recorded the activities and the information-handling activities of the subjects. The framework for this experiment is depicted in Figure 1. All design sessions lasted two hours and a half. The results of these sessions are described in van Breemen [11]. We also have data from three design sessions that were performed after a change was introduced. The analysis of these data is under way.

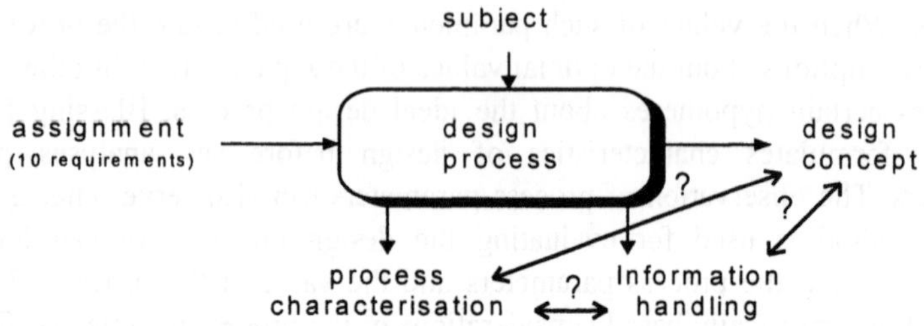

Figure 1: The framework for our experiment.

Existing observation methods did not fulfil our demands, therefore we have developed our own observation method [12, 13]. The method is focused on observable information (drawings, written or spoken words) rather than the thoughts in the designer's head. The subjects had to fill out questionnaires containing, among other things, questions to check for specific domain knowledge. The subjects could request information from an experimenter located in the same room. The experimenter provided simple answers. No hints, advice or judgements were given. The subject was asked to briefly describe the start of any new activity, and any result obtained. In addition, at regular time intervals, the experimenter asked the subject to briefly describe his or her current activity, possible results, and possible needs for information. All questions, answers, activities and results (either pronounced or directly observable) were listed during the session by a second experimenter, watching the session on a video monitor in a separate room. This list, together with the subject's writings and drawings, was translated into a structured protocol file, by following a well-defined procedure [14, 15]. Based on these protocols, we did several types of analyses. We also presented the results of the subjects to seven experts and asked for their opinions. The four methods of evaluation as described in earlier, are applied to our data and discussed in the following paragraphs.

3.2 The Designer's Opinion

After the design assignment was finished, the subject was asked to complete a questionnaire. Most of the questions concerned the setting of the experiment or the backgrounds of the subject. They are beyond the scope of this paper. Two of the questions were about the designer's performance. One concerned the designed concept, the other was about the process.

The designed concept. The question was: Are you satisfied with the result?
Five subjects answered yes, three subjects answered no, and two subjects answered both yes and no. One subject was unsure about the working of a subsolution (subject 3), and one subject found the concept insufficiently detailed (subject 5).

The design process. The question was: Are you satisfied with the way you worked? Six subjects answered yes, three subjects answered no, and one subject ticked 'yes', but annotated 'but not quite'. He noted some requirements, but too late in the process.

Three subjects (s) thought their work was too unstructured (s7, s8, s9), and three subjects had time problems (s3–s5). Some subjects noted problems in their design process, but were still satisfied with it (s2, s3, s5, s8). Table 1 summarises the results.

Table 1: Satisfaction of the designers with their concepts and processes.

Subject:	1	2	3	4	5	6	7	8	9	10
Satisfied with the design *concept*	yes/ no	yes	no	yes/ no	no	yes	no	yes	no	yes
Satisfied with the design *process*	yes	yes/ no	yes	no	yes	yes	no	yes	no	yes

3.3 The Experts' Opinions

3.3.1 Introduction

Following contacts with professional designers in the bicycle industry and with expert teachers of generator design, we selected seven jurors to express their opinion on the design results from our experiments. Because we are interested in the relative performance of the designers, we asked the experts to express the order of perceived quality, by relating the designs to each other, while using whatever arguments they deemed necessary. In addition, we asked the experts to evaluate 10 aspects of the designs.

The method used to interview the experts had some constraints:

- The method must ensure that the experts give their opinion in an unprejudiced manner.
- To facilitate study of the differences and similarities of the criteria that the experts use, and the criteria we used in our experiment, the aspects to judge must be related to the requirements used in the experiment.
- The weight of the arguments used should be estimated.

In our experiment we observed a lack of information in the documentation that the designers made. This limited us in evaluating the requirements. In a considerable number of cases we could not say with certainty whether or not a design could meet a certain requirement. Therefore we asked the experts in this study to express the 'ease of judgement' that they experienced. The general set-up of the interview with the experts is described in van Breemen [16].

3.3.2 Placing the Designs in Order of Perceived Quality

After evaluating and criticising the ten design results, the experts were asked to arrange the designs in order from best to worst. It was stressed that they could use whatever means or criteria they needed to place the designs in whatever order they wanted. The design judged as the best is ranked with 10 points, the next design with 9 points, and so on. The seven rankings, thus obtained, agreed to a reasonable extent. (The Kendall Coefficient of Concordance is 0.6069, where 0 stands for no agreement and 1 means full agreement, see McPherson [17]).

Table 2: The order in which the ten designs were placed by the seven experts.

exper / ranking	1	2	3	4	5	6	7
10 (best)	8	1	8	2	5	8	8
9	2	8	2	7	8	2	2
8	3	3	6	3	6	3	6
7	7	7	3	9	7	7	9
6	9	2	7	1	9	6	3
5	6	6	9	6	2	9	1
4	4	9	1	5	1	1	5
3	1	5	5	8	3	10	7
2	10	4	4	4	4	5	4
1 (worst)	5	10	10	10	10	4	10

Each column contains the judgement of one expert. The numbers in the columns are the rankings given to the different designs. The design judged as the best is ranked with 10 points, the next design with 9 points, and so on.

Table 2 indicates that there is a high consensus about which designs belong to the top two: numbers 8 and 2. And 6 out of 7 experts agree on the two worst designs: numbers 10 and 4. But, on the other hand, some designs are placed on very different positions by the individual experts. We see this for designs 1, 5, 7 and 8, with design number 5 as the extreme. Figure 2 shows the total rankings of each design. The ranking is presented as the percentage of the maximum possible. For example, design 8 has a total ranking of 10 + 9 + 10 + 3 + 9 + 10 + 10 = 61 points, out of 70, which works out at 87%.

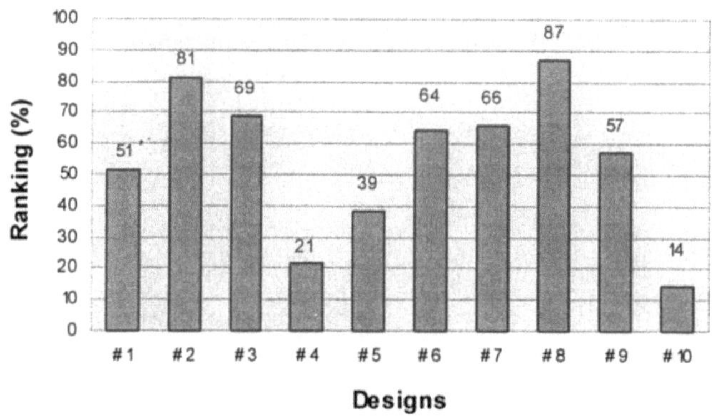

Figure 2: The mean ranking of the 10 designs.

3.3.3 Scoring the Designs on 10 Aspects

The experts were asked to express their opinion on 10 aspects of each design. Every aspect was characterised by two antonyms, expressing contrasting extremes of the aspect, see Table 3.

Table 3: The aspects on which the experts were asked to express their opinion.

Aspect no.	Description (positive)	Description (negative)
1	Fully integrated	Hotchpotch
2	Position well chosen	Position absolutely wrong
3	Universally applicable	Very specific for one brand and typ of bicycle
4	Easily mountable	Mountable only with great effort
5	Outstanding appearance	Poor appearance
6	Is fully functional	Is not functional at all
7	Pioneering	Imitation
8	Weatherproof	Fine-weather-only product
9	Efficient	Wastes energy
10	Correct basic principle of functioning	Wrong basic principle of functioning

The experts could choose from five check boxes between those extremes. The five check boxes were given the values 0%, 25%, 50%, 75% and 100% respectively, and we calculated for each aspect the average of the scores of the seven experts. Unfortunately, from these average values we cannot deduce the scores given by the individual experts. Their judgements were markedly different. The average scores of the 10 designs on all aspects are presented in Figure 3.

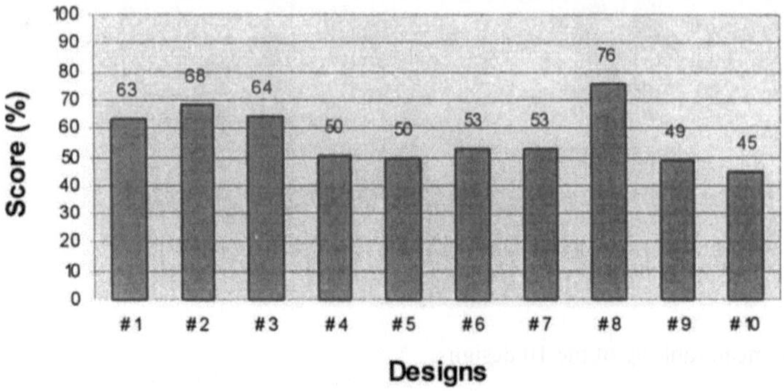

Figure 3: The mean scores of the 10 designs, when scored on the aspects of Table 3.

When we compare these scores with the ratings of the placing in perceived order, in Figure 2, we see that in both cases designs number 8, 2, and 3 score best, and design number 10 scores worst. However, there is not much similarity for the other designs, and design number 4 scores particularly differently.

3.4 Fulfilment of the Requirements

Table 4 shows which requirements had to be fulfilled by the designs. To express the mutual importance of the requirements, they were provided by the researchers with weights which together amount to 100%. To determine of the weights we applied the method of pairwise comparison, described by Zangemeister [18].

Table 4: Requirements and their weights.

	Requirement	weight (%)
1	Single product	5.5
2	On rear axle	5.5
3	No on/off mechanism	1.8
4	General applicability	14.5
5	Easily mountable	12.7
6	Powered lights	16.4
7	Light at any speed	9.1
8	Light at zero speed	7.3
9	Max. 12% extra power	18.2
10	Water resistance	9.1

The weighted totals of the 10 designs are presented in Figure 4. Generally, these values are not very high. This is partly because subjects often produced no proper descriptions or sketches that could be used to evaluate whether the requirements had been met. This happened in spite of the fact that we stressed the need for clear descriptions before the experiment, as well as in the design brief. Another reason is that the available time for the design assignment was only two and a half hours. Many subjects found this period too short to make a sufficiently detailed design concept.

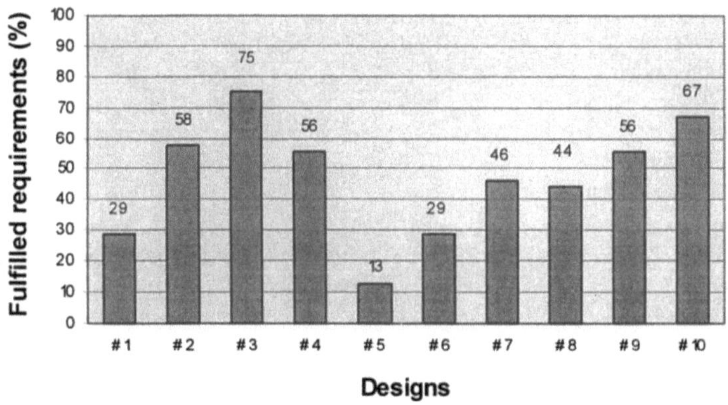

Figure 4: The percentage of fulfilled requirements of the 10 designs.

3.5 Internal Process Parameters

We analysed the collections of information addressed by the individual subjects during their design assignment [19]. An independent expert designer identified what information was needed to perform the design assignment. We distinguished between engineering issues and engineering aspects.

An engineering issue is an overall description of an engineering item. The expert was given the assignment and asked to list all engineering issues that should have been taken into account to solve the problem. In this way 22 engineering issues were indicated. In addition, two other issues were addressed by the subjects. The issues are listed in Table 5.

Table 5: Engineering issues and the number of subjects that gathered information about them.

	Engineering issues	No. of subjects used to gather information about the issue
1	Rear axle area	10
2	Generator types	10
3	Generator output	9
4	Calculation models	8
5	Batteries	8
6	Magnets	7
7	Maximum energy	7
8	Stand light performance	6
9	Charging	4
10	Connections	3
11	Mounting	3
12	Materials	2
13	Tooling	2
14	Tests	1
15	Patents	1
16	Splash and drip resistance	1
17	Bearings	0
18	Cost price	0
19	Bicycle type	0
20	Quantities	0
21	Environment	0
22	Assortment	0
23	On/off switching	1
24	Wear	1

Within the engineering issues we have described the engineering aspects. They can be interpreted as a subdivision of the issues. No single aspect was addressed by all the subjects. We found only seven aspects that were addressed by at least seven subjects. They concern generator type, lighting consumption and the maximum extra energy that the cyclist must deliver. All subjects requested information about the bicycle's rear axle area, but individual subjects requested different aspects of this issue. The number of aspects that were addressed by each of the ten subjects is presented in Figure 5. The number of aspects addressed correlates well with the percentage of fulfilled requirements, as presented in Figure 4, see [19]. However, if the subject addresses a certain group of aspects continuously, and for a long time, this is an indication of a serious block in the design process [20].

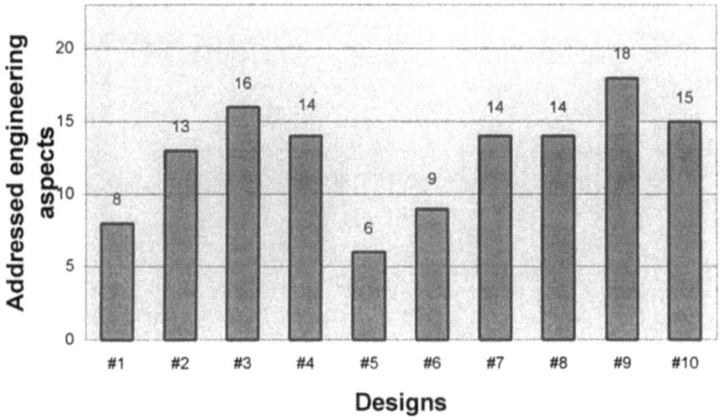

Figure 5: The number of aspects addressed by the 10 subjects.

4 Evaluation Methods Compared

Now we compare the results of the different evaluation methods. For this purpose we plotted of all the previous results on one graph, see Figure 6. In spite of the large amount of data, we can easily recognise two pairs of data sets that show some similarity.

At first we see similarity between the number of engineering aspects addressed and the percentage of requirements fulfilled. To make this similarity more obvious we scaled up the number of engineering aspects that were addressed, see the right vertical axis in Figure 6.

Apparently, in our case, the process parameters measured were a good indicator for the percentage of requirements fulfilled; but not for the opinions of the experts or designers.

The second similarity is between the experts' ranking of perceived quality, and the scores given on the 10 aspects of the design. Surprisingly, this similarity can not be found when we look at individual experts, see [16]. It only occurs in the average values of all experts. Here we see (just as in scoring the designs on ten aspects) that the final judgements of the experts correspond to a certain extent, but differ markedly on the individual aspects.

There is no indication that subjects who judge their own concepts or processes as good do indeed get higher values in the experts' judgements, or in the rating of fulfilment of requirements.

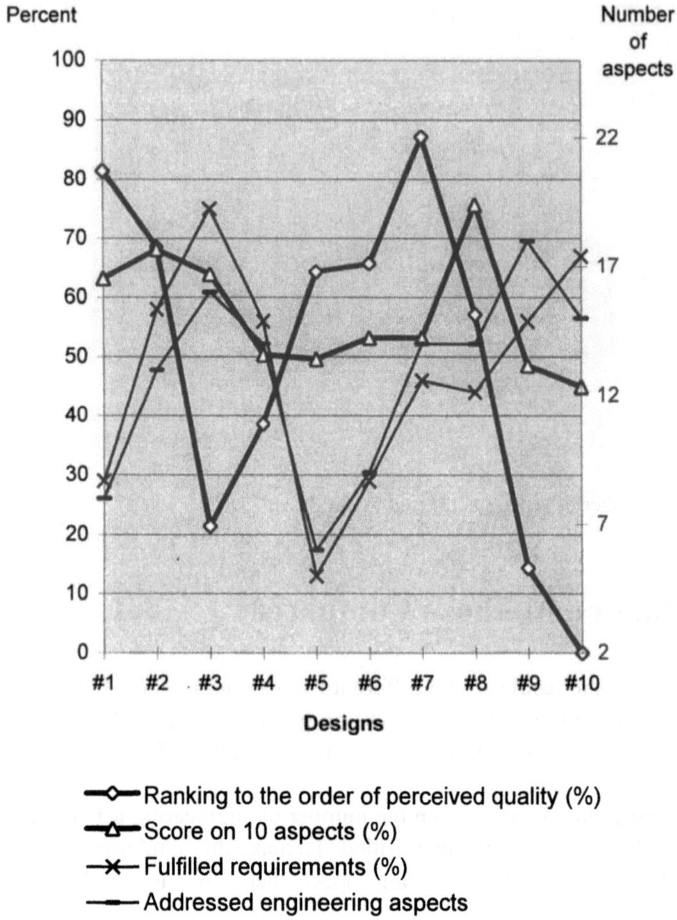

Figure 6: All previous results combined into one plot.

5 Discussion

We have looked at four methods of evaluating designs. We applied these different methods to the same design case and compared the results. Different methods can lead to very different results. It is important to match the method to be applied to the influences that are investigated. If an unbiased comparison of design processes is the goal, then evaluation based on the fulfilment of requirements is an applicable criterion. But this does not guarantee the agreement of an expert from industry. Experts often take aspects into account that were not in the list of requirements. This can be overcome if the requirements themselves are set by experts (see the discussion in [21]). Furthermore, experts may differ in the requirements that they focus on. They especially differ in the way they value aspects that are not listed in the requirements. We give a few examples:

- Some of the designs did not work because the generator was placed on an axle that did not turn. Most of the experts considered this as an unforgivable error, and ranked these design as the worst ones. But one expert considered this as a fixable problem, so he ranked these designs significantly higher.
- One design was placed on the front wheel, although one of the requirements stated that the rear axle must be used. Some experts discounted the design because of this discrepancy. But one expert thought the front wheel was indeed a better place, and the design allowed for placement on the rear wheel as well.
- Nearly all experts emphasised the visual appearance of the product, more than the electrical details. The latter, however, are essential for the operation and working life of the product.

In several cases, the documentation of the subjects was ambiguous or incomplete. This was a handicap for the proper evaluation of the designs. It may also have influenced the differences in the results that stemmed from the different methods.

Our interest in evaluation methods stems from the need to measure the effect of a specific influence on the design process. Our goal is to observe design processes with and without a specific influence, and evaluate them. Then we can make statements about the effect of the influence. However, we cannot yet present empirically based statements. The following should be considered as our initial impressions of the influences that we noticed during our observations.

We noted several subjects who spent much time in detailing part of a design that contained a fundamental error. Typically, in those cases the subjects wasted a lot of time because they were unaware of the error. That awareness could be made apparent by the application of an evaluation method or tool.

However, we also noted on several occasions that a subject delayed evaluation, as if to postpone the possible need for change should the result of evaluation be poor.

Sketching often helped subjects to get more insight into the design problem [22]. Some subjects generated a number of concepts from which they selected the most appropriate. During the generation of later concepts, they often applied knowledge that they had gained from earlier ones. Apparently their early concepts contributed not so much to a possible solution, but more to an exploration of the design problem, its constraints, and its dominant aspects. We also observed that subjects differed in their estimation of the importance of aspects.

An effect seen time after time was that subjects, when they received new information, immediately tried to apply it. Perhaps they did this to get a feeling of what the role of the information received might be. However, it is also possible that they hoped the new information to be the miraculous missing link that would solve their problem as soon as they applied it.

When a subject received a lot of information within a short time, this often seemed to block the process. This influence may be related to the one above, the one of immediate application of received information, because this attitude is only practicable if the information is administered in appropriate doses.

Subjects were very pleased that there was an experimenter who could immediately answer their questions. Possibly this is because continuing while waiting for an answer forces a designer to keep track of the context in which the question occurred. This is confirmed by our observations that subjects did not start another activity even if finding the answer to their question took some time. Exceptions occurred in cases where a subject posed a question while generating the sketch of an idea. The explanation may be that generating the sketch of an idea is already the recording of the context itself. It needs no separate effort to remember the context of the question, because the reason for the question is already in the sketch. Christiaans and Van Andel [23] report similar findings. Instead of an experimenter, they used a hypertext information system to support information retrieval by the subjects. The system was extensively used and it was appreciated by the subjects.

During our design experiment, as described in 3.1, at regular time intervals the experimenter asked the subject to briefly describe her or his current activity, possible results, and possible needs for information. We call these regular questions *samples*. On a number of occasions these samples provided information about activities and results that were not expressed by the subjects on their own initiative.In several instances a subject took a decision and terminated current activity immediately after giving a sample. These findings may indicate that regularly describing the current activity helps the subject to remain aware of what he or she is aiming at, to stick at the job and to generate results quickly before moving to other areas.

Lack of domain knowledge was another important influence for many subjects. They had difficulties within the following domains:

- generation of electricity
- charging and decharging batteries
- energy balance and effect of energy losses
- construction and geometry of a bicycle rear wheel.

Many designs did not work because of errors in these domains. Moreover, subjects spent so much time on aspects of these domains that they neglected other important aspects. Subjects did not readily make assumptions for values or solutions in a domain they were not familiar with. Instead of making such an assumption, they tried to clarify all the details before addressing other aspects of the design. Just as a chain is no stronger than its weakest link, it seems that a designer's understanding is restricted by the aspect he understands the least. That makes lack of domain knowledge a real bottleneck. The designer has to obtain the information lacking, understand it and become familiar with its application, before he can attain a balanced understanding of the design as a whole.

References

1. Blessing L.T.M. 1994: *A Process-Based Approach to Computer-Supported Engineering Design.* Thesis, University of Twente, Enschede, The Netherlands.
2. Vergeest J.S.M. 1996: The Position of CAD Development in Design Research. In: I. Horváth & K. Váradi (eds) *Proceedings of the International Symposium on the Tools and Methods for Concurrent Engineering (TMCE'96).* Technical University of Budapest, Budapest, Hungary.
3. Christiaans H.H.C.M. 1992: *Creativity in Design – the Role of Domain Knowledge in Designing.* Thesis, Lemma, Utrecht.
4. Fricke G. 1993: *Konstruieren als flexibler Problemlöseprozeß – Empirische Untersuchung über erfolgreiche Strategien und methodische Vorgehensweisen beim Konstruieren.* Fortschritt-Berichte VDI Reihe 1 Nr. 227, VDI-Verlag, Düsseldorf.
5. Elsas P.A. van 1997: *Free-form Displacement Features in Conceptual Shape Design.* Thesis, Delft University Press, Delft.
6. Pahl G., Beitz W. 1996: *Engineering Design – A Systematic Approach.* Springer, Berlin, London.
7. VDI-Richtlinie 2225 1977: *Technisch-wirtschaftliches Konstruieren.* VDI-Verlag, Düsseldorf.
8. Archer L.B. 1979: Whatever Became of Design Methodology? *Design Studies* 1, 17–20.
9. Franke H.J. 1975: Methodische Schritte beim Klären konstruktiever Aufgabenstellungen. *Konstruktion* 27, 395–402.
10. Lewis W.P. 1981: The Role of Intelligence in the Design of Mechanical Components, Man-Machine Communications. In: T. Sata & E. Warman (eds) *CAD/CAM*, North-Holland Publishing Company, IFIP.
11. Breemen E.J.J. van 1996: The DESYS D_1 Experiment: Empirical Results of Observing Product Development Processes. *Technical report TUD-IO-K329 Delft University of Technology*, Faculty of Industrial Design Engineering, Engineering Department, Delft, The Netherlands.
12. Knoop W.G., Breemen E.J.J. van, Vergeest J.S.M., Wiegers T. 1996: Towards More Effective Capturing of Empirical Data from Design Processes. In: O. Akin & G. Saglamar (eds) *Proceedings of the Conference on Descriptive Models in Design*, Department of Architecture, Carnegie Mellon University, Pittsburgh, PA, USA.
13. Vergeest J.S.M., Breemen E.J.J. van, Knoop W.G., Wiegers T. 1995: An Effective Method to Analyze Chronological Information Aspects in Actual Engineering Processes. In: Q. Sun et al. (eds) *Proceedings of the 5th International Conference on Computer Applications in Production and Engineering (CAPE'95)* Chapman & Hall, London.
14. Knoop W.G., Wiegers T. 1995: The DESYS D_1 Experimental Procedure: A Method to Observe Information Handling in Actual Product Development Processes. *Technical Report TUD-IO-K328, Delft University of Technology*, Faculty of Industrial Design Engineering, Engineering Department, Delft, The Netherlands.
15. Knoop W.G. Wiegers T. 1995: The DESYS D_1 Experimental procedure: Appendices. *Technical Report TUD-IO-K340, Delft University of Technology*, Faculty of Industrial Design Engineering, Engineering Department, Delft, The Netherlands.
16. Breemen E.J.J. van 1998: Quality Assessment of Design Results from Experiments, Report on Quality Assessment, Performed by Industry Experts, of the Results of 10 Design Processes Performed in Experiments. *Technical report TUD-IO-K363, Delft*

University of Technology, Faculty of Industrial Design Engineering, Engineering Department, Delft, The Netherlands (in preparation).

17. McPherson G. 1990: *Statistics in Scientific Investigation.* Springer, Berlin.
18. Zangemeister C. 1973: *Nutzwertanalyse in der Systemtechnik* (in German), 3rd edition, pp. 158–163. Zippel-Druck, Berlin.
19. Wiegers T., Breemen E.J.J. van, Knoop W.G., Vergeest J.S.M., 1996: Required and Requested Information During Conceptual Engineering Design as Found by Categorization of Empirical Data. In: J.M. McCarthy (ed.) *Proceedings of the ASME 1996 Design Engineering Technical Conferences* (published on CD-ROM) American Society of Mechanical Engineers, New York, index 1533.
20. Wiegers T., Knoop W.G. 1997: Discovering Information Patterns During Conceptual – Design. In: A. Riitahuhta (ed.) *Proceedings of the 11th International Conference on Engineering Design (ICED'97),* Tampere University of Technology, Tampere, Finland, 2, pp. 433–436.
21. Knoop W.G. 1997: The Emperical Research Discussion Platform. *Technical Report TUD-IO-K366, Delft University of Technology*, Faculty of Industrial Design Engineering, Engineering Department, Delft, The Netherlands.
22. Hacker W., Sachse P., Schroda F. 1998: *This volume,* pp. 205–216.
23. Christiaans H.H.C.M., Andel J. van 1997: *This volume,* pp. 233–248.

Successful Industrial Product Development

Gerd Fricke

1. Introduction

Industrial development projects involve such complexity that an individual action of organisation will only be one aspect relating to the success of the product development. Frequently the goals are not clearly defined and will change during the development process. The proceedings and the methods of problem-solving are not known in detail and coordination of the parties involved is a complicated matter. Neither performance-related nor uncoordinated proceedings will automatically result in complete failure.

The sales figures are decreasing. A new product is to be developed and produced at a low price.

Relying on the many years of experience and the intuition of the engineering manager, the entire design department enthusiastically starts working on the development of a totally new product that their manager has some knowledge of. The sales manager is informed about this some weeks later and, together with the production department, strongly advocates that an already existing product has initially to be developed further, in order to reduce its costs. Now the design department starts working in parallel: ideas and first proposals for both goals are worked out. The marketing department is sceptical and, after some time has passed, starts to analyse the market situation on its own. Then, a sudden increase in customer complaints requires that a well-established product has to be modified immediately. The head of the design department now tries to reconstruct the product in cooperation with a new engineering design firm working on very favourable cost terms. However, after some time it is found that their CAD systems are not completely compatible with those of the main firm. Working under great pressure, a software company cannot find a solution to this problem. And although first drafts of the new development are now available and tests of the first prototypes are being run, most of the design staff is spending time on trying to eliminate the reason for the complaints. However, after 2 months have passed the sales figures are beginning to rise again. The board of directors enquires about the state of the new product's development. The marketing study carried out has identified good potential in particular for the high price segment not yet considered. The head of the design department is under more and more pressure and entrusts ...... [1].

The successful development of industrial products requires mastery of a complexity of goals, products, processes and interfaces, as well as the collaboration of the great number of different persons and institutions. In this respect, development is considered to be the process from the first idea about a product up to its introduction to the market. With the background of joint scientific research of individual designers [2, 3] in mind, some realities of industry will be analysed. These occurred in the course of several developments that Norbert Dylla and myself were supervising personally at three market leaders in different sectors. The goal is now to derive performance enhancing measures from the results.

2. Complexity of Industrial Product Development

The complexity of industrial development results from the type, number and networking of goals, from product elements, processes and interfaces, together with the method of collaboration (Figure 1). Normally, this far exceeds the complexity of those tasks that are examined in detail at universities or covered during training. Accordingly, it will, of course, not be possible to consider all important aspects that contribute significantly to a positive outcome in industry.

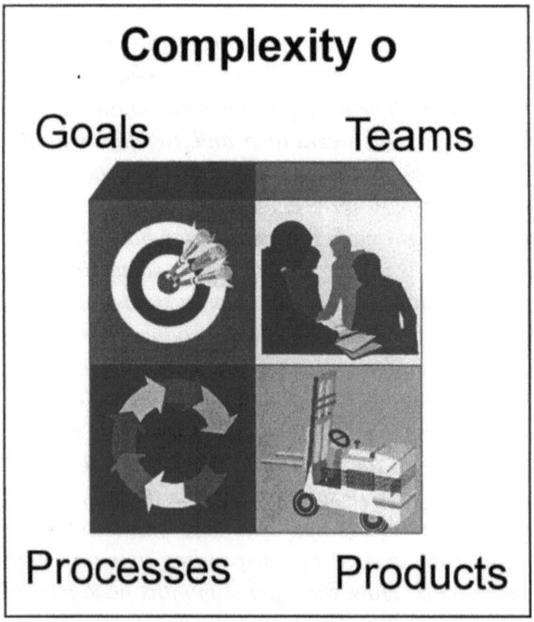

Figure 1: Complexity of product development.

2.1 Complexity of the Goals

Functional goals

A great number of frequently unclear objectives (definition, formulation) are often for the product and its 'handling' (i.e. production and marketing). These are connected to each other in various ways (e.g. specific functions can only be realised in connection with other functions) and may unexpectedly change in the course of development (dynamics).

Time and cost goals
As an aggravating circumstance, limited and clearly defined time and cost requirements will have to be met in addition to the functional goals mostly established in a fuzzy way at the beginning.

To furnish an example of the complexity of the goals, consider a so-called 'innovative development' [1]: a new apparatus is to be developed for a market of which nothing is known so far. A proposal has been presented to a large manufacturer for gas heating units, for example, to develop a gas heating system for vehicles in Eastern Europe. Future customers will comprise both the users (individual final consumers) and, on the other, the producers of vehicles (major customers). The behaviour of the two customer groups differs greatly from what we are used to in Western Europe. For example, the final consumers do not require hot water as a standard feature, and the production process of the rather small-scale vehicle manufacturer involves a lot of craftsmanship. Though only very rough concepts have yet been established, the behaviour of the customers has to be determined in a comprehensive and realistic way before any feasibility studies are prepared. Frequently, unrealistic proposals and requests (marketing) or unnecessary limitations on the product requirements involving inapplicable emphases (technical engineering) are made. Therefore analyses beyond the limits of the individual departments and the fixing of development goals have to be performed by the marketing department (i.e. marketing, field services, service) in cooperation with the technical department (i.e. development, purchasing, production, repair). Only then will it be possible to identify the requirements of the future customer in the most realistic and comprehensive way and thus, to satisfy them. Only then will orientation toward customer demand be more than a catchword.

In the early phases of product planning, only the selected use of certain methods is considered practical, for example 'benchmarking', 'product portfolio', 'house of quality' and the systematic definition of performance specifications. Moreover, product planning must essentially be based on internal business and development goals and strategies, as well as on clearly defined responsibilities and operations. To sum up, such a base must be 'lived'. To reach this goal, various individuals will have to collaborate adequately, oriented toward the company's overall success and the satisfaction of the customer. Such collaboration and personal exchange of information are characterised by misunderstandings and mental leaps in the early stages of development. These initial developments will require input from individuals and free thinkers, though all efforts will have to be directed toward the primary goals.

Concerning the definition of the goal, many uncertainties will remain, irrespective of a systematic procedure in which customer requests and product requirements are defined in specific, goal-oriented discussions. Moreover, far-reaching and clear decisions will have to be taken in the early phases. If, for example, any increase of performance is requested or the use of the heating units in other countries is specified, further time-consuming feasibility studies (constructions, tests, approval tests) might have to be planned and more expensive overall solutions found.

In addition there is also the requirement to establish the product in the market as quickly as possible. This means that activities will have to run simultaneously (Simultaneous Engineering) that normally would only be carried out after the results of some had become available. The higher the degree of overlapping, the higher the risk (Figure 2). Irrespective of the uncertainties outlined above, specific proceedings and decisions will have to be interlocked, coordinated, moved, revised and speeded up beyond the confines of individual departments. The purchasing department, for example, will need to buy the tool steel required for a plastic operating panel even though no final clarification of the operating functions has yet been obtained and the temperature resistance tests are still running.

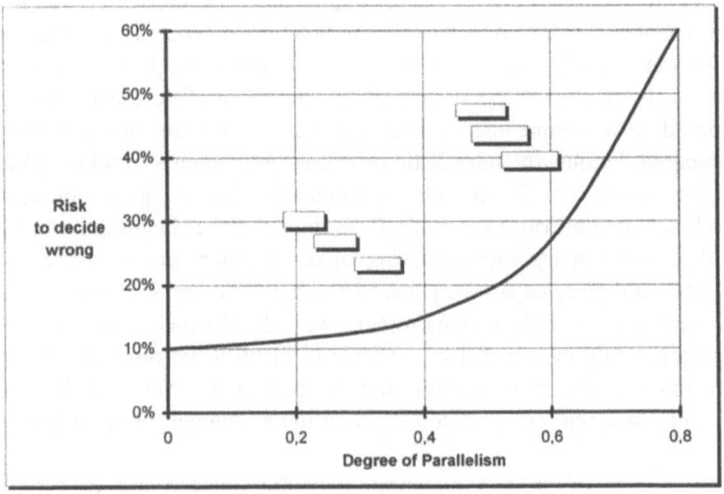

Figure 2: Risk of simultaneous engineering.

Finally, if a competitor places a product on the market with additional functions that have not been thought about, even more changes will have to take place. Dynamic planning will be necessary, flexible enough to be permanently updated and directed toward the overall goal, over and over again.

Taking the example of the heating units for Eastern Europe, the cost pressure would be increased further by the region's moderate spending power – that is why cost optimisation as an accompanying feature will have to be taken into consideration permanently throughout the entire process, from the first idea, to design and production, up to service and/or disposal. Intensive permanent collaboration is essential by those responsible for marketing and development, for purchasing (including both suppliers and production) and for overall control and sales. In addition, the involvement of management at regular intervals throughout the entire development process is an absolutely essential prerequisite for a successful project. To encourage that success, a transparent optimisation of costs and target dates is necessary.

2.2 Complexity of the Product

The complexity of many products, especially in plant engineering, is great; i.e. the requirements (functional goals) established by the customer can only be met by using a great number of components that are frequently highly dependent on each other. A machine for the production of newspapers, for example, comprises components for unwinding the paper webs, printing units, elements to guide, cut and turn the paper webs and to fold the product. Some of these components are, by themselves, very complex assemblies of modules requiring constant interaction. A unit for four-colour printing of paper on both sides consists of several thousand parts, many of which will have to be designed in accordance with special customer requests. And all these adaptations will have to be done after consideration of the dependencies within the component itself and of the marginal conditions of its neighbouring components. All mechanical parts of the assembly with its electric drives will also have to work in harmony with the decentralised and superior electronic controls.

In addition to that, the newspaper printing machine is only one part of a production plant which, for example, also includes components for the transportation, storage and preparation of the paper webs, as well as for transportation and further processing of the actual newspapers. This increases the complexity even more.

The more components – parts of which will have to met extreme requirements – that have to be combined in modules, the more difficult the coordination of product development will be, and the more complex the realisation of the product being ready for immediate use will be. This is particularily the case with plant engineering.

2.3 Complexity of the Process

Complexity of the process is defined by the number of sequences and product elements (for example parts, module) that will require permanent coordination via different interfaces, in particular for cases when modifications that influence the process take place over and over again. This, for example, may apply to series products when sequences involving a great number of single elements in purchasing, logistics, production and marketing have to be anticipated during various product development phases (i.e. various trial, test, pilot and initial series).

This situation is aggravated further when different electronic, mechanical and other working principles are involved and have to interact, requiring several specialists to find an overall optimum solution, and when new results obtained from more recent investigations demand basic modifications in other areas.

2.4 Complexity of the Team

In international corporations, development projects can no longer be carried out at isolated locations. Irrespective of the type of production – single-item or series production – international coordination is required between internal and external individuals and institutions with different motivations, qualifications and languages.

Should requirements become more and more demanding (more functions at less cost, available sooner, global use, safe in quality), further complex development projects cannot be realised effectively by departments operating in isolation. Powerful teams need to cooperate in a goal-oriented manner beyond the limits of single departments. Successful management of a development project demands basic technical knowledge and product as well as market experience. On the other hand, the social competence of the project manager is of the utmost importance to successful cooperation in the development team. Sensitivity in specific situations will be necessary: positive attitudes in the search for new ideas are to be promoted, critical reflection furthered in analysis and evaluation, discordances among the staff quickly eliminated, and a clear flow of information guaranteed to management and other persons concerned. Commitment and initiatives, in addition to the ability to express criticism, must be adequately supported. These many different activities can only be successfully coordinated if collaboration is effected in such an open and responsive way.

To master these complex dependencies in daily industrial operations in order to bring developments to the market successfully, overall methodical procedures have to be followed involving high social and professional competence on the part of the project team and the project management. This includes more than systematic procedures by the individual designers; it also requires goal-oriented systemising of the organisation structure, overall optimisation of the development process, as well as the consistent use of adequate tools.

3. Measures for Successful Industrial Product Development

The above statements allow us to conclude that a clear orientation toward the goals by all persons involved in the development process will be an essential factor. The basis for this will be clear company and development goals, as well as strategies that define the basic procedures but leave enough latitude for ideas, initiative and commitment.

Many innovative corporations who have been successful in their markets over many years make an effort to become or remain the market leaders in their respective sectors.

Several empirical studies exist on the economic advantages of a market leader [4]: their return on investment (ROI) is higher than that of their competitors (Figure 3). Frequently innovative, flexible and open corporate cultures concerning organisation and technology can be found in such enterprises, a fact that makes them predestined for long-term success in their market.

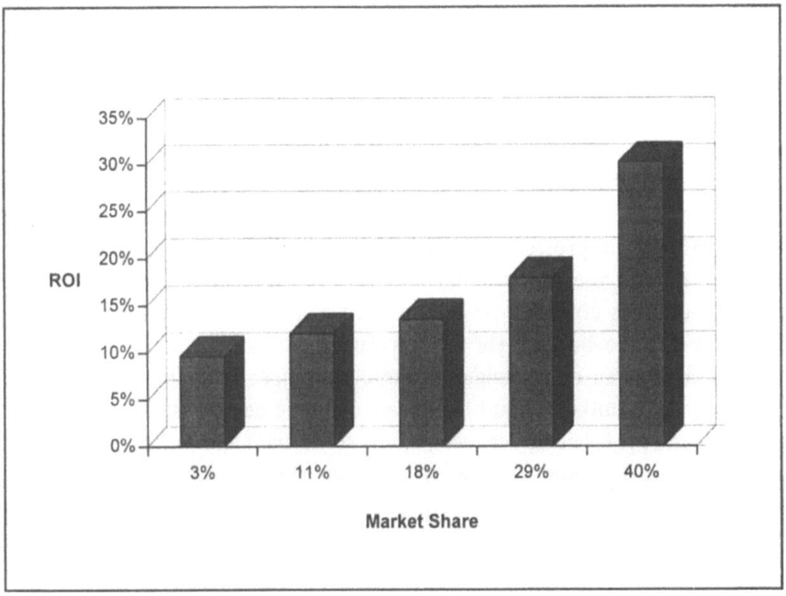

Figure 3: Profitability of market leaders.

In addition, successful corporations deal actively and directly with their customers, with the following consequence: their development processes are not primarily directed toward technical novelty (new development, improvement or product variation) but more toward their importance in the market [1]. Therefore, development processes should be distinguished by their importance for the market (see Table 1).

Table 1: Market driven development processes.

Market Importance	⇒	Development Process
open up a new market segment	⇒	innovation projects
increase market share	⇒	novelty projects
improve market margin	⇒	optimisation projects

Company organisation structure, control of the processes and the use of adequate tools must correspond to the development process. Some examples of success-enhancing measures will be presented in this respect.

3.1 Effective Organisation

Consistent leadership, goal-oriented project work and flexible, committed cooperation are essential for successful development of a product:

- Major basic decisions should be taken by a committee that involves all sectors of a corporation; in medium-scale companies this will normally be an extended board of directors. With regard to some major projects, for example, the start of the project, the performance specification and the milestones will be announced via documents presented for decision to each project manager.
- Depending on the culture of the corporation and the complexity of the project, developments are organised as pure, impacting or matrix [1]. For moderately complex innovation or novelty projects, a matrix organisation would be the organisation of choice.
- The project manager greatly influences the future success of the product. In particular for the realisation of major projects, the project manager should have several years of project experience, responsibility beyond a single department, and at least the professional authority to issue directives to the core team. His activities comprise product-related management tasks including responsibility for the budget. He reports directly to the deciding committee. Ideally he acts as an entrepreneur in the corporation, is distinguished by the capability to take initiatives and technical/economic decisions, is competent in both social and professional respects, and is self-reliant and independent.
- The project team should consist of the core team of up to 5–7 persons, supplemented by specialists as required for the respective phases. Every important department should participate. Innovations should, for example, require the marketing department to take part. Upon completion of the project, the project team is dissolved. Project teams are formed irrespective of the complexity of the project and the development process. Project teams should always act in innovative and major novelty projects. Less complex novelty projects and all optimisations could, however, be handled within the linear organisation.
- Allocation of the tasks depends on the product and project structure. Design and development of complex systems, e.g. newspaper printing machines, will require adequate organisation. On one hand this organisation should have only a minimum of simple interfaces, but on the other it must take into consideration department-overlapping tasks. It is practical, for example, to organise design departments by components such as printing unit or folding equipment. Yet it must be possible to overcome the limits of this organisation when similar functions are needed for different components. Thus it does not make much sense establishing a separate electronics group for every single component. The primary goal here should be to use uniform modules and software. A mixture of component-related and overlapping organisational units have proved reliable. A purely product-oriented organisation structure is not very suitable for handling complex customer orders or innovative and novelty projects. It must be supplemented by a project organisation that is responsible for the smooth realisation of customer requests and that coordinates

the contributions that the individual component departments make to order.

- In order to allow the project team to concentrate on its important tasks, project overlapping functions can be handled by support units, such as development control, approvals, patents and technical calculations (Figure 4). Furthermore, 'Task Forces' and groups of experts can be employed flexibly to provide additional support in difficult situations, for example to increase the design capacity over a short period of time.

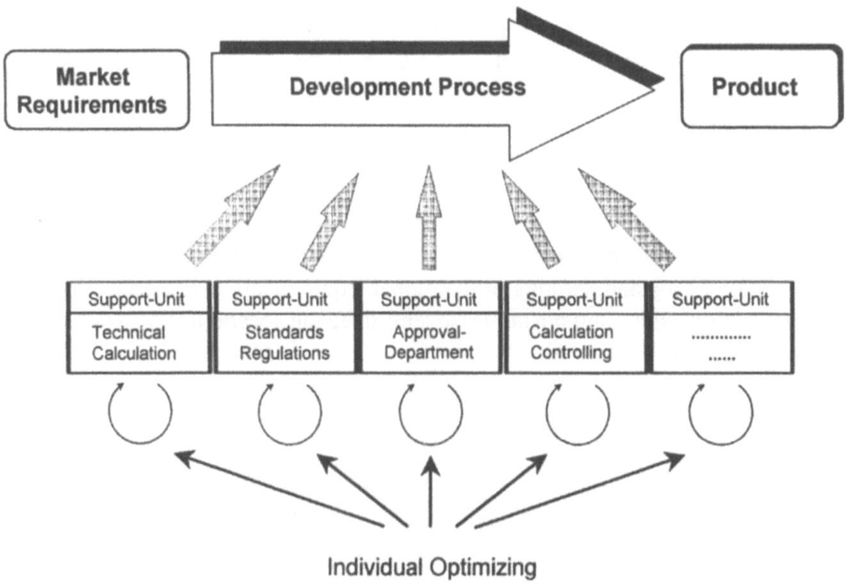

Figure 4: Main development process and support units.

3.2 Optimisation of the Development Process

The organisation provides the roof under which product development with its great variety of processes is actually effected. These processes will have to be harmonised and optimised in their entirety.

- Concerning the overall development management, distinction is to be made between research, development and handling of standard products. Intensive interchange of information and interface optimisation are essential. In addition, farsighted planning and a clear definition of the priorities shall prevent permanent helping out in favour of development of new products without problems being generated for current projects or subsequent developments. Clear goals and budgets will also promote the successful realisation of advance development, e.g. in the form of action lists [1].

- Planning of the development process requires permanent updating and coordination. Summarising the planning for all projects gives an overall survey, an essential factor to take into account before making basic decisions, for example, concerning development strategy or the long-term planning of resources. Decentralised planning of every single project furthers their efficient control. Depending on the requirements, exact planning of subsequent phases is based on the respective outline plan. Certain overlapping of the phases (see Figure 2), so-called 'Simultaneous Engineering' should be considered to allow for quick product introduction to the market. However, early introduction of product planning and the feasibility study are also essential features.
- A clear start to the project promotes its success. The use of defined corporate and development goals, market analyses (knowing the customers, the competitors and their products) and feasibility studies should lead to realistic requirements. These should be defined in a performance specification to be signed by the deciding committee. Complex development projects will need clear structuring and defining of the global requirements so that a development organisation with components in several different locations can work with it (see Figure 5).

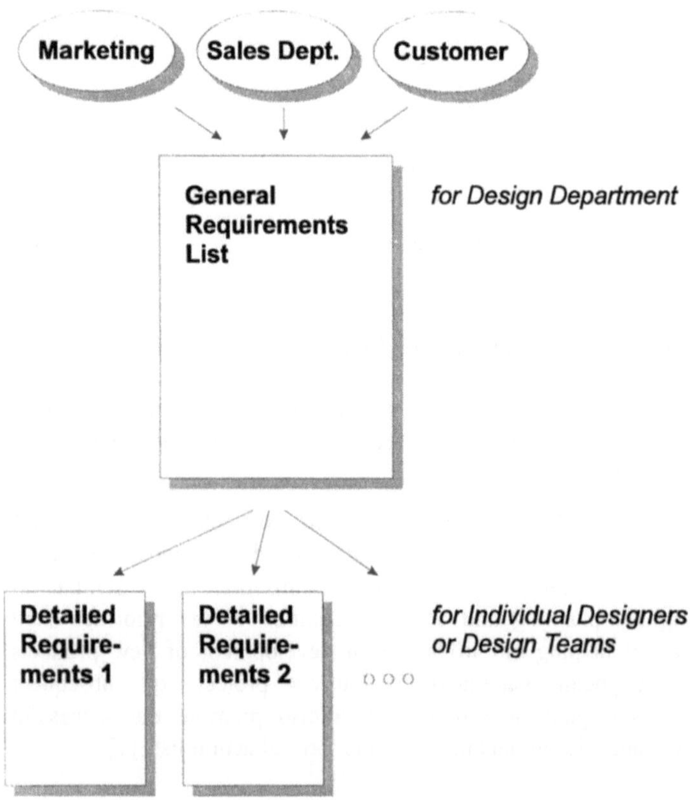

Figure 5: Specifying general requirements.

According to a survey of more than 100 corporations in the mechanical engineering sector, most problems involved in development processes arise from the realisation of the projects [5]. To avoid such a situation, target dates, costs, resources and technical quality must be monitored constantly. In addition, methodical search for solutions, analysing them in practical tests and evaluating them systematically will all be very helpful [6]. Important aspects of project management must be taken into consideration, for example regular meetings of the core team, layoff of the team members from normal business, and records of the allocation of tasks (responsibilities, dates).

In huge development projects involving a great number of tasks, the development progress in one organisational unit will originate interface requirements to be met by others, e.g. support units. Care shall be taken that these requirements are permanently recorded for the remaining project participants and exchanged ('duty to deliver'). A product-oriented flow of information is effected by the project manager (information management). All project-relevant information should be available to all parts of the corporation.

3.3 Use of Development-specific Tools

Highly developed software facilitates work on the most complex problems. On the debit side, high non-recurring and recurring costs, commitment of resources and start-up difficulties all have to be taken into consideration. The expense and use of such tools should be analysed in the corporation following the principle of 'not all that is possible will be necessary'.

- Planning support
  Depending on the complexity of the project and the corportation-specific marginal conditions, it is advisable to use the most simple and extendable programs. However, many corporations still use manual planning systems or planning systems with partial computer support only [7].
- Support for preparation of product data
  CAD/CAE/CAM systems are being used more and more, particularly for a comprehensive product description such as a three-dimensional geometric description (3D-CAD). However, models of complex plant systems via 3D-CAD cannot be applied to the completed detailed design, since the amount of data involved cannot be handled by today's hardware and software components. Most applications and advantages are at present to be found in engineering, calculation, rapid prototyping and tool optimisation. Further support for tool design, assembly optimisation, quality control and automatic calculation, for example, has been realised in single cases. The same applies to systems on the basis of virtual reality.
- Support of methods
  Using methods to find and analyse solutions will have considerable effect on the product development process, even those without computer support. Therefore, many methods can be used effectively for conventional applications. With the support of computers, representation, analysis and flow

of information about the results can all be improved, for example by obtaining clarification of the goals (QFD), risk analyses (FMEA) or comprehensive test programs (data-bases).

The methods used within the organisation and for the optimisation of development processes, together with the use of individual development-specific tools, do not constitute an absolute condition for, but are in clear support of, the development and corporative success. All considerations in this respect must focus on the people who are participating in the actual process.

4. How Big is the Influence of the Individual Person?

The project manager, as an integrating professional leader, and the members of the core team, collaborating closely and skilled in a goal-oriented manner, are the prime determinants of success. Therefore the general goals have to be correct at the beginning of the development process. The more complex a product is and the greater the number of persons and organisational units participating, the more important is the role of competent support units and an effective and flexible organisation structure. In this respect, methods are 'only' a helpful support for the proceedings of the individual or of groups, and they are expressed in the selection of an adequate organisation form. Methodical procedures are not a *sine qua non* condition for successful development, though they will significantly support the attainment of such success.

References

1. Fricke G., Lohse G. 1997: *Entwicklungsmanagement*. Springer-Verlag, Berlin.
2. Dylla N. 1991: *Denk- und Handlungsabläufe beim Konstruieren*. Hanser-Verlag, München.
3. Fricke G. 1993: *Konstruieren als flexibler Problemlöseprozeß*. VDI-Verlag, Düsseldorf.
4. Buzell R., Gale B. 1987: *The PIMS Principles. Linking Strategy to Performance*. The Free Press, New York.
5. Diebold 1997: *Produktinnovation – Von der Idee zum Erfolg*. Diebold, Frankfurt.
6. Pahl G., Beitz W. 1993: *Konstruktionslehre 3.Aufl.* Springer-Verlag, Berlin.
7. Ehrlenspiel K. 1995: *Integrierte Produktentwicklung*. Hanser-Verlag, München.

Résumé of Section VI

Considering the multitude of factors that determine business success, such as economic conditions or marketing, it is extremely difficult to attribute the overall success of a product to characteristics of a particular design process. Nevertheless, the problem of evaluating complex, creative and therefore unique design processes is often approached using the superficial level of a business process with the two dimensions of 'input' and 'output'. Obviously, this 'black-box approach' is not sufficient, because it does not allow a direct causal link between the concrete design activity and the parameters of the result. For example, the question of *which* individual prerequisite of a designer is important, and *when,* for a specific parameter result cannot be answered on this superficial level. Consequently, the overall process has to be subdivided into design steps and then broken down into design activities (which have to be defined) in order to encompass the real design work. Moreover, this process of defined description is important for a generalisation of the findings.

Another point made is that theoretical considerations still fail to give a clear definition of the input and output variables. The key question is still not answered: how does one label and examine the claimed variables using 'characteristic' and 'value'? Thus the practical use of theoretical models for design productivity remains unclear. If these basic questions are not addressed, the measurement of successful product development becomes an issue that is treated on a highly general and superficial level with no relation to empirical examination or design practice.

Reflecting on empirical investigations on the evaluation of design results illustrates the various sources of the problems. For example, investigations of the different approaches for evaluating a design product show that the most important question is that of choosing criteria. Different criteria can radically change the outcome of an assessment.

In consequence, the question of design performance measurement remains one of the most important unsolved questions in design research and practice. It becomes clear that many of the key problems can only be solved by concrete empirical research in design practice, including the human problem-solving process with its individual and group-related influences.

Conclusion and Outlook

Design Research: Approaching a Complex Field

The contributions of this book provide a survey on a wide field of research from various viewpoints: engineering design processes in their diversity are investigated with different aims and methods and therefore with a great variety of results. This variety of engineering design in practice and research is encouraging because it illustrates the importance of engineering design research; however, it is becoming more and more difficult to explore and to compare the multifarious results of empirical and theoretical design research.

The principle underlying any idea of comparison is to evaluate the different results, methods or investigations according to specific criteria. By doing this, the biggest problem seems to be the selection of criteria, because motives for comparison may be very different, and the system of categorisation will vary according to the goal of the evaluation. Additionally, the selection of criteria has to proceed on a relative widespread level if it is to form the basis for comparison of systems of widely different contents.

For example, this volume is subdivided into the various fields of research according to the influences on the design process and the result, i.e. individual, team, organisational, task and process influences. Even this simplistic classification is difficult because there are several contributions that consider more than one group of influences.

Another attempt to compare the contributions of this volume applies the categorisation of papers according to the various research aims proposed by Blessing *et al.* (this volume). Blessing *et al.* set four criteria: ‘success-criteria’, ‘description and analysis’, ‘prescriptions of design methods or tools’ and ‘interventions and analysis’, and these are used to categorise the contributions, as Table 1 illustrates. The table is based only on the information contained in the papers.

Table 1. Classification of the papers in this volume in relation to the proposed design research methodology described in the paper of Blessing *et al.* (page 44).

Crit: Product success criteria
DesI: Descriptive study of type I
Pre: Prescriptive study (statements that are direct translations of descriptive findings are not considered)
DesII: Descriptive study of type II
++: Main focus of the paper
+: Short description (DesI), indication of method or tool (Pre), or initial study (DesII)

Name	Crit	DesI	Pre	DesII	Comment
Dörner					Individual level *Research method development*
Hubka & Eder		+	++		Individual level *Theoretical approach*
Roozenburg & Dorst		++			Individual level
Schregenberger					*Research agenda*
Günther & Ehrlenspiel	+	++	+		Individual level
Cross & Clayburn Cross	+	++			Individual level
von der Weth	+	++			Individual level
Lindemann, Kleedörfer & Gerst			+	+	Company and project level
Andreasen & Hein	+		+	+	Company level
Ottosson	+		+	+	Project level
Lloyd		++			Team level
Leifer & Mabogunje	+	++	+	+	Team level
Hales	+	++			Multiple levels
Frankenberger & Badke-Schaub	+	++			Multiple levels
Hacker, Sachse & Schroda	+	++			Individual level
Radcliffe		+			Multiple levels *Research method development*
Christiaans & van Andel		+	+	++	Team level
Meerkamm	+		++		Team level
Wiegers, v. Breemen, Knoop & Vergeest	+	+			Individual level *Research method comparison*
Duffy & O'Donnell	+				Company level
Fricke	+	++			Multiple levels

This categorisation of research goals illustrates that the majority of studies are concerned with the observation and analysis of design processes (DesI), while some contributions entail prescriptions (Pre) and prove them (DesII). A few contributions explicitly discuss approaches to research methods for design research (e.g. Dörner).

It seems interesting that studies, although attempting to analyse and observe design processes, do not link their findings to product success criteria (Crit), although other papers do make this link, but without explicitly defining product success.

Some of the papers only implicitly give additional details of the categories addressed, or even details of the way in which the study was executed. Obviously there is a need for more transparency in terms of a more precise description of aims, approaches and results.

Some Conclusions

As a concurring result, engineering design appears as a complex activity in a network of influences related to the *individual*, the *team*, the *organisation* and the *task*. Of course, the bringing together of selected results concerning different influencing factors provides deeper insight into the complexity of design activities. We learn about interrelations of factors in regard to particular situations or requirements, and about the importance of human factors in the course of the design work.

Naturally, the compilation and analysis of design activities may be approached in various ways, and in doing so, the design *process* becomes more and more the crucial point of interest. This volume introduces different views and documentation of design work, from structured descriptions in protocols and time-schedule diagrams over the whole process, including design steps, sub-functions and collaboration between colleagues (see for example, Günther & Ehrlenspiel and Frankenberger & Badke-Schaub, in this volume), to ethnologic descriptions of communication that are independent of the overall problem-solving context (see Lloyd, in this volume).

For further research, it is useful to apply this knowledge of different methods to the compilation and evaluation of design processes. Possibly common standards and better documentation of these methods would enlarge the acceptance and application of common procedures and increase mutual interest.

However, in spite of various investigations with different methods, many questions concerning factors and influences on design success are still unanswered. How should we proceed? There are two main trends observable.

First, there is a move to *open* the viewpoint and to integrate other fields, such as business sciences. This aim is in line with the claim for a need to put design research into the context of business objectives. In this regard, the gap between theory and practice of design is obvious.

Second, widening the perspective also reveals the problem of losing touch with concrete design work. There are still many open questions about design-related problems. Should further research focus more on the design aspects?

In any case, what we need are more descriptive studies of design processes from different perspectives on the detailed resolution level in order to gain more information about situation-specific influences in actual design processes. The difficulties of generalisation may be reduced by deeper knowledge about influences in particular situations. Many interesting questions remain, for example, the huge importance of language and 'group language' in design processes, which is involved with psychological motives and processes and may be related to individual as well as to group strategies to a considerable degree.

All in all, there are no golden rules to guarantee successful design results, but there are a number of trends and results derived from the contributions of this book that should encourage further investigations and the unveiling of the 'mystery of designing'.

Visions, Challenges and Trends

Any investigation of design processes and results, as well as engineering design research itself, aims to optimise the connection between requirements and performance and to improve the clarity of demands and results. This means that vision of the future is the optimal company providing optimal conditions for the best designers in an optimal group setting, with complete knowledge of the facts and methods. This ideal world would inevitably lead to correct decisions and successful innovative products.

But reality is different and the future will be more challenging than ever. The European perspective, as well as world-wide globalisation, will be the crucial milestones over the next 5 years. Organisations will develop into new forms, and new markets will increase competition and foster diversity. Because of this, opportunities will arise to distribute the various partial processes of design, manufacturing and sales etc., as well as to disperse parts of design processes themselves as can be observed in many German companies, known by the term 'outsourcing'. On a macroscopic view, social and cultural changes also lead to the development of values and ethical norms for evaluating technologies. For example, the ecological need to anticipate the complete life cycle of a product in the design phase broadens the responsibility of the engineering designer.

Taking all these facts into account, this situation enhances the role of the designer and enhance the magic triangle of quality, time and cost. Along with these pressures on design we will experience less transparency but escalating uncertainty.

Of course, new information technologies aim to provide complete information on a subject. But facing the increasing amount of data, the designer often experiences even more uncertainty and a strong desire to reduce this information overload. Such tools should therefore allow the designer to set priorities in his search and to adjust the information system to his actual needs. Without a doubt, knowledge and information will increase, but their availability as well as the management of information will be restricted more and more to a very few skilled people, and the individual designer will work in continuously changing work teams and under widely varying working conditions.

Consequently, progress in future will bring its own new requirements, not just in the technical sector but also in the social environment, e.g. concerning communication and collaboration between various cultures and through new media (e.g. 24-hour development for different places around the world).

To summarise, designers will increasingly face situations that are more complex, more dynamic, more unstructured and thus more uncertain than ever before. Therefore, the question remains: How can future success be supported in engineering design theory and practice?

Knowing that designing is a complex and heterogeneous process, it is important to formulate a general theory about designing in practice that includes the aspects of thinking, remembering and imaging in the course of the design work, as well as individual parameters that concern personality, and parameters that encompass the complex social interactions (see also Dörner, in this volume). The aim of theory is to combine the high number of phenomena in design practice into a composite of main general mechanism containing a limited number of variables. This theory should be based on observations, analyses and predictions of designers in real working situations. From such a general theory we could derive methods, as well as educational principles and rules that should enable us to face the various social and engineering requirements for different working conditions, branches and products.

According to the stated view, the acting *designer* will become the 'key' variable to be supported in his work. In consequence, *human factors* should be more emphasised in design research and later on in education. In this field of 'soft factors' especially, interdisciplinary research (e.g. with psychologists) can help to answer the many unanswered questions.

And, of course, the 'crucial data' for answering these questions and to build a general theory of designing are the designers themselves. Being aware and sensitive of influences on their work, designers should be able to inform us about the importance of potential design-related factors on their work. In this regard, *designers* are not only to be seen as the key to successful product development but also *as the key to successful design research.*

MIX
Papier aus verantwortungsvollen Quellen
Paper from responsible sources
FSC® C105338

If you have any concerns about our products,
you can contact us on
ProductSafety@springernature.com

In case Publisher is established outside the EU,
the EU authorized representative is:
Springer Nature Customer Service Center GmbH
Europaplatz 3, 69115 Heidelberg, Germany

Printed by Libri Plureos GmbH
in Hamburg, Germany